S267 How the Earth Works: The Earth's Interior

Block 1

The Physical and Chemical Properties of the Earth

Prepared for the Course Team by

Peter J. Smith and Hazel Rymer

The S267 Course Team

Chairman

Peter J Smith

C J Hawkesworth

Course Coordinator

Veronica M E Barnes

Course Manager

Val Russell

Authors

Andrew Bell

Stephen Blake

Nigel Harris

David A Rothery

Hazel Rymer

Editors

David Tillotson

Gerry Bearman

Sue Glover

Designer

Caroline Husher

Graphic Artist

Alison George

BBC

David Jackson

The Open University, Walton Hall, Milton Keynes, MK7 6AA.

First published 1993.

Edited, designed and typeset in the United Kingdom by the Open University.

Printed in the United Kingdom by Thanet Press Ltd, Margate, Kent.

ISBN 0 7492 8162 6

This text forms part of an Open University Second Level Course. If you would like a copy of *Studying with the Open University*, please write to the Central Enquiry Service, PO Box 200, The Open University, Walton Hall, Milton Keynes, MK7 6YZ. If you have not already enrolled on the Course and would like to buy this or other Open University material, please write to Open University Educational Enterprises Ltd, 12 Cofferidge Close, Stony Stratford, Milton Keynes, MK11 1BY, United Kingdom.

3.1

S267b1i3.1

S267
HOW THE EARTH WORKS: *The Earth's Interior*

BLOCK 1
THE PHYSICAL AND CHEMICAL PROPERTIES OF THE EARTH

1.1 INTRODUCTION AND STUDY GUIDE

Just as it would be hard to imagine a book about how the human body works that didn't first describe body parts, it would be difficult to envisage a course on how the Earth works that didn't first explain what we know of the planet's structure and composition. This, the first substantive Block of the Course *How The Earth Works: The Earth's Interior*, is therefore largely (but not entirely) concerned with what the Earth *is*. It would be wrong, however, to see it merely as a service Block for those that follow. It's true that subsequent Blocks will be more overtly concerned with the workings of the Earth (and other planets), and it's equally the case that they will draw on some of the information in this one, but Block 1 also has its own justification. Part of the aim of this Course is to introduce you to the state of the Earth from crust to core as being of importance in its own right. In short, the Course's subtitle, *The Earth's Interior*, has a point.

We begin with a look at how the Earth formed as part of the Solar System. This is a field in which there have recently (from the late 1980s onwards) been some revolutionary developments in thought, overthrowing some long-standing ideas about just how the Earth grew from the original solar nebula. There is more than merely incidental background here, however, for, surprising as it may seem, some of the best information we have about the Earth's overall composition comes not from the Earth itself but from the Sun and meteorites, bodies from elsewhere in the Solar System. Moreover, the Earth's current thermal state is still very much under the influence of the heat that the planet acquired during its formation. A study of the Solar System and its origin has therefore much to tell us about the Earth of today and, in particular, enables us to construct a model of the Earth's chemical constitution, which, because the Earth's deep interior is inaccessible, is the best we have.

We next turn to the Earth's physical structure, the most important indication of which comes from seismology, the study of the behaviour of earthquake/explosion (seismic) waves as they pass through the Earth. As a preliminary, we examine the characteristics of seismic waves and of the earthquakes and non-earthquake sources that generate them, but we then go on to the more important matter of how, by noting the times at which seismic waves arrive at recording instruments, we can determine how the velocity of the waves varies with depth in the Earth and identify subsurface boundaries, or discontinuities. This is then followed by a description and discussion of those boundaries — the primary division of the Earth into crust, mantle and core; the finer structure observed within these major layers; and the alternative division of the Earth into lithosphere, asthenosphere, mesosphere and core.

Important though seismic studies are, they are not the only way of investigating the Earth's interior; measurements of gravity also provide structural information. We therefore go on to look at what gravity anomalies are, how they are measured, how they are corrected and how they are interpreted, a topic that leads naturally into an examination of how density varies with depth in the Earth. Armed with knowledge of how the Earth is structured, we then ask what the individual layers consist of and how they differ chemically. This, in turn, leads to a re-evaluation of the model of the Earth's chemical composition developed from Solar-System studies in the light of what we can deduce about the chemistry of the real Earth.

The Earth's interior is not just a matter of physical structure and chemical constitution; it also contains energy, particularly thermal energy. We therefore look next at the Earth's heat. Study of the Earth's thermal characteristics poses some tricky problems because, although we can easily measure the heat flowing outwards through the Earth's surface, it is much more difficult to deduce from that how the heat is distributed at depth. Moreover, although we know that substantial heat is still being produced in the Earth (mainly by the decay of radioactive isotopes), we also know that some of the heat now in the Earth has been present since the planet's formation. What we don't know is the precise balance between the two sources.

Our ignorance of thermal details is unfortunate, because one thing that we can be very sure of is that it is heat that makes the Earth 'work', particularly by generating convection currents within it, currents that drive, or at least influence, the plate tectonics, volcanism, seismicity and other phenomena observed at the surface. Nevertheless, Earth scientists have made considerable progress in understanding convection in the Earth, the topic with which we end this Block and which forms an appropriate introduction to Block 2, on plate tectonics.

Block 1 contains, apart from this introduction, 13 main Sections with the following approximate Course-unit equivalents (CUEs):

Section	CUE
1.2	0.3
1.3	0.5
1.4	0.5
1.5	0.6
1.6	0.5
1.7	0.5
1.8	0.5
1.9	0.1
1.10	0.2
1.11	0.5
1.12	0.7
1.13	0.1
1.14	0.5

Before proceeding further with this Block, you should, if you have not already done so, read Block 0, which provides general information on the Course and gives details of (and help with, where required) the skills necessary to study it. You should also by now have listened to AV01, 'Skills'.

1.2 THE EARTH IN CONTEXT

The Earth may be very important to humanity, but in the context of the Universe as a whole it is an insignificant speck. So, indeed, are the Sun and the planets and other bodies that revolve around it — our **Solar System**. The Sun is just one of many billions of stars in the **Universe** (the whole of existing matter, energy and space), and a very average one at that. Some stars are much hotter than the Sun and glow white and blue; others are much cooler and appear orange or orange-red. The Sun, which looks yellow under normal Earth-atmospheric conditions, comes somewhere in between, with a surface temperature of about 6 000 K (although its interior is much hotter). Some stars have diameters greater than the distance of the Earth from the Sun (about 1.5×10^8 km), whereas others are smaller than the Earth itself (average diameter 12 742 km). Again, the Sun lies between these extremes, with a diameter of about 1.4×10^6 km.

S1 S2

Stars cluster together in **galaxies**, of which there are at least a billion throughout the known (observable) Universe. But galaxies are no more averse to their own company than are stars, and themselves cluster in groups of from 2–3 to 1 000 or more. There is even some evidence that clusters of galaxies group together to form so-called superclusters. Galaxy clusters and superclusters can measure tens to hundreds of millions of light years across. (A **light year** is the distance travelled by light in one year; and as the speed of light is 3.0×10^5 km s^{-1}, 1 light year is equivalent to a distance of 9.46×10^{12} km.)

ITQ 1

Proxima Centauri, the nearest star to the Sun, is 4.2 light years away. How far is that in kilometres?

ITQ 2

How long does it take light from the Sun to reach the Earth when the Earth is at its average (mean) distance from the Sun?

Something like 10^9 galaxies lie within the range of modern optical and radio telescopes, some spiral in shape (loosely or tightly coiled), some elliptical (long and narrow to almost circular) and some irregular. Plate 1.1, Block 1 Colour Plate Section, shows just one of these, our near-neighbour the Andromeda Galaxy (M31), a spiral form. Each galaxy, in turn, contains billions of stars, some, but not all, of which, like our Sun, are thought to have planets orbiting them (although no such planets have yet been observed). The star system to which the Sun belongs is known as the Milky Way Galaxy, or simply the **Galaxy**, a member of a cluster of about 30 galaxies known as the **Local Group**. The Galaxy is about 10^5 light years in diameter, contains at least 10^{11} stars, and is a flattened system with a central bulge and fairly loose spiral arms. Its centre is about 30 000 light years away from the Sun, which, with its planets, rotates about the galactic centre once every 225 million years. The progression Universe → Local Group → Galaxy → Sun and its neighbouring stars is illustrated in Figure 1.1.

The Sun lies at the centre of the Solar System, which, besides the Sun, comprises nine planets (in order of increasing distance from the Sun: Mercury, Venus, Earth, Mars, Jupiter, Saturn, Uranus, Neptune and Pluto), satellites (or moons) of the planets, many minor bodies (asteroids, comets and meteoroids) and some interplanetary dust and gas. Like most stars, the Sun is a ball of hot gases producing energy by nuclear reactions. Because of these reactions, the Sun has a core temperature probably in excess of 1.4×10^7 K; and because it is gaseous, it has a low density (about 1 400 kg m^{-3}). But the low density should not deceive anyone into believing that the Sun is of insignificant mass, for it is also very big, having a diameter 109 times

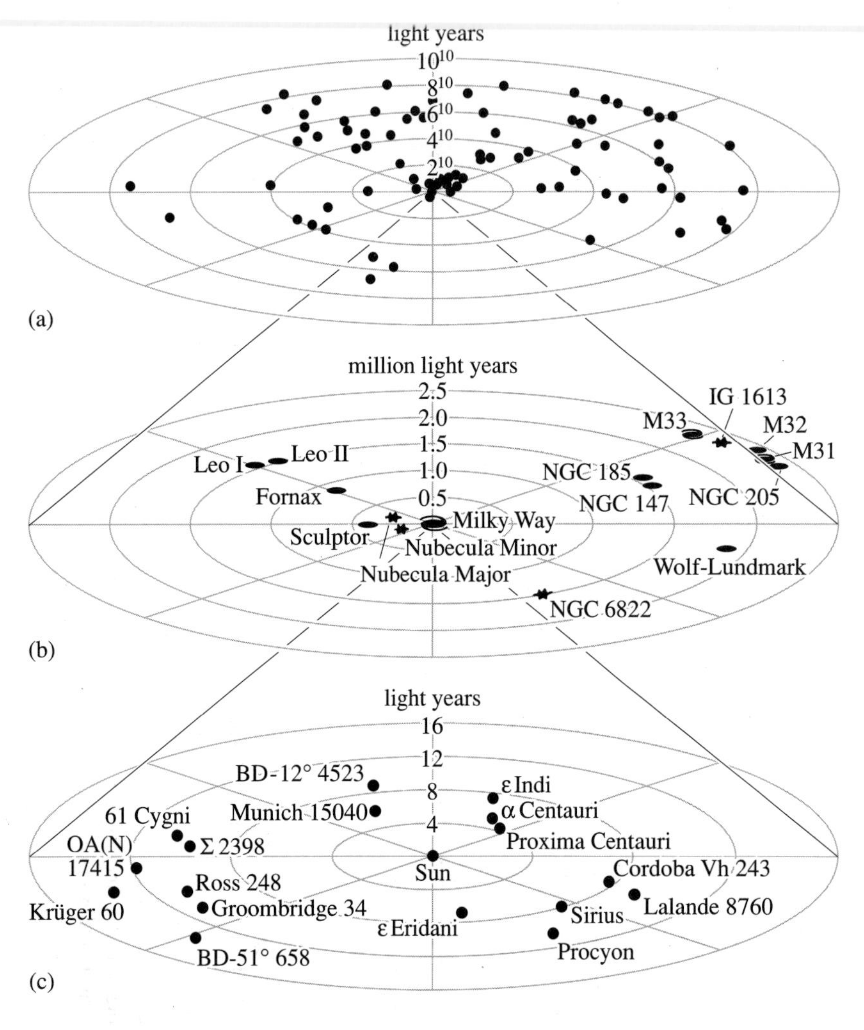

Figure 1.1 Galaxies are observed out to the limits of the currently known Universe (a), about 10^{10} light years away. Much closer to us is the Local Group of galaxies (b), which lies within about 2.5 million light years. The (Milky Way) Galaxy is a flattened system with a central bulge and fairly loose spiral arms. Within the Galaxy, which is about 10^5 light years across, lie the Sun's near-neighbour stars (c), those shown here being those out to about 16 light years.

that of the Earth and almost 10 times that of Jupiter, the largest planet in the Solar System. Its mass, about 2×10^{30} kg, is more than 330 000 times that of the Earth and more than 1 000 times that of Jupiter, which means that it contains most of the mass in the Solar System.

ITQ 3

Table 1.1 lists some basic planetary, solar and lunar data. Use the information provided on the masses of the bodies in the Solar System to determine the proportion of the total Solar System mass accounted for by the Sun.

Most planetary orbits are roughly circular, but not exactly so. Each of the planets in the Solar System orbits the Sun in an ellipse, with the Sun at one focus of the ellipse and nothing at the other. As a result, none of the planets maintains a constant distance from the Sun and none has a constant orbital speed. The Earth–Sun distance, for example, varies from 147×10^6 km to 152×10^6 km within a single orbit. There is uniformity in all this, however, in that all the planets orbit the Sun in the same direction. The plane formed by the orbit of the Earth around the Sun (which, of course, includes the Sun) is called the **ecliptic.** The other planetary orbits also form planes that include the Sun, and all such planes are inclined to the ecliptic. However, with the exception of that of Pluto (and, to a lesser extent, Mercury), the angle of inclination is very small, which means that the Solar System is essentially 'flat', or disc-like, as Figure 1.2 illustrates. This is really quite a remarkable phenomenon; and any hypothesis of the origin of the Solar System must take it into account, as well as the fact that all planets orbit in the same sense.

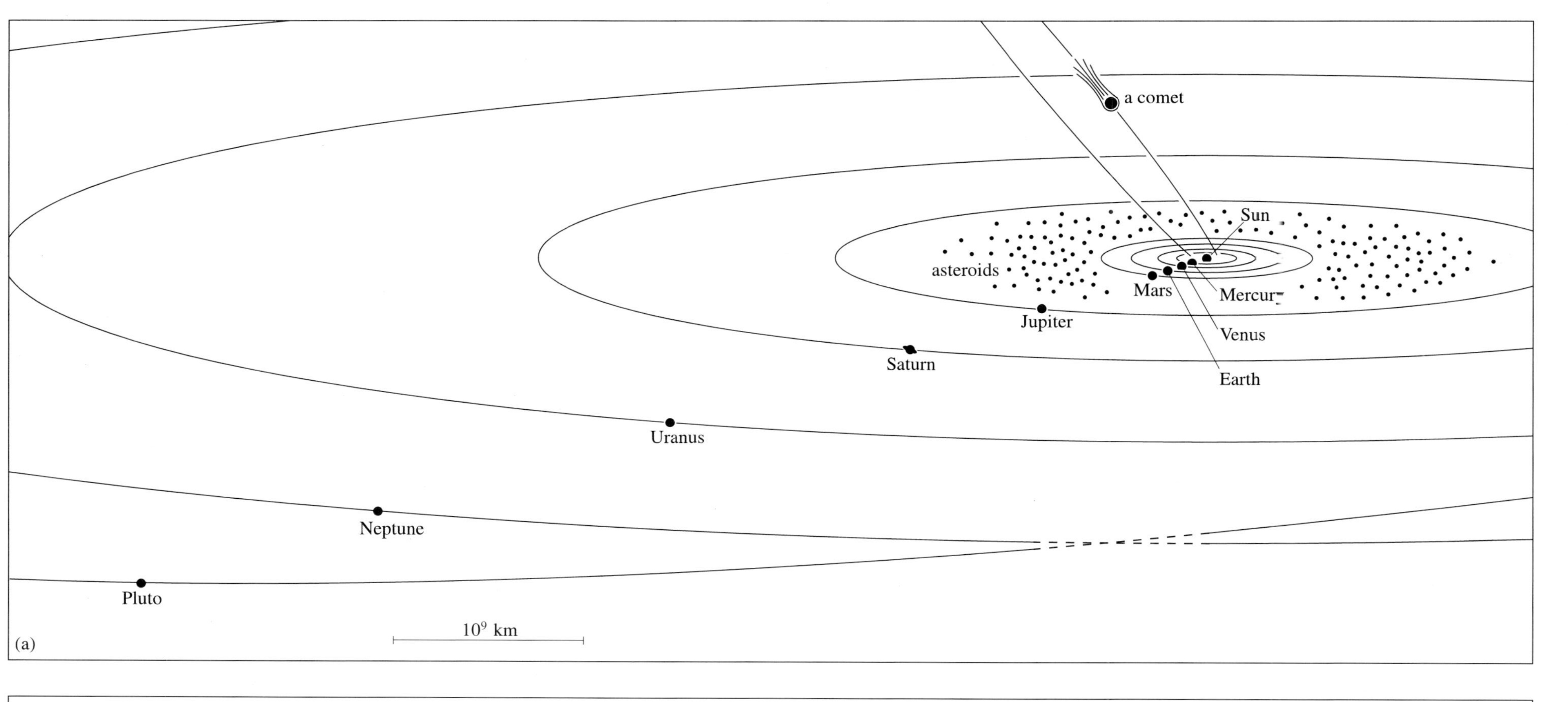

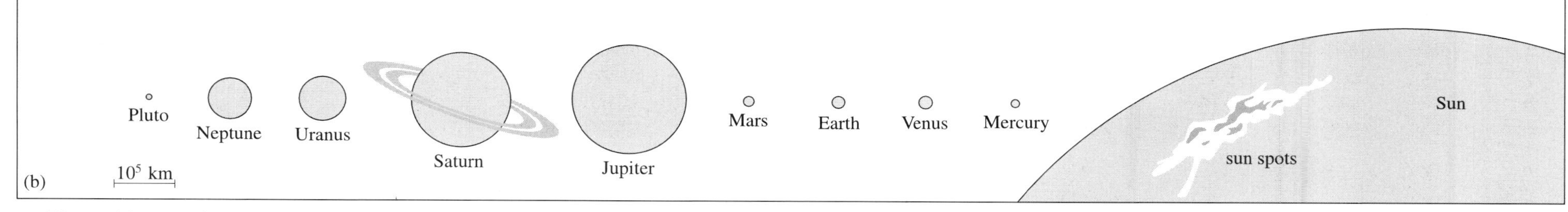

Figure 1.2 The orbital paths of the planets (a) and the relative sizes of the planets and the Sun (b). Seen in this conventional representation — i.e. viewed from above and with the planetary north poles uppermost — all the planets orbit the Sun in an anticlockwise direction.

Table 1.1 Planetary, solar and lunar data

Name	Mean distance from Sun (10^6 km)	Mean distance from Sun (AU)*	Orbital period (Earth days or years)	Inclination of orbital plane to ecliptic (°)	Mean orbital speed (km s^{-1})	Period of rotation about axis (Earth days)	Inclination of equator to orbital plane (°)	Equatorial diameter (km)	Equatorial diameter relative to Earth	Mass (kg)	Mass relative to Earth	Mean density (10^3 kg m^{-3})	Number of known satellites (early 1991)
Sun	–	–	–	–	–	25.4	–	1 392 530	109	2.0×10^{30}	332 848	1.4	–
Mercury	57.9	0.39	88 days	7.00	47.89	58.65	~2	4 878	0.38	3.3×10^{23}	0.06	5.43	0
Venus	108.2	0.72	225 days	3.39	35.03	243.01	177.3+	12 102	0.95	4.9×10^{24}	0.81	5.25	0
Earth	149.6	1.00	365 days	0.00	29.79	1.00	23.44	12 756***	1.00	6.0×10^{24}	1.00	5.52	1
Mars	227.9	1.52	687 days	1.85	24.13	1.03	25.19	6786	0.53	6.4×10^{23}	0.11	3.95	2
Jupiter	778.3	5.20	11.9 years	1.31	13.06	0.41	3.12	142 984++	11.21++	1.9×10^{27}	318	1.33	16
Saturn	1 427.0	9.54	29.5 years	2.49	9.64	0.43	26.73	120 536++	9.45++	5.7×10^{26}	95	0.69	17+++
Uranus	2 871.0	19.19	84.0 years	0.77	6.81	0.75	97.86+	51 118++	4.01++	8.7×10^{25}	14.5	1.29	15
Neptune	4 497.1	30.06	164.8 years	1.77	5.43	0.66	29.6	49 528++	3.88++	1.0×10^{26}	17.1	1.64	8
Pluto	5 913.5	39.53	248.5 years	17.15	4.74	6.39	122.46+	2 300	0.18	1.3×10^{22}	0.002	2.03	1
Moon	149.6	1.00	27.3 days**	5.15**		27.3	6.68**	3476	0.27	7.3×10^{22}	0.012	3.34	–

* AU = astronomical unit, the mean distance of the Earth from the Sun: 1.496×10^8 km.

** Refers to orbit about the Earth. (The mean distance of the Moon from the Earth is 384 392 km.)

*** The equatorial radius of the Earth is 6 378.137 km, the polar radius is 6 356.752 km, and the radius of the sphere having the same volume as the Earth is 6 371.000 km. In this Course, the last figure will be referred to as 'the average radius of the Earth'.

+ These angles are greater than 90° because the planets have retrograde rotation (see text).

++ As the large outer planets have no solid surfaces, these are the diameters at the 10^5 N m^{-2} pressure level in their atmospheres. (10^5 N m^{-2} is the approximate atmospheric pressure at the Earth's surface.)

+++ Plus seven very small ones.

Note 1: Some of the figures in this table may appear to be slightly inconsistent with each other because they have been rounded (e.g. corrected to one decimal place).

Note 2: Because astronomical data are continually being revised, any table such as this is bound to be out of date almost as soon as it is published. This table was compiled in mid-1991 and contains the most authoritative figures available at that time.

The planets not only revolve around the Sun, they simultaneously rotate about their own axes. This has a number of important consequences. One is that the planets bulge slightly at their equators (i.e. they are not perfectly spherical) because of the centrifugal forces set up by the rotation (balanced by the internal gravitational attraction which tends to draw the shape into a sphere). Another, perhaps more profound, effect is that where a planetary interior (e.g. that of the Earth) contains fluid layers, differential movements can arise between the fluid and solid parts of the planet. One curiosity of the Solar System's rotating planets is that the axes of rotation are never perpendicular to the orbital planes of their respective planets. The Earth's axis, for example, is tilted at 23.44° from the perpendicular; or to put it another way, the Earth's equator lies at an angle of 23.44° to the ecliptic. These and other data for the Sun, Moon and planets are listed for reference in Table 1.1.

1.2.1 BODIES IN THE SOLAR SYSTEM

One of the most striking features of the Solar System is its diversity; in addition to the Sun, there are planets, satellites of planets, asteroids, comets, meteoroids and interplanetary dust and gas. Moreover, there are considerable differences even among the planets and their satellites.

❑ Look at the 'mean density' column of Table 1.1. What is the most obvious pattern among the densities listed there?

■ There is a clear group of higher-density bodies (Mercury, Venus, Earth, Mars and the Moon) and a distinct group of lower-density bodies (Jupiter, Saturn, Uranus, Neptune, Pluto and the Sun).

The higher-density planets (Mercury, Venus, Earth and Mars) are known as the **inner planets**, or terrestrial planets, because they are the planets closest to the Sun (i.e. in the inner Solar System). They are essentially rocky bodies comprising mainly silicates (compounds based on SiO_4) and iron. The lower-density **outer planets** (Jupiter, Saturn, Uranus and Neptune, but not Pluto), by contrast, consist largely of gases such as hydrogen, helium, methane, ammonia and water vapour, although some or all may have small rocky cores (the meaning of 'small' in this context will become clear later — Section 1.3). With the exception of Pluto, the outer planets are also much larger than the inner planets (see Figure 1.2).

Briefly, the Solar System comprises the following bodies and groups of bodies:

Mercury is the smallest planet in the inner Solar System and the nearest to the Sun. Its orbit is more eccentric (i.e. it is less closely circular) than that of any other planet except Pluto, a phenomenon that has a marked effect on its surface temperature, which varies from as low as 90 K at night to as high as 720 K during the day (average about 400 K). The surface of Mercury is heavily cratered (the result of meteorite impacts), which has resulted in its often being described as 'Moon-like'. In fact, the planet's surface is more varied than that of the Moon; there is evidence of more extensive early volcanism; and there are much more extensive cliff systems, thought to be the result of global compression during early cooling. Mercury has a very tenuous atmosphere (the pressure being only 10^{-12} of that of the Earth), but what atmosphere it does have consists largely of helium, sodium (very surprisingly), and oxygen. The planet has a very high density for its size, suggesting that it has an unusually large iron core, accounting for perhaps 60–70% of its total mass. We shall return to this point later.

Venus, the second planet from the Sun and the most brilliant as viewed from the Earth, has the most closely circular orbit of all the planets. With mass, diameter and (probably) internal structure very similar to those of the Earth, it is often described as the Earth's 'sister planet', but there are some crucial differences between the two bodies. One is that the surface of Venus is much less mobile than that of the Earth — i.e. there is no evidence of widespread plate tectonics — although there is evidence of extensive volcanism. Secondly, Venus has a thick atmosphere that comprises mainly carbon dioxide. This means that the planet is in a perpetual 'greenhouse', giving it an average surface temperature of about 730 K. A peculiarity of Venus is that its rotation is **retrograde** — i.e. the planet rotates in the sense opposite to that of the Sun and most other planets (it rotates clockwise in the perspective of Figure 1.2).

Earth, the third planet from the Sun, is the largest of the inner planets and has the highest mean density. Uniquely among the planets (as far as we know), it has oceans and continents; its surface has long been subject to plate-tectonic and other geological processes; and it has an atmosphere comprising largely nitrogen and oxygen. We shall say no more about the Earth here as it is the chief subject of this Course.

The **Moon**, Earth's only satellite, is unusually large in relation to its primary planet (mass of Moon/mass of Earth ~ 1/82). Most planetary satellites have masses of only hundredths of those of their primaries; and they include three satellites of Jupiter and one of Saturn that are, in absolute terms, more massive than the Moon*. Despite the relatively close matching of Earth and Moon in terms of size, however, the two bodies are quite different in nature. The Moon has a highly cratered surface, has been geologically inactive for several billion years, and has a significantly lower mean density. The Moon's orbital period about the Earth and its rotational period are equal; so the same face of the Moon is always directed towards the Earth. (This is the most stable configuration gravitationally.) The Moon has no atmosphere, but it does have a tail of mainly sodium gas extending outwards from its dark side by about 24 000 km.

S3

Mars, the first planet beyond the Earth's orbit, is both smaller and less dense than the Earth. The surface is cratered and shows evidence of volcanic, tectonic and erosional activity. There is an atmosphere consisting mainly of carbon dioxide. One of the planet's most conspicuous features is its polar ice caps, although, curiously, whereas that at the north pole is thought to comprise water ice, that at the south pole is believed to consist of carbon-dioxide ice. Mars's average surface temperature is about 215 K — i.e. the planet is cold.

Asteroids are minor planetary bodies orbiting the Sun, all the better-known ones being in the region between the orbits of Mars and Jupiter. All are small, the largest, Ceres, measuring about 930 km across. By 1990, more than 4 000 of the larger asteroids had been named and the number was increasing by 150–200 a year; but there are many hundreds of thousands of other, unnamed examples, ranging in size down to mere boulders. Asteroids range from roughly spherical (e.g. Ceres) through elongated to very irregular in shape (see Plate 1.2), and their orbits range from circular to very highly eccentric. Hidalgo, the asteroid with the most eccentric orbit known, comes to within about 300 million kilometres of the Sun at one extreme (close to Mars orbit) and moves out to 1 440 million kilometres at the other (close to Saturn orbit). Although most known asteroids lie within the **main belt** between Mars and Jupiter, individual asteroids can be found almost anywhere within the inner parts of the Solar System. Asteroids with orbits overlapping (but

* The only satellite–planet pair in which the satellite–planet mass ratio is greater than (and, in fact, much greater than) that of Moon–Earth is Pluto and its moon Charon.

larger than) that of the Earth are known as **Apollo asteroids**; those with orbits lying mainly within the Earth's are called **Aten asteroids**. More than 40 Apollos/Atens had been discovered by 1990. There is always a chance that an Apollo or Aten will collide with the Earth; and, indeed, such collisions have almost certainly happened in the past, perhaps with profound effects. Since 1980, for example, a popular (and controversial) hypothesis among Earth scientists has been that an asteroid impact with the Earth led to the mass extinction of dinosaurs and other species about 65 million years ago. Asteroids vary widely in composition; they are probably fragments of material that elsewhere in the Solar System accreted into planets, but which in this case failed to coalesce for one reason or another. They almost certainly hold important clues to the origin and evolution of the Solar System. This may prove to be especially so in the case of a family of icy asteroids in the **Kuiper belt**, extending from Neptune's orbit out to about 80 AU (defined in Table 1.1). The first Kuiper belt object was discovered in 1992 and several others followed. By mid-1995 it was thought likely that the belt contained about 10^8 bodies more than about 30 km across.

Jupiter, the fifth planet from the Sun and the first of the outer planets, is the largest and most massive planet in the Solar System, accounting for more that two-thirds of the total mass of the planets. It consists mainly of light elements such as hydrogen and helium in gaseous and liquid form, although there is some evidence that it has a small rocky core. It rotates much more rapidly than does the Earth, giving it an even more pronounced equatorial bulge. Moreover, because of its largely fluid nature, Jupiter exhibits rotational effects not shared by the rocky, inner planets; for example, surface features at different latitudes rotate at different speeds. The motions of Jupiter's visible surface are thus very complex.

Saturn, **Uranus** and **Neptune** are broadly similar to Jupiter, also being 'giants' and having rapid rotations. Saturn, of course, is well known for its 'rings', consisting largely of ice and frozen gases, although Uranus, Neptune and Jupiter also have rings, albeit less conspicuous ones. Uranus is unique in that its rotational axis lies more or less in its orbital plane. In fact, the planet is tilted at rather more than 90°, so that its rotation, like that of Venus, is considered to be retrograde. Saturn, Uranus and Neptune probably all have small rocky cores. There is some evidence that, beyond their core regions, whereas Jupiter and Saturn consist largely of hydrogen and helium, Uranus and Neptune also contain heavier elements such as oxygen and nitrogen.

Pluto, though technically an outer planet, is not a gaseous giant. On the contrary, it is much smaller than the Earth and even than Mercury, although its density is only marginally higher than those of the giants. The mean distance of Pluto from the Sun is greater than that of Neptune, but the orbit is so eccentric that at times Pluto comes in closer to the Sun (e.g. over the period 1979–99). The origin of Pluto is uncertain, but the planet may turn out to be no more than the nearest (and largest) Kuiper-belt object.

Planetary satellites (natural ones, that is) are common, only Mercury and Venus being deficient in this respect. By 1990, at least 60 satellites had been discovered, ranging from icy bodies to rocky ones. We shall say no more about them here, because we propose to examine some of them in more detail in Block 5.

Comets are bodies which, when moving through the inner Solar System, generate spectacularly illuminated tails of dust or plasma (ions and electrons), or both. An example is shown in Plate 1.3. Although cometary nuclei are typically only a kilometre or so in diameter and apparently consist largely of ice but with some rocky particles ('dirty iceballs'), their tails frequently stretch for tens of millions of kilometres. In 1950, J. H. Oort concluded from an analysis of cometary orbits (which are generally very elliptical) that comets usually reside in a zone some 10 000–100 000 AU from the Sun — now called the **Oort cloud** — from which some are from time to time dislodged (e.g. by the gravitational disturbances caused by passing stars) and sent on their way towards the inner Solar System. In origin, Oort-cloud comets are probably Kuiper-belt-like bodies that formed closer to the Sun than the surviving part of the belt, and which were flung outwards by gravitational interactions with Jupiter or another giant planet. The Oort cloud is thought to contain trillions of comets with a total mass 25 times that of the Earth, but has never actually been seen. The difficulty is that the spectacular effects produced by comets result from their interaction with the Sun and so are only observed in the inner Solar System. The traditional view is that comets formed early in the Solar System's history and are thus potentially capable of providing valuable information on its formation. However, it is also clear that comets in the Oort cloud must be altered by collisions, cosmic radiation and the heat of passing stars; so they cannot be entirely pristine. Incidentally, comets could, in principle, and may well have in practice, hit the Earth.

Very small bodies orbiting the Sun — i.e. bodies ranging in size from smaller than a grain of sand to several kilometres — are known collectively as **meteoroids**. Meteoroids that enter the Earth's atmosphere, producing a luminous tail as they burn up, are known as **meteors**, although many meteors, particularly those occurring in showers, are produced by material thrown off from comets or by the breakup of comets. (Technically, comets are meteoroids, but they are not usually thought of as such.) **Meteorites** are what meteoroids become known as once they hit the surfaces of planets or planetary satellites — they are fragments of rock of any size. Most meteorites are fragments of asteroids (which means that generally there is no difference between a large meteoroid and a small asteroid), although a few are thought to come from the Moon and Mars. Unlike comets (because of their nature and general inaccessibility), meteorites have provided immensely important information about the origin of the Solar System and the Earth, for which reason we shall look at them in much more detail in Section 1.4.

1.2.2 Planetary density and compositional comparison

❑ Suppose that there were several rocky planets of identical composition and all chemically homogeneous, but each having a different diameter. Would all these planets have the same mean density?

■ No, because although different materials have different densities, density (= mass per unit volume = mass/volume) does not depend *only* on chemical composition. Consider, for example, two chemically identical homogeneous spheres, A and B, of the same diameter under 'natural' conditions at the Earth's surface (i.e. where the pressure is one atmosphere). Now suppose that sphere A is subjected to a huge external pressure such that its volume is reduced to one half of its original volume, whereas sphere B is not subjected to any additional pressure at all. The density of sphere B will not have changed; but because density = mass/volume and because its volume has been reduced by half, sphere A will have doubled its density. In short, spheres A and B will still be chemically identical but will now have different densities. Planets are not, of course, subjected to large external pressures, but the same principle applies. Pressure within a planet — even a chemically homogeneous one — will increase with depth because of the weight of overlying matter.

At any depth, therefore, the material there will have a higher density than the same material would have had were it at the planet's surface. This phenomenon is known as **self-compression**. Generally, the bigger the planet, the greater will be the pressure at depth, the greater will be the density at any given depth, and hence the greater will be the planet's mean density.

The method of calculating the effects of self-compression is mathematically complicated and beyond the scope of this Course. However, one result of the calculations is that if the density of the Earth were to increase with depth as a consequence of self-compression alone, the pressure at the centre of the planet would be $3.18 \times 10^{11}\,\mathrm{N\,m^{-2}}$. Even more complicated calculations for the real Earth show, by contrast, that the actual pressure at the centre is probably $3.64 \times 10^{11}\,\mathrm{N\,m^{-2}}$. The difference arises because the real Earth not only undergoes self-compression but also has denser materials towards its centre. The Earth has an iron core overlain by a silicate mantle; and iron is denser than silicates. The density in the Earth thus increases with depth not only because of self-compression but also because of chemical variations. (We shall be looking at the internal structure of the Earth later in the Course. For the time being it is sufficient to be familiar with the gross structure, which is summarized in Figure 1.3.)

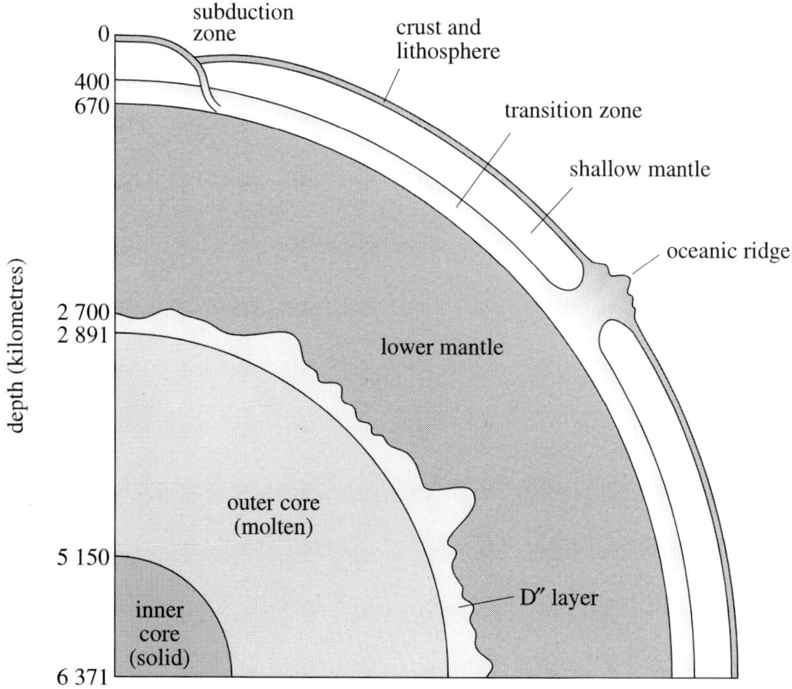

Figure 1.3 The gross structure of the Earth. The mantle as a whole comprises the shallow mantle, the transition zone, the lower mantle and the D″ layer (the significance of which will become clear later in the Course). The radius of the inner core is 1 221 km and that of the outer core is 3 480 km; and these figures have been subtracted from the average radius of the Earth (6 371 km) to give depths to, respectively, the inner core–outer core boundary and the (outer) core–mantle boundary. The mantle comprises iron/magnesium silicates, and the core is mainly iron.

The fact that the mean density of a planet depends partly on self-compression and partly on actual chemical variations provides an interesting way of comparing the chemical compositions of the terrestrial planets and the Moon, at least to the extent of trying to determine whether they are all chemically the same or quite different. As all five bodies formed in the same (inner) region of the Solar System and are all rocky, a case could be made for supposing that their bulk compositions are rather similar, but it would be useful to have some evidence other than the merely circumstantial.

Suppose, then, that the terrestrial planets and the Moon were chemically identical and homogeneous. In that case, self-compression would be the only consideration; and the bigger the planet, the higher would be its average density. The relationship between increasing planetary size (e.g. radius) and mean density might or might not be linear, but at least it would presumably be regular in some way. However, if the bodies differed greatly in overall chemical composition, that regularity would be disrupted, because there would be greater variation in mean density; and any attempt to plot mean density against radius would be likely to result in scattered points. So what actually happens?

ITQ 4

Plot on a piece of graph paper the radii of the terrestrial planets and the Moon against those bodies' mean densities. What pattern, if any, do you observe? (Note: It's simpler to plot not the radii but the (equatorial) diameters relative to that of the Earth, which are proportional to the radii.)

S4

We have dwelt on the question of self-compression, mean planetary density and the bulk compositions of the inner planets at some length because it demonstrates rather well the difficulties faced by Earth and planetary scientists studying planetary interiors (including the Earth's) and illustrates the sorts of procedures they must often adopt. We would like you to think about what immediately follows as you work your way through the whole of this Course, for the Course is not just about how the Earth works but also about the type of skills Earth scientists use to infer how it works.

The obvious fact is that the Earth's interior is almost completely inaccessible. During the 1980s, the Russians drilled a hole in the Earth's crust to a depth of about 14 km; and so in one small area of the world, scientists have access to the Earth's interior down to a level equivalent to 0.2% of the planet's radius. It's not much. Elsewhere, however, scientists are even less privileged; they have direct access only to the extent of the deepest mine and access to Earth materials only to the bottom of the deepest borehole, which, if they're lucky, might be a few kilometres. Of course, they have other resources. They have astronomical data of the type discussed in this Section; they can make certain observations (e.g. of gravity) and measurements (e.g. using seismic waves generated by explosions) at the Earth's surface; they are able to perform laboratory experiments on the sort of materials they might suspect lie in the Earth's interior; they can devise hypotheses, or speculations as some people might call them; and, with direct access only to the uppermost crust, they can sometimes find and examine there material that the Earth itself has chosen to raise from the depths (e.g. rock fragments brought up by volcanoes). But, unlike surface geologists, they cannot visit their field area.

The result is that we know little, if anything, about the Earth's interior with absolute certainty. Deep-Earth and deep-planet scientists can make inferences, but they can seldom be sure. Moreover, despite the vast amount of data now available bearing on the Earth's interior, information is still lacking in so many areas and, where it exists, is often ambiguous. The inference about the possible similarity between the bulk chemical compositions of the Earth, Venus, Mars and the Moon, obtained through ITQ 4, illustrates these points. The context may be planetary rather than purely terrestrial, but the principles are no different.

As regards the topic covered by ITQ 4: Scientists *hypothesized* (objectively, of course) that the inner planets and the Moon might have similar compositions. Then, having already *theorized* that even a chemically homogeneous planet would experience an increase in density with depth because of self-compression and having *determined* certain planetary data astronomically, they *perceived* that, in principle at least, theory and measurement might together offer a way of comparing planetary bulk compositions. They then *tested* their idea by plotting the relevant data and found that, despite some obvious problems (e.g. the possible influence of internal planetary structure), the result was a strong *inference* that the inner planets and the Moon do indeed have broadly similar overall compositions. This is a far cry from, say, going into the laboratory and measuring the properties of a pure chemical element. Earth and planetary scientists have to deal with their worlds as they are, warts an' all, and must take their meagre data where they find them.

We stress again, however, that what emerges from this sort of exercise is not a fact or proof but an inference or possibility. Some inferences may be stronger than others; but however strong they might appear, they are ever likely to be overthrown by some newly discovered data that inconveniently contradict them. As it happens, the particular inference about the similarity of the compositions of the Moon and the inner planets has not been disproved, or at least not entirely. There is now evidence from other sources to suggest that perhaps Mars and the Moon fall rather too neatly into line as far as the result of ITQ 4 is concerned; Mars and, more so, the Moon probably contain rather less iron (and thus have too low a mean density for their size) than ITQ 4 would suggest. The procedure used in ITQ 4 was evidently rather too coarse, but it's interesting that it worked at all.

SUMMARY OF SECTION 1.2

The Earth is a planetary member of the Solar System, which is part of the Galaxy, itself a member of the Local Group of galaxies and just one of about 10^9 galaxies that populate the observable Universe. The Solar System is highly diverse, comprising planets, planetary satellites, asteroids, comets, meteoroids and interplanetary dust and gas, and has certain conspicuous properties (such as its disc-like shape and the common orbiting direction of its main constituents) with which any hypothesis for the origin of the system must conform. The planets and planetary satellites are also very varied, ranging from hard rocky objects to largely liquid/gaseous bodies. The Sun, the planets and the planetary satellites rotate about their own axes, mostly, but not always, in the same sense.

Planets undergo self-compression resulting from the pressure of overlying matter, which raises their mean densities above those that the same materials would have had under normal (one atmospheric pressure) conditions. Earth and planetary scientists must frequently make inferences about terrestrial and planetary interiors on the basis of fairly, sometimes very, tenuous data. One such inference is that the Earth, Venus, Mars and the Moon are of broadly similar bulk composition, although Mars and the Moon deviate somewhat from that and Mercury deviates much more severely.

OBJECTIVES FOR SECTION 1.2

When you have completed this Section, you should be able to:

1.1 Recognize and use definitions and applications of each of the terms printed in the text in bold.

1.2 Outline the main features of the Universe in general and the Solar System in particular as the context in which the Earth exists.

1.3 Perform simple calculations using, and draw general conclusions from, astronomically derived data on the Earth, Solar System and Universe.

1.4 Explain the general principles of self-compression in planets and draw general conclusions therefrom.

1.5 Recognize and apply the sort of reasoning that enables inferences to be drawn about planetary and terrestrial interiors in the absence of firm, unambiguous data.

Apart from Objective 1.1, to which they all relate, the four ITQs in this Section test the Objectives as follows: ITQ 1, Objective 1.3; ITQ 2, Objective 1.3; ITQ 3, Objective 1.3; ITQ 4, Objectives 1.4 and 1.5.

You should now do the following SAQs, which test other aspects of the Objectives.

SAQs for Section 1.2

SAQ 1 (*Objectives 1.1, 1.2, 1.3, 1.4 and 1.5*)

State, giving reasons where appropriate, whether each of the following statements is true or false:

(a) As seen from the Earth, there is a hierarchy in the Universe, in order of decreasing size of the largest member of each group: galactic superclusters (if they exist); galactic clusters, including the Local Group; the Galaxy; stars with or without orbiting planets, including the Sun/Solar System; individual planets within the Solar System; planetary satellites; asteroids; comets; interplanetary dust and gas.

(b) All the planets except Pluto orbit the Sun in the same direction.

(c) All planetary orbital planes lie within 18° of the ecliptic.

(d) There are three planets with retrograde rotation: Mercury, Venus and Uranus.

(e) All planetary axes lie at angles of at least 10° to their respective orbital planes.

(f) There are two distinct groups of planets — the higher-density inner planets and the lower-density outer planets.

(g) Pluto is a gaseous giant.

(h) Of the terrestrial planets, Mercury seems to be least like the Earth in overall composition.

(i) The surface of Neptune is heavily cratered as a result of meteorite impact.

(j) Venus can be regarded as the Earth's 'sister planet' because its mass and diameter are close to those of the Earth and because in all major respects it appears Earth-like.

(k) The Earth–Moon size ratio is smaller than for any other primary-satellite pair in the Solar System, except for Pluto–Charon.

(l) Most meteorites are probably fragments of asteroids.

(m) All planets undergo self-compression to a greater or lesser extent.

SAQ 2 (*Objective 1.3*)

How long does it take sunlight reflected from Pluto to reach the Earth when Pluto is at its mean distance from the Sun and the Sun–Pluto and Earth–Pluto distances are the same?

SAQ 3 (*Objective 1.3*)

What is the approximate distance in kilometres to the limit of the currently observable Universe?

SAQ 4 (*Objective 1.3*)

Calculate the proportion of the total mass of the planets represented by Jupiter.

SAQ 5 (*Objective 1.4*)

Explain the implications, in terms of the core, of a planet lying well to the left of the line in Figure 1.123 (see answer to ITQ 4). (Note: There is no such planet; this is just an exercise in reasoning.)

SAQ 6 (*Objective 1.2*)

Before moving on to Section 1.3, make a list of the main features of the Solar System that you think should be 'explained' by any hypothesis for the Solar System's origin.

1.3 THE ORIGIN OF THE SOLAR SYSTEM

No one will ever know for certain how the Solar System — and hence the Earth — originated. All that one can aspire to in this context is to devise plausible hypotheses that (a) conform to the known laws of nature and (b) do not conflict with any of the observable phenomena (some of which were listed in the answer to SAQ 6). Many hypotheses for the origin of the Solar System have been put forward in the past, most of which have apparently been consistent with the laws of nature but many of which have been discarded as new, incompatible observations have been made or as new discoveries about the behaviour of materials have rendered the relevant hypotheses implausible, either in whole or in part. What has emerged in their place since the mid-1970s is a consensus view of the general characteristics of Solar System formation, based on the work of planetary scientists around the world. This has come to be known as the **standard model** for the formation of the Sun and planets. However, the model is fairly flexible; alternative routes occur at several points within it and the detail of most of it is vague.

According to the standard model, the Sun formed about 5 000 million years (Ma) ago as a result of gravitational instability in a dense, rotating interstellar molecular cloud. 'Gravitational instability' here simply means that the gravitational (i.e. force of attraction) interactions between particles keeping the cloud in existence were destabilized in some way such that a new, more stable set of interactions evolved — one that included the presence of a star, the Sun. The end result of this process, however, was not just the Sun but also a flattened rotating disc surrounding it and containing 1–2% of the original gas and dust. This disc is termed the **solar nebula**.

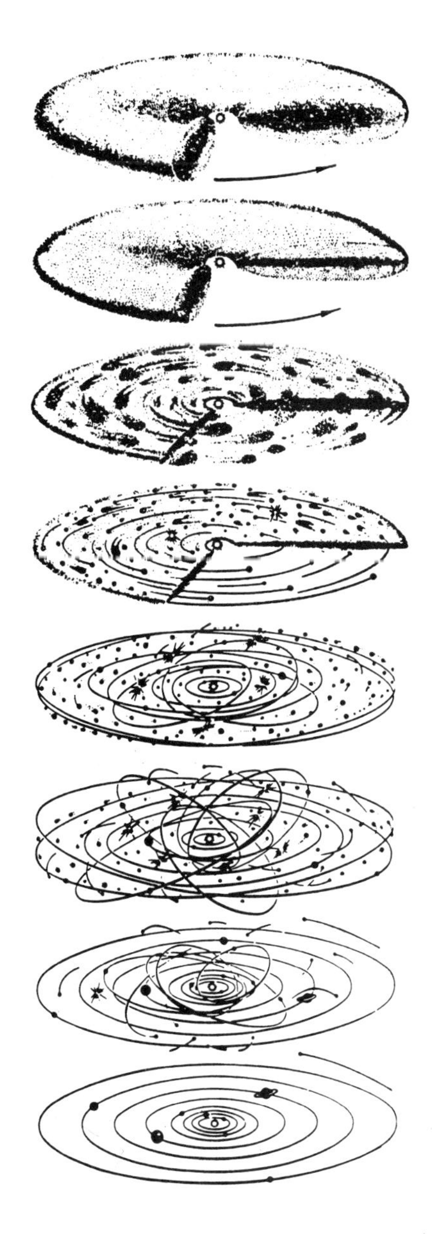

Figure 1.4 A schematic view of the formation of the planets from the dust component of the solar nebula. Because of the onset of gravitational instability, the nebula gradually breaks up to form planetesimals. The planetesimals then grow by collision and coalescence until all that remains are the planets, planetary satellites, asteroids and small pieces of debris. The initially flat disc-like system thickens as the planetesimals grow, because of the gravitational influences these bodies exert on each other; but in the final stages, as the bodies become large and reduce in number, the system becomes almost flat again.

The planets subsequently formed by the coalescence of material within the solar nebula. In the inner part of the embryo Solar System, temperatures evidently became low enough (less than 1 800 K) to permit the condensation of magnesium and iron silicates and metallic iron. The particles thus formed gradually combined with others (and, as cooling of the nebula continued, with other types condensing out at lower temperatures) to build up rocky bodies with diameters of up to about 10 km (mass ~ 10^{15} kg), known as **planetesimals**. As the accretion process continued, the small planetesimals grew and merged gradually to form **planetary embryos** in the mass range 10^{22}–10^{23} kg (i.e. bodies of Moon/Mercury size); and they, in turn, continued to collide and coalesce into fully formed planets. A very schematic view of the transition from solar nebula to planets is shown in Figure 1.4.

A crucial question arises here, namely, how long did the planetary accretion process take? Among the important factors governing the rate at which collisions and mergers took place would have been the density of the original solar nebula, its temperature and the temperature gradients within it, the number density of the subsequently formed planetesimals (i.e. the number of planetesimals per unit volume of space), and the speeds of the planetesimals. Such things are difficult to determine, but estimates are not impossible.

ITQ 5

Assuming that the inner solar nebula (i.e. that corresponding to the terrestrial planets) had a radius of 300×10^6 km and a thickness equal to the diameter of one large planetesimal (10 km), it would have had a volume of 2.8×10^{18} km^3. Estimate the number density (per km^3) of planetesimals of average mass 10^{15} kg within this volume.

Estimates such as that made in ITQ 5 have been used to show that planetesimals could combine into planetary embryos of mass about 10^{23} kg in about 10^5 years, which is remarkably rapid on the geological time-scale (the age of the Earth being 4 550 Ma). By the time the solar nebula had coalesced into planetary embryos, however, there would have been fewer bodies flying around (i.e. the number density would have been much reduced), collisions would have been less frequent, and the rate of growth of the planets would have been lower. It would therefore have taken about 10^7–10^8 years for the planetary embryos to combine into almost fully grown planets.

These timings are crucial in a number of ways, one of which concerns the fate of the gases in the solar nebula. If the 'dust' in the inner nebula gradually accreted into the planets, whatever happened to the gas component? Here's where one of the alternative pathways comes in. Having observed that gas clouds around new stars dissipate very rapidly, some planetary scientists argue that gas in the inner solar nebula must have been lost within about 3×10^6 years, in which case no gas would have been present during the Earth's final stages of formation and the planets would have been left with no major **primordial atmosphere**. (It has been widely agreed for a long time that the Earth's present oxygen-rich (oxidizing) atmosphere is a 'secondary' one, having formed later, possibly by the release of gases from the Earth's interior, albeit modified subsequently by interactions with life forms. By contrast, the primordial atmosphere, if there was one, would have been a hydrogen-rich (reducing) atmosphere attracted gravitationally from the solar nebula.) Whether or not the Earth had a primordial atmosphere is perhaps neither here nor there. The problem is that Jupiter and the other gaseous giants are hydrogen-rich, which means that they must have formed before the solar-nebula gas had been dissipated. This, in turn, means that the rocky cores around which the hydrogen-rich gases accreted must also have been formed by that time. But the rocky cores of Jupiter and the other gaseous giants, which in Section 1.2.1 we described as 'small' and which are indeed small compared with the gas giants' overall diameters, actually have masses at least several times that of the Earth (up to 15 times by one estimate). Thus if we insist that the gas of the solar nebula had disappeared within 3×10^6 years, we are left to explain how and why the cores of the gaseous giants could have been formed in 3×10^6 years, whereas it took the Earth, with a much smaller mass, 10^7–10^8 years to form. Clearly, there is something wrong here, demonstrating (a) that hypotheses for the origin of the Solar System are far from perfect and (b) that the formation of the inner planets cannot be considered separately from that of the outer.

The alternative (minority) view, notwithstanding the contrary evidence from stars, is that the solar-nebula gas was not dissipated until *after* the Earth was fully formed, in which case the planet would have formed entirely in a hydrogen-rich environment and would therefore have gravitationally attracted a hydrogen-rich primordial atmosphere. Calculations show that this atmosphere would have had a mass of about 10^{23} kg, or about 10^5 times that of the present atmosphere, and would thus have exerted a huge drag on the young orbiting Earth. The dynamic implications of this have not been worked out; so it's not yet possible to say whether or not it is consistent with astronomical data. Another consequence of the presence of the gas during the late stages of the Earth's formation, if by that time the solar nebula had cooled to about 300 K, is that the Earth may have ended up with more **volatiles** than it would otherwise have done, having absorbed them from the gaseous nebula. This effect is difficult to quantify, but clearly it would have set up the young Earth with a somewhat different chemical mix. (Volatiles are substances — elements or compounds — that are easily vaporized — i.e. at low temperatures. Clearly, they include those substances that are gaseous at normal Earth-surface temperatures — hydrogen, helium, carbon dioxide, etc. — but they also include such elements as sodium, potassium and lead. The opposite of volatile is **refractory**, referring to substances that have very high vaporization temperatures. A well-known refractory substance, for example, is aluminium oxide, Al_2O_3, which is so refractory that it is used to line blast furnaces. Unfortunately, Earth scientists often use the terms 'volatile' and 'refractory' rather loosely, and it is not always clear which particular substances they are talking about. In fact, there are no clear-cut divisions; substances range from 'very refractory' to 'very volatile' and some of those in between are more volatile/refractory than others. Moreover, the same substance can change its volatility as conditions (e.g. temperature, the chemical environment) change. As we shall need to refer to volatile and refractory materials later in the Course, we are providing Figure 1.5 as a guide to the elements at least.)

1 H																	2 He
3 Li	4 Be											5 B	6 C	7 N	8 O	9 F	10 Ne
11 Na	12 Mg											13 Al	14 Si	15 P	16 S	17 Cl	18 Ar
19 K	20 Ca	21 Sc	22 Ti	23 V	24 Cr	25 Mn	26 Fe	27 Co	28 Ni	29 Cu	30 Zn	31 Ga	32 Ge	33 As	34 Se	35 Br	36 Kr
37 Rb	38 Sr	39 Y	40 Zr	41 Nb	42 Mo	*43 Tc*	44 Ru	45 Rh	46 Pd	47 Ag	48 Cd	49 In	50 Sn	51 Sb	52 Te	53 I	54 Xe
55 Cs	56 Ba	*57 La	72 Hf	73 Ta	74 W	75 Re	76 Os	77 Ir	78 Pt	79 Au	80 Hg	81 Tl	82 Pb	83 Bi	*84 Po*	*85 At*	*86 Rn*
87 Fr	*88 Ra*	†*89 Ac*	*104*	*105*	*106*	*107*	*108*	*109*	*110*	*111*	*112*	*113*	*114*	*115*	*116*	*117*	*118*

*	58 Ce	59 Pr	60 Nd	*61 Pm*	62 Sm	63 Eu	64 Gd	65 Tb	66 Dy	67 Ho	68 Er	69 Tm	70 Yb	71 Lu
†	90 Th	*91 Pa*	92 U	*93 Np*	*94 Pu*	*95 Am*	*96 Cm*	*97 Bk*	*98 Cf*	*99 Es*	*100 Fm*	*101 Md*	*102 No*	*103 Lw*

refractories

volatiles 1300–600 K

volatiles < 600 K

Figure 1.5 The periodic table of the elements, showing which elements are refractory and which are volatile. Those elements marked with an asterisk (La and the upper row of the lower block) are known as the lanthanides or, more commonly these days, rare-earth elements (REEs for short); those marked with a dagger (Ac and the lower row of the lower block) are known as the actinides. Those elements in italics are short-lived radioactive elements. You do not need to understand the whole of this diagram at this stage; just concentrate on those elements marked as volatiles (two temperature ranges) and those marked as refractories. As a general working rule, the boundary between refractories and volatiles can be taken as 1 300 K (vaporization/condensation temperature). The number against each element is the atomic number, which equals the number of protons in the nucleus.

There is an important change in perspective embodied in the standard model (irrespective of the gas problem) that needs to be emphasized, because it greatly affects the initial state of the planet on which, 4 550 Ma later, we now live. The present standard model was not the first to involve planetesimals as the building-blocks of planets. On the contrary, planetesimals had been involved for decades, if not longer. However, previous hypotheses do not appear to have admitted the possibility of planetesimal growth, apart, that is, from the five destined to become the terrestrial planets and the Moon. It was almost as if particular planetesimals had been designated 'Earth', 'Venus', etc. right from the start and had then grown by attracting other planetesimals, the implication being that throughout their growing lives (even in the final stages) they were still absorbing planetesimals (i.e. small bodies up to about 10 km in diameter). The new standard model rejects that as being unlikely. It claims, apparently more reasonably, that while the bodies that were to become the planets were growing, so were the other planetesimals within the inner Solar System. The position at the planetary-embryo stage of growth was therefore not one of five largish terrestrial bodies in a field of remaining planetesimals but many planetary embryos surrounded by comparatively few, if any, planetesimals. Precisely which embryos were to become Earth, Venus, Mars, etc. would not have been clear until a comparatively late stage — until some sort of dominance had been achieved (apparently at about 5×10^6 years).

ITQ 6

Estimate how many planetary embryos of mass 10^{23} kg each it would have taken to form the four terrestrial planets.

The formation of the Earth must have been a violent affair by any standards; but as long as it was envisaged in terms of millions of incoming planetesimals, the end result, the Earth, could have been regarded as, if not the product of a 'gentle rain', at least as that of some sort of regular, continuous process. The standard model postulates something very different, however. It sees each of the planetary embryos (mass range 10^{22}–10^{23} kg) as a body still due to receive several tens of giant impacts at highly irregular intervals. The implications of this are profound, both for the initial state of the Earth about to be sent on its evolutionary course and for the way in which we must think about the origin of our planet, which now appears to have had more randomness about it than has hitherto been envisaged.

Before looking at some of these implications, we should perhaps make it clear that the standard model represents a consensus but is not universally accepted. Perhaps the most important rival hypothesis is that which claims that the first product of the solar nebula was, besides the Sun, a series of 'giant gaseous protoplanets' each of which then condensed into an individual planet. In any case, the standard model should not be taken as representing anything resembling certainty. Understanding of the steps within it is rudimentary at best, even for the terrestrial planets; for the gaseous ones it's practically non-existent. On the other hand, there is evidence. The young star β Pictoris, for example, has just the sort of flattened nebular disc postulated for the Sun 5 000 Ma ago. In 1990, George W. Wetherill, a major contributor to Solar-System studies over the years wrote: 'At present, when compared with what needs to be known, the [standard] model tells us rather little that can be considered at all certain. Some specific problems have been explored rather thoroughly and quantitatively, but most are in the qualitative "scenario" stage. At present the model can be regarded as a very useful working hypothesis, one that can provide a focus for workers of varying backgrounds and experience, facilitate communication between them, and permit a rational division of labour, rather than requiring everyone to make up a personal grand model. Only time will tell if it will succeed as a unifying theory.'

1.3.1 GIANT IMPACTS

If the standard model for the origin of the Solar System is accepted, the terrestrial planets would, during their later stages of formation, have been subjected to impacts by bodies as massive as 10^{23} kg. Earth scientists have not been slow to invoke such giant impacts to explain a number of phenomena that have been difficult (but sometimes not impossible) to account for otherwise, among them the following:

(1) As we noted in Sections 1.2.1 and 1.2.2, Mercury appears to have an anomalously large iron core in relation to its diameter. Or to put it the other way round, it appears to have an anomalously small silicate mantle. Given the occurrence of giant impacts, it is not implausible to suggest that part of the planet's mantle could have been vaporized away by just such an impact. Certainly, energy calculations show that to be possible.

(2) One of the longest-standing and most vexing problems in the planetary sciences has been that of the origin of the Moon. Hypotheses put forward over the years have envisaged the Moon as a passing body captured by the Earth's gravity field, as part of a two-body Earth–Moon

system generated from the solar nebula by whatever process that produced all the terrestrial planets, and as material thrown off from the Earth by rotation at a time during the Earth's formation when the rotation rate was perhaps much higher than it is now. However, all of these hypotheses have failed because they cannot explain the dynamic (orbital and rotational) characteristics of the Earth–Moon system. It would appear that this major difficulty can now be resolved by hypothesizing that a Mars-size impactor hit the Earth during the later stages of its formation, vaporizing material from the Earth's mantle, which then rose into Earth orbit and subsequently condensed to form the Moon (see Figure 1.6). But while this solves the dynamic problem, it may pose geochemical difficulties. You will recall that we pointed out in Section 1.2.2 that the answer to ITQ 4 had painted an over-rosy picture of the Moon's bulk geochemical similarity to the Earth and Venus and that, in particular, the Moon was relatively deficient in iron. In fact, there are also some other geochemical dissimilarities.

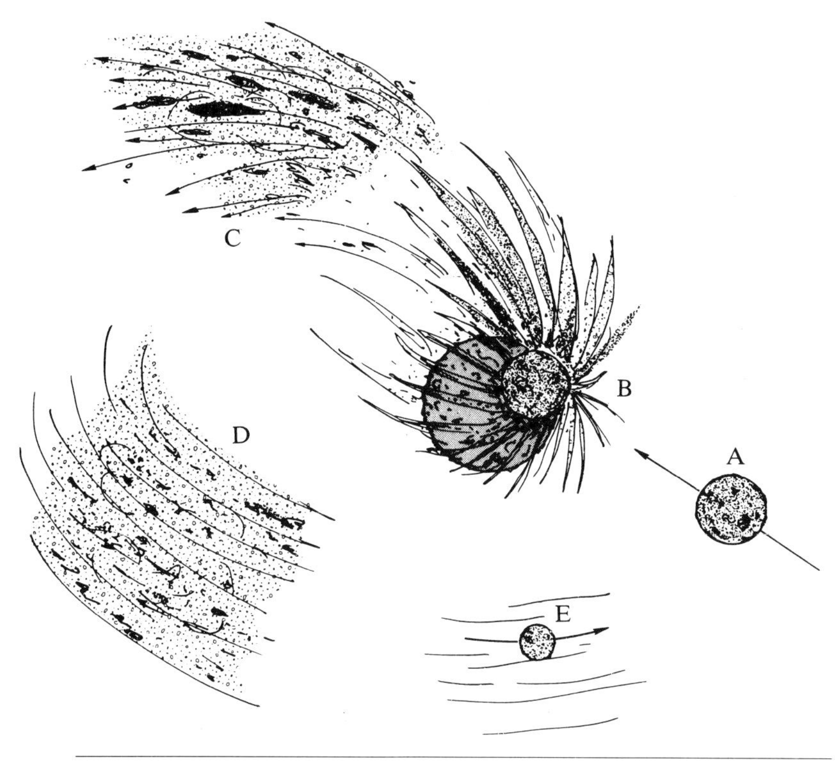

Figure 1.6 Cartoon illustrating the giant-impact hypothesis for formation of the Moon. A Mars-sized body (A) collides with the growing Earth (B), ejecting debris (C) which enters Earth-orbit (D), where it later re-accumulates into the Moon (E).

ITQ 7

Table 1.2 contains some comparisons between selected volatile and refractory elements in terrestrial and lunar basalts. Use the figures there in conjunction with Figure 1.5 to draw two conclusions about geochemical differences between the Moon and the Earth.

Such differences are hard to explain on the basis of a single giant impact. As the impactor would also be partially or completely vaporized on impact, it might be thought that the resulting Moon would contain material both from the Earth's mantle and the impactor (as, indeed, it probably does) and thus have a composition reflecting a mixture of the two. However, this is unlikely to be an answer, for the impactor, having come from the same region of the solar nebula as the Earth, would presumably have a composition not much different from it. One possible solution is that the giant impact may only have resulted in a less-than-full-sized Moon, which would then have gone on to attract to itself, and combine with, debris from a wider scavenging zone. There are other possible solutions. Given the difficulty of being sure about the compositions of planetary interiors, Earth scientists seem to think that the geochemical

Table 1.2 The relative abundances of selected volatile and refractory elements in terrestrial and lunar basalts.

Element	Ratio of abundances Moon : Earth	
bismuth (Bi)	11.5×10^{-3}	V
zinc (Zn)	8.5×10^{-3}	V
titanium (Ti)	2.0	R
rubidium (Rb)	3.5×10^{-2}	V
indium (In)	3.8×10^{-2}	V
potassium (K)	6.5×10^{-2}	V
thorium (Th)	4.2	R
sulphur (S)	4.4	V
germanium (Ge)	6.9×10^{-2}	V
uranium (U)	5.0	R
sodium (Na)	8.0×10^{-2}	V
iridium (Ir)	1.1×10^{-1}	R
lead (Pb)	9.0×10^{-2}	V
copper (Cu)	1.1×10^{-1}	V
barium (Ba)	6.0	R
gallium (Ga)	3.0×10^{-1}	V

problem with the giant impact hypothesis of the Moon's origin is less serious than the dynamic one. Or maybe they're so astonished and overjoyed to have solved the dynamic problem after several centuries that they are underestimating the geochemical one. Time may tell.

(3) Another long-standing mystery of the Solar System is why all the planetary equators are tilted away from their respective orbital planes at diverse angles (see Section 1.2 and Table 1.1). There seems to be no way that this situation could arise from orderly condensation from the solar nebula.

❑ In this respect the question of equatorial tilts differs from that of orbital direction. Can you see why? Or to be more specific, can you see how the fact that the planets all orbit the Sun in the same direction is 'explained' by the standard model?

■ The standard model begins with a solar nebula in the form of a rotating flattened disc. The direction in which the planets now orbit the Sun is simply a legacy of the direction in which the nebula was rotating in the first place. There is no such antecedent for the tilt directions.

An explanation of the tilts follows naturally from the giant impact hypothesis. The tilts are simply the net, random result of the planets' having experienced one or more giant impacts which knocked them out of 'true'.

(4) You will recall from Section 1.3 that the Earth may or may not have had a primordial hydrogen-rich atmosphere, depending on whether or not the gas of the solar nebula was present during the Earth's late stages of formation. However, if the Earth did have such an atmosphere, it becomes necessary to explain what happened to it, given that it must have been lost before the present oxygen-rich atmosphere appeared. It could have been dissipated by a giant impact, which could easily have provided the required amount of energy for the job.

(5) A giant impact could equally easily have melted, or even vaporized, all or part of the Earth's mantle, just as it appears to have vaporized part of Mercury's. This would have been an important event in the Earth's early history, having an influence on today's planet. We shall therefore look at it more closely in Section 1.3.3.

1.3.2 THE DIFFERENTIATION OF THE EARTH

Most hypotheses for the origin of the Solar System assume that, irrespective of the precise processes involved, the planets were derived from the mixture of materials in the solar nebula. The end result, at least in the case of the Earth, is a planet divided into crust, mantle and core (see Figure 1.3). The question inevitably arises, therefore, as to how and when the **differentiation** into distinct layers took place. There are two conflicting hypotheses about this, known, respectively, as the homogeneous and heterogeneous accretion models. To understand them, however, it's useful to know a little about what would have happened as the solar nebula condensed.

The pressure in the solar nebula would have decreased with radial distance from the Sun and is thought to have been very low in the region in which the Earth was to form, perhaps about $10\,N\,m^{-2}$ (i.e. about 10^{-4} of current atmospheric pressure at the Earth's surface). At such a pressure, elements and compounds would vaporize (and hence, with decreasing temperature, condense) at lower temperatures than they would at the present Earth's surface. The refractory compound Al_2O_3, for example, has a vaporization/condensation temperature under current Earth-surface conditions of about 3 500 K, but at $10\,N\,m^{-2}$ and in the presence of other solar-nebula material it would have been below 1 700 K. All the material in the solar nebula would have been in vapour form (i.e. the nebula would have been entirely gaseous) had the temperature there been in excess of about 1 900 K. The most recent estimates, however, suggest that the original temperature was more like 1 500 K, at which level some of the more refractory materials would already have condensed to form tiny dust particles (less than 0.1 μm in size) that were destined to become the basic building blocks of the terrestrial planets and the cores of the gaseous giants. Incidentally, the dust component of the solar nebula would only have been about 2% by mass, the rest being gas (chiefly hydrogen and helium).

The order in which the various materials would have condensed out of the solar nebula as temperature fell with time is illustrated in Figure 1.7. At first sight this diagram may appear formidably complicated, but if you examine it and its caption carefully it should soon become clear. The very first materials to condense out of the solar nebula would have been highly refractory (but not very abundant) metals such as osmium (Os), tungsten (W), zirconium (Zr) and rhenium (Re). Below about 1 750 K, there would have followed, rather more abundantly, the only slightly less refractory Al_2O_3, the platinum-group metals (platinum, Pt; rhodium, Rh; palladium, Pd; and iridium, Ir), the REEs (see caption to Figure 1.5), uranium (U), thorium (Th) and, later, the silicates and titanates of calcium (Ca), aluminium (Al) and magnesium (Mg). Then at about 1 400 K, metallic iron would have started to appear, to be followed, with only a small temperature drop, by the bulk of the silicate material and, later, elements such as copper (Cu), silver (Ag) and zinc (Zn). By this time some of the less volatile volatiles would have appeared, reaching FeS (iron sulphide) at about 750 K, which would have been closely followed by FeO (one of the oxides of iron). Still later, off the bottom of the diagram (below 400 K), would have come ammonia (NH_3), methane (CH_4) and water. The temperature would never have fallen low enough to condense out such gases as hydrogen and helium.

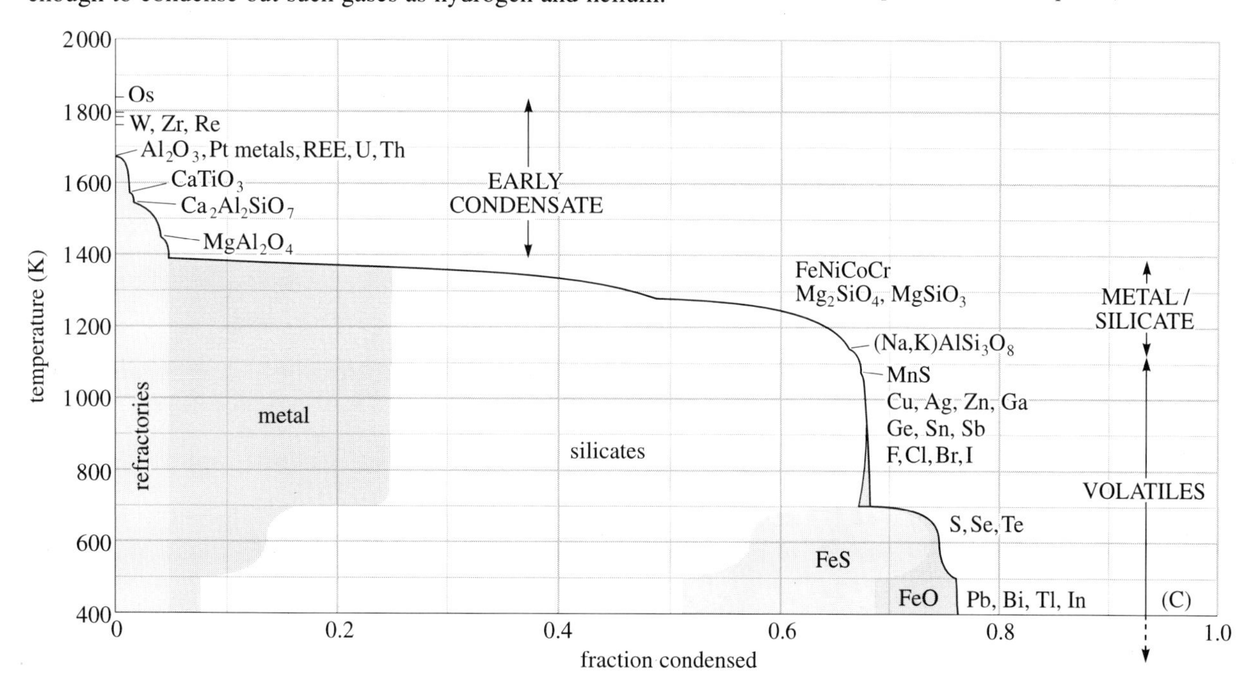

Figure 1.7 The condensation sequence of the solar nebula at a pressure of about $10\,N\,m^{-2}$. Temperature is on the left-hand axis; so moving down that axis represents a fall in temperature. The order in which the materials condense can be read off the diagram from the top downwards, starting at the top left-hand corner. The horizontal axis (see top) simply represents the fraction of the material that has condensed. Thus when, for example, 0.4 (40%) of the solar-nebula materials has condensed, the growing planet will comprise those materials to the left of a vertical line through 0.4. (Note: Under the conditions of pressure and temperature in the solar nebula, materials would have condensed direct from the vapour phase to the solid phase.)

Homogeneous accretion model

This assumes from the start that all the above condensation processes took place before the formation of the planets began. Immediately before the accretion process started, therefore, the solar nebula — or at least the inner part of it — would have been populated by small, cold fragments of rock, each containing one, some, or all of the materials shown in Figure 1.7. These fragments would then have gradually coalesced in random arrangements to form bodies in which the full complement of refractories and volatiles would be distributed throughout their whole volumes, *assuming, that is, that the bodies remained cold.* Thus the Earth, in particular, would have ended up fully assembled as a chemically uniform, or homogeneous, body with no segregation into crust, mantle and core.

It follows from this that the differentiation of the Earth would have taken place after the planet's initial assembly. But how? And when? As to the how, the story generally put about since the 1960s is that large 'blobs' of liquid iron (perhaps hundreds of kilometres in size) must have accumulated within the largely silicate (solid) matrix and then have gradually sunk towards the centre of the Earth to produce the core. The blobs would have sunk because iron is denser than silicates and they would have been *able* to sink through the solid silicates because the latter were deformable. However, analyses published as recently as 1990 clearly demonstrate that liquid iron cannot percolate through a solid silicate matrix to enable blobs to form in the first place, thus scotching a hypothesis in which many Earth scientists have had faith (in general terms at least) for several decades.

But the failure of the **blob model** may not matter, because its environmental context — liquid iron attempting to percolate through solid silicates — was evidently wrong anyway. As we shall see in Section 1.3.3, an accreting Earth could not have remained cold but would have been molten or partially molten. For all or much of its growing life, therefore, the Earth would have comprised an emulsion of liquid iron and *liquid* silicates. (An emulsion is a mixture in which one substance is suspended in another.) Analyses suggest that iron blobs would not have formed under these conditions either, but that the iron would have accumulated into clouds of droplets, each of the latter having a dimension of about 1 cm. These droplets would then have rained down towards the centre of the Earth (again, because iron is denser than silicates) to form the core. This **rainout model** is illustrated in Figure 1.8. The process would have been extremely rapid; the droplets fall so fast that they would combine to form the core in only a few tens of thousands of years.

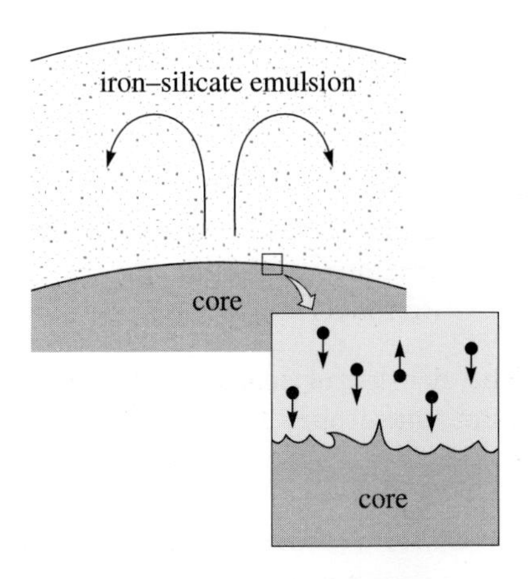

Figure 1.8 The rainout hypothesis for the formation of the Earth's core, in which droplets of iron fall towards the centre of the Earth from an iron/silicate emulsion produced by melting. The theory predicts that the rainout will proceed even if the emulsion is convecting (curved arrows). During core growth, the surface of the core becomes agitated as the droplets impinge upon it, producing 'waves' and even throwing a few iron droplets upwards ('surf').

❑ What conclusion would you draw from this about the timing of core formation?

■ The time-scale for the descent of the iron droplets is so short that core formation must have begun long before the Earth was fully formed. On the other hand, the core itself couldn't have fully formed until all the iron due to reach it via impacting bodies had arrived. So core formation must have been going on from the time that the Earth became molten enough until it reached its full planetary size.

Prior to 1990, it had been widely agreed that the core must have formed early in the Earth's life, but it wasn't usually envisaged as having formed simultaneously with the accretion of the Earth, at least in the context of the homogeneous accretion model.

Heterogeneous accretion model

The broad concept of the homogeneous accretion model (not including the rainout hypothesis) was devised long before the standard model of the formation of the Solar System, but clearly it is consistent both with the standard model (with its giant impacts) and with the earlier impact model involving only planetesimals. The heterogeneous accretion model, on the other hand, is more applicable to the Solar System formation hypothesis mentioned briefly at the end of Section 1.3 — i.e. the one that assumes that the solar nebula separated into protoplanets *before* cooling and condensing.

The heterogeneous accretion model assumes that the Earth went through the condensation sequence as a unit rather than having been constructed from smaller bodies that had already resulted from condensation; in other words, the Earth formed during the cooling and condensation of the solar nebula.

❑ Can you envisage what the consequences of this would be?

■ The Earth would have been built up layer by layer in the order in which materials condense from the solar nebula; differentiation into crust, mantle and core would be concurrent with accretion.

As Figure 1.7 shows, the first major set of materials to condense (i.e. apart from the refractory elements such as Os, W, Zr and Re) would have been the refractory oxides, silicates and titanates of Ca and Al, the Pt group metals, the REEs, and so on. The Earth would therefore have had a refractory core. Following the scheme of Figure 1.7, the next layer would have been of iron, and then would have come the layer of major silicates. The volatiles would have condensed last.

❑ There is a problem here right away. Have you spotted what it is?

■ The Earth actually has an iron core, not a refractory one. There therefore has to be an additional mechanism by which the refractory core predicted by the heterogeneous accretion model rises above the iron layer.

The density of many of the materials in the primary refractory core would have been lower than that of iron; so, in principle, the materials could have risen through the iron layer. However, it would have been rather a traumatic event in the early history of the growing planet. There are also several other less-than-desirable difficulties with heterogeneous accretion, one of them being that of the condensation temperatures of metallic iron and the major silicates (centre block of Figure 1.7).

❑ Can you deduce what that difficulty is?

■ The silicates begin to condense out of the solar nebula at only a slightly lower temperature than does the iron. The difference is so small, in fact, that the two materials would probably have condensed out more or less together, forming an iron–silicate mixture rather than two distinct layers.

All in all, the heterogeneous accretion model is as much out of fashion as the planetary formation model with which it is closely associated, but that doesn't mean that either is necessarily wrong.

1.3.3 Heat in the Earth

Every geological process taking place on or within the Earth involves the transfer of energy; and most of that energy is eventually discharged into space as heat. In this sense, therefore, the Earth can be envisaged as a huge thermal ball from which heat is forever leaking away. But that makes geothermal heat sound altogether too passive, as a by-product of geological processes almost. In fact, the reverse is true. The Earth's internal heat is the prime mover. It is the processes by which heat escapes from the Earth that *cause* most of the major geological phenomena (e.g. plate tectonics); and in so far as these phenomena involve movement of rock, the Earth can be regarded as a heat engine (i.e. a 'machine' that converts heat into mechanical work). Of course, internal heat is not the Earth's only prime mover. Certain geological processes at the surface (e.g. the weathering of rocks) are caused largely by external energy (from the Sun); some internal processes (the generation of the Earth's magnetic field in the core is one possibility) are motivated, or at least modified, by energy provided by the Earth's rotation; and, in the past at least, geological landforms have been produced by the energetic impact of large meteorites. But some other planetary bodies rotate, have been subjected to meteorite impact and have their surfaces modified by solar energy, and yet are geologically dead, or, at least, far less active than the Earth. As Plate 1.4 vividly illustrates, the Earth, unlike any other terrestrial planet, is a churning mass — and it is churning because of heat.

To the extent that almost everything we perceive about the Earth — including its very structure — is a consequence of heat, heat must be regarded as the Earth's most fundamental property. Unfortunately, however, it's very difficult to pin down with numbers. As you will see later in this Course, some characteristics of geothermal heat can be measured quite accurately (e.g. the rate of heat flow outward through the Earth's surface) and some can be plausibly estimated (e.g. the amount of heat currently being generated within the Earth by the decay of radioactive elements); but determining how much heat there is in the Earth, how it is distributed and redistributed and where it all comes from is extremely difficult. It is even hard to discover how temperature varies with depth throughout the Earth now — a problem not made any easier by the fact that some of Earth's current heat content was almost certainly generated when the Earth was formed, by processes barely understood.

The thermal properties of rocks are such that any heat present in the Earth at its formation could only have been conducted a distance of a few hundred kilometres throughout the planet's whole lifetime. (**Conduction** is the process by which heat in a material is transferred from a region of higher temperature to a region of lower temperature by the interaction of atoms and molecules within the material.)

❑ Is conduction likely to be the only process by which heat in the Earth is transferred?

■ No. As Plate 1.4 shows, there are mass movements of material in the Earth, and the moving masses carry heat with them. This process of heat transfer is called **convection**.

Convection is very much faster than conduction in transferring heat. So is it possible that convection throughout the Earth's lifetime could have effectively transferred to space all the heat, if any, present at the Earth's formation? No one knows the answer to that, but geophysicists suspect not. If they are right, it means that the thermal characteristics of the Earth today must be, in part, the result of heat generated long in the past; and that, in turn, directs attention to the means by which the Earth acquired heat at its formation.

During the 19th century, it was widely assumed that the Earth had formed as a molten ball and had been cooling ever since. In other words, the Earth possessed **original heat**. With the standard model, which envisages planetary bodies as having accreted from what were originally cold rock fragments (i.e. they had no original heat), you might suppose that by the time the Earth fully formed it was as cold as its constituent bodies. But this would be quite wrong. During its 10^7–10^8 years of growth the Earth would have become very hot indeed, having acquired heat from a number of different sources and thus having **initial heat** with which to begin its subsequent evolution. Indeed, the problem soon becomes not one of explaining how a coldly accreting Earth could end up with heat but one of explaining how the Earth managed to avoid being vaporized by the huge amount of heat that was produced.

There are various processes by which a body accreting from cold fragments will become hot. Unfortunately, in trying to discuss these processes in the context of an Earth formed by the standard model, we face an insuperable difficulty. If the Earth was a product of giant impacts, its formation would have been so violent and unpredictable, especially in the later stages, that it would be impossible to describe in terms of known thermal processes. We therefore propose to cheat a little. Purely for the sake of argument, and to enable us to distinguish the various heat-generating processes involved, we shall consider an Earth that accreted from planetesimal-sized bodies. We shall then return to the implications of giant impacts.

Heat of accretion

Any planetesimal or larger body falling towards the Earth will acquire a velocity directed towards the centre of the Earth because of the gravitational attraction on the body by the Earth. The body will thus have **kinetic energy** (the energy possessed by a body by virtue of its motion). The kinetic energy in this case is a type of **gravitational energy**, so called because it arises from the effect of gravity. If we take the velocity of the body immediately before impacting the Earth as v, the body's kinetic energy (E) at that point will be given by

$$E = \tfrac{1}{2}mv^2$$

where m is the mass of the body.

If we now suppose that the body of mass m impacts with the Earth (mass M) and somehow gets embedded in it, and if we further suppose that *all* the body's kinetic energy is converted into heat, we can write

$$\tfrac{1}{2}mv^2 = \Delta T(m + M)C \qquad \text{(Equation 1.1)}$$

where ΔT is the temperature rise produced in the combined body of mass $(m + M)$ and C is the **specific heat capacity** of Earth material (i.e. the amount of heat required to raise the temperature of 1 kg of the material through 1 K). The right-hand side of Equation 1.1 is the heat content added to the combined body by the impact. For an impact by a planetesimal (i.e. a body much smaller than the Earth), $(m + M)$ will be effectively equal to M, so we can write

$$\tfrac{1}{2}mv^2 = \Delta T\, MC \qquad \text{(Equation 1.2)}$$

or

$$\Delta T = \frac{mv^2}{2MC}. \qquad \text{(Equation 1.3)}$$

S5

Note that in these four equations, for SI units, m and M are in kg, v is in $\mathrm{m\,s^{-1}}$, E is in joules (J), ΔT is in K and C is in $\mathrm{J\,kg^{-1}\,K^{-1}}$.

ITQ 8

S6

A planetesimal of mass 10^{15} kg impacts the Earth with a velocity of $10\,000\,\mathrm{m\,s^{-1}}$. Calculate the rise in temperature in the Earth assuming that the heat generated by the impact spreads rapidly and uniformly throughout the whole Earth. (Take the average specific heat capacity of Earth material to be $750\,\mathrm{J\,kg^{-1}K^{-1}}$.)

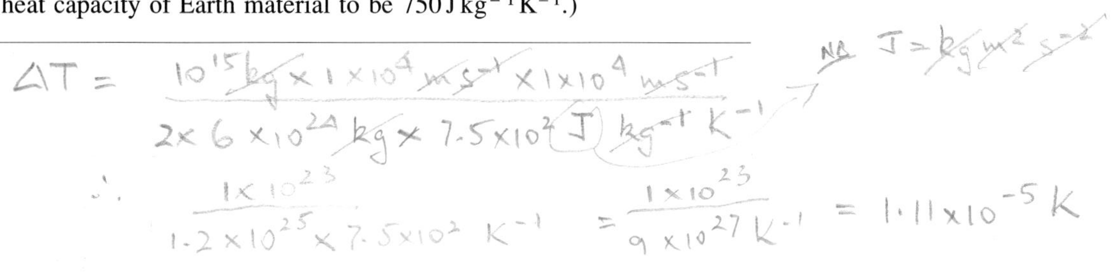

ITQ 9

Suppose that the Earth were constructed entirely of 10^{15} kg planetesimals, each of which generated the temperature rise obtained in ITQ 8. What would the total temperature rise be?

The calculations in ITQs 8 and 9 are, of course, based on quite unrealistic assumptions. For one thing, the heat generated by a planetesimal impacting the Earth would not heat the Earth throughout but only a small region around the impact site. Moreover, even if a large number of planetesimals were to hit the surface in a uniform distribution, they would not generate heat beyond a certain depth. Secondly, not all the kinetic energy of a planetesimal would be converted into heat in the first place. Some, for example, would be used to excavate a crater (by ejecting rock fragments which, by the time they had fallen back to the Earth's surface, would have cooled). In fact, the best estimates suggest that only 5–15% of the original kinetic energy of a planetesimal would ultimately be retained by the Earth in the form of heat.

The various uncertainties involved make it very difficult to determine exactly how much energy would have been acquired by the Earth during accretion. Nevertheless, by making estimates of the masses and velocities of Earth-accreting planetesimals, one geophysicist has concluded that the amount of gravitational energy imparted to the growing Earth would have been about 10^{32} J.

ITQ 10

Calculate the temperature rise that the proportion of this energy retained by the Earth as heat (say, 10%) would produce in an Earth of present size, again assuming rapid transmission throughout the Earth and assuming that all the energy is converted into heat.

The energy imparted by impactors would have been sufficient to raise the temperature of the Earth well above the vaporization temperatures of its constituents, and hence, one might imagine, cause the Earth to evaporate back into space as rapidly as it tried to form. As the Earth is here now, however, that clearly didn't happen. One reason is that vaporization temperature increases with pressure. For materials deep in the early Earth, therefore, vaporization temperatures may *not* have been exceeded as the result of heat generated by giant impactors. Moreover, some (perhaps most) of the heat of impact would have been radiated into space as fast as it was released. Even so, the energy available from impactors would have been so great that there is a possibility that *some* of the Earth would have been vaporized, although much of the vapour thus produced may have chilled in the near-Earth environment and, being still gravitationally bound to the Earth, would have re-accreted.

Heat of compression

You will recall from Section 1.2.2 that as a planet grows, it undergoes self-compression. This process, too, converts gravitational energy into heat. (That compression generates heat can easily be demonstrated by feeling the valve of a newly pumped-up bicycle tyre.) The amount of heat generated by compression in the Earth is difficult to calculate for the same reason that it is difficult to calculate the effect of compression on density, but one widely quoted estimate is 2.5×10^{32} J. This poses a more serious problem for the possible evaporative fate of the early Earth than does the heat of accretion, and not only because the energy is greater.

❑ Can you see why?

- Because whereas accretional heat caused by planetesimals is generated near the Earth's surface, from which it can easily be radiated into space, the heat of compression is buried deeply, where the compression is greatest and from where is it difficult to escape.

The heat of compression could in theory have raised the temperature of the Earth by tens of thousands of degrees (40 000 K by one estimate).

Core formation

If the Earth formed by homogeneous accretion, the core must have formed as the result of the descent of iron towards the Earth's centre. The gravitational energy lost by iron in falling towards the centre would, according to one estimate, have generated about 10^{31} J of heat. This is small by accretional or compressional standards, but would nevertheless have been enough by itself to raise the temperature of the Earth by about 1 500 K. Moreover, the heat released by core formation would, like the heat of compression, have been deeply buried.

Short-lived radioactive isotopes

Among the materials going into the newly forming Earth would almost certainly have been short-lived radioactive isotopes, which, like all radioactive isotopes, decay to stable isotopes, emitting heat as they do so. The **half-life** of a radioactive isotope is the time it takes for half of any quantity of it to decay; and by 'short-lived' here we mean radioactive isotopes with half-lives of less than about a million years. Short-lived radioactive isotopes would therefore have been effective heat producers for several million years just as the Earth was accreting.

A number of short-lived radioactive isotopes would have been incorporated into the Earth, including ^{26}Al (aluminium), ^{36}Cl (chlorine), ^{60}Fe (iron), ^{244}Pu (plutonium) and ^{129}I (iodine). Of these, ^{26}Al would have been the most important from the heat production point of view, having a half-life of 0.88 Ma as compared with less that 0.5 Ma for each of the others (which would therefore have been effective heat sources for a much shorter period of time). Estimating the heat likely to have been produced by ^{26}Al before it decayed away completely is difficult, but even crude attempts made it clear that total heat production comparable to that from accretion (i.e. of the order of 10^{32} J) can by no means be ruled out.

Long-lived radioactive isotopes

Some radioactive isotopes have such long half-lives (e.g. ^{87}Rb, rubidium, at 48.8×10^{3} Ma) and have such low abundances in the Earth that they are not significant heat generators. Others, however, cannot be so neglected. In fact, there are four thermally important long-lived radioactive isotopes — ^{235}U, ^{238}U (uranium), ^{232}Th (thorium) and ^{40}K (potassium). As these are important heat producers *today*, we shall be looking at them in more detail later in the Course. The point here is to ask whether they could have produced a significant quantity of heat during the 10^{7}–10^{8} years it took the Earth to accrete, especially as their abundances would have been much higher then than now. In fact, fairly plausible calculations suggest that they would have generated only 10^{28}–10^{29} J, which is small compared to the amount of heat from other sources.

Tidal dissipation

It is well known that the Moon and Sun interact gravitationally with the Earth to generate tides in the Earth's oceans. It is perhaps less well known that tides, albeit much smaller ones, are produced in the same way in the solid Earth (and Moon). The effect of both solid and fluid tides is to slow down the Earth's rotation and increase the distance from the Earth to the Moon, both of which processes have been going on since the Earth–Moon system formed. The energy lost from the Earth's rotation is dissipated through friction by the tidal motions, and thus ends up as heat capable of adding to the Earth's heat store. The current rate of loss of rotational energy is put by one estimate at about 10^{25} J per million years, which is small compared to the heat generated by the Earth's accretion, etc. However, in the past it would have been much higher, because the Moon would have been much closer to the Earth. Unfortunately, it's impossible to say how much higher.

Two broad conclusions may be drawn from all this. The first is that we have only a very crude idea of just how hot the Earth could have become during its formation, partly because we don't know the details of the processes involved and partly because, even if we did, it would be impossible to put figures on them all. The second conclusion is that, notwithstanding the crudeness of the thermal estimates, there seems to be a fairly firm indication that there would have been an embarrassingly large amount of heat available — enough, indeed, to melt the Earth, and perhaps even vaporize it, several times over.

Under these circumstances it is difficult to imagine that anyone could have estimated temperature distributions in the growing Earth, but several people have tried by one means or another. One recent (1989) example is shown in Figure 1.9, although this can be little more than educated guesswork. There are two basic constraints on the temperature distribution in the growing Earth: (a) that the temperature could not have been so high that the Earth was vaporized out of existence, and (b) that, given the large amount of heat apparently available, the temperature could not have been much lower than that required to satisfy (a). The temperature distributions in Figure 1.9 therefore echo these constraints.

The evidently high temperatures within the Earth during its later stages of formation have led some scientists to suggest that during this period the outer layer of the planet (perhaps to a depth of several hundred kilometres) was permanently molten, resulting in what has come to be known as a **magma ocean** (magma is molten rock). This is still a matter of active debate. Although some geophysicists argue that the high temperatures dictate the presence of a magma ocean, others claim that they need not do so. At the same time, there are geochemists who argue that, because a magma ocean would have left geochemical imprints that aren't actually observed today, there couldn't have been one; but there are other geochemists who claim that such imprints wouldn't necessarily have been produced, so their absence proves nothing. The issue has not yet been resolved.

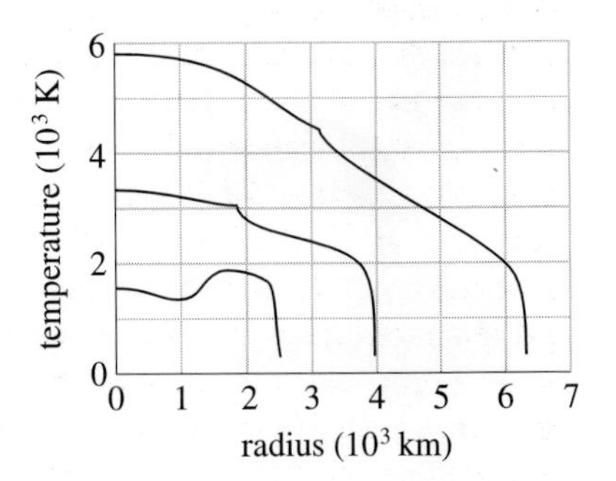

Figure 1.9 Schematic temperature distributions in the Earth at three stages in the planet's accretion — i.e. when the radius is about 2 500 km, 4 000 km and 6 371 km (Earth fully formed). Zero radius on the horizontal axis represents the centre of the Earth.

What everyone does agree on, however, is that if there had been one or more giant impacts late in the Earth's growth, there would certainly have been a magma ocean and perhaps even a completely molten mantle — which brings us to a phenomenon not yet considered in this Section. Figure 1.9 is a result of thinking of the accretion of the Earth as an ordered, albeit violent, process — i.e. as gradual growth through the acquisition of planetesimals. But it would be meaningless if giant impacts took place, for to accept the concept of giant impacts is inevitably to accept that any conclusions based on ordered growth are likely to be invalid. A single giant impact would have imparted to the Earth so much energy that it would have violently changed the planet's thermal characteristics in one random, catastrophic stroke. Scientists hate this sort of thing because it invalidates their carefully constructed hypotheses based on orderly progression, but that doesn't mean to say that such events as giant impacts did not occur (or even that they did).

What we want you to appreciate, however, is that even if there were giant impacts, the above heat-generating processes would still have taken place. There would have been collisions in which kinetic energy was converted into heat, the growing planetary bodies would have undergone self-compression, core formation would have occurred, short-lived and long-lived radioactive isotopes would have decayed and released heat, and there would have been tidal dissipation. The circumstances in which these processes took place would not have been much like those described above, and the numerical values quoted (dubious even under the conditions assumed) would certainly not be valid; but the conclusion that the early Earth apparently experienced a thermal crisis would stand.

Giant impacts during the Earth's later formation would simply have accentuated the thermal crisis already in evidence from more conventional heat sources, for several reasons. First, having a large mass, a giant impactor would have had a high kinetic energy and thus produced a great deal of heat. Second, having a high kinetic energy, an impactor would have thoroughly disrupted both its target and itself, thereby making it more difficult for the resulting heat to escape. (Planetesimals, by contrast, would not only have imparted heat just to the Earth's surface layer but, by agitating that layer, would have encouraged heat to escape.) Third, the impactor would have induced throughout both its target and itself seismic waves, which, having decayed away and lost energy as heat, would thereby have deeply buried the heat.

❑ The giant impactor would also have introduced the embryo Earth to yet another source of heat. Can you deduce what that is?

■ The impactor would presumably have been hot for the very same reasons that the target planetary body was hot. Planetesimals, by contrast, would have been comparatively cold.

At this point, the thermal problems of the early Earth really begin to get out of hand; so we had better turn to something more amenable to analysis — the Earth's initial chemical conditions.

SUMMARY OF SECTION 1.3

According to the current standard model for the formation of the Solar System, the Sun originated about 5 000 Ma ago as the result of gravitational instability in a dense, rotating, interstellar molecular cloud. Surrounding the Sun was a disc-like nebula of dust and gas which also subsequently became gravitationally unstable, leading to the formation of the Earth and the other planets. As the solar nebula cooled, the constituents of the nebula condensed out to form particles which accreted into small planetesimals (mass ~ 10^{15} kg), then into larger planetary embryos (mass ~ 10^{22}–10^{23} kg), and finally into the terrestrial planets (mass ~ 10^{23}–10^{24} kg) and the rocky cores of the gaseous giants. Planetary embryos took about 10^5 years to form; full-sized terrestrial planets took about 10^7–10^8 years. There may or may not have been solar-nebula gas present throughout the whole of the terrestrial planets' growth; so the Earth may or may not have had a primordial hydrogen-rich atmosphere. A particular feature of the standard model is that it admits the possibility of planetesimal growth and hence of giant impacts during late stages of the terrestrial planets' accretion. An alternative (albeit now a minority) view of the origin of the Solar System claims that giant protoplanets formed before the solar nebula cooled, and then condensed out of the nebula individually. In any case, the standard model is just the current consensus; only time will (may?) tell if it's true.

The giant-impact hypothesis has the advantage that it can explain certain phenomena otherwise difficult (but not necessarily impossible) to account for. These include the anomalously small mantle of Mercury, the origin of the Moon, the diverse tilts of planetary equators from their orbital planes, and the disappearance of the Earth's primordial atmosphere (if there was one). Giant impacts could also have melted the Earth's mantle.

The components of the solar nebula would have condensed out in known order — refractories first and volatiles last. According to the homogeneous accretion model, the solar-nebula constituents would have condensed out before planetary accretion began. This is consistent with the standard model for the origin of the Solar System, but it means that the core was not a primary feature of the Earth. The most satisfactory core-formation model appears to be that, during the Earth's accretion, droplets of iron fell through an iron–silicate emulsion. The heterogeneous model, by contrast, assumes that the individual planets were designated before the solar nebula cooled and that the layering of the Earth was a natural consequence of the condensation sequence.

Although the Earth formed from a cold solar nebula (according to the standard model), it would have become very hot during accretion. Possible sources of heat include the kinetic energy of impactors, self-compression, core formation, short-lived radioactive isotopes, long-lived radioactive isotopes and tidal dissipation. It's very difficult to quantify these sources, but all the indications are that, in its late stages of accretion, the Earth was lucky to avoid vaporization. It could well have been largely molten. The existence of a magma ocean is currently a matter of dispute, but giant impacts would have made one, and a mantle-deep one at that, inevitable.

Objectives for Section 1.3

When you have completed this Section, you should be able to:

1.1 Recognize and use definitions and applications of each of the terms printed in the text in bold.

1.6 List the main tenets of the standard model for the origin of the Solar System, including those relating to the time-scale of accretion and to the issue of whether or not the Earth had a primordial atmosphere; and recognize that the standard model represents the current consensus but is not the only possible model.

1.7 Explain how giant impacts can account for particular features of the Solar System and its individual bodies.

1.8 Recognize the condensation sequence in the solar nebula.

1.9 Describe the main features of the homogeneous and heterogeneous accretion models for the Earth with particular reference to the formation of the Earth's iron core and the time-scale of core formation.

1.10 Summarize the possible sources of heat that would have enabled an Earth accreting from cold fragments to reach a high temperature, and perhaps even become molten, by the time it was fully formed.

1.11 Perform simple calculations relating to the material covered in Objectives 1.6–1.10.

Apart from Objective 1.1, to which they all relate, the six ITQs in this Section test the Objectives as follows: ITQ 5, Objectives 1.6 and 1.11; ITQ 6, Objectives 1.6 and 1.11; ITQ 7, Objective 1.7; ITQ 8, Objectives 1.10 and 1.11; ITQ 9, Objectives 1.10 and 1.11; ITQ 10, Objectives 1.10 and 1.11.

You should now do the following SAQs, which test other aspects of the Objectives.

SAQS FOR SECTION 1.3

SAQ 7 (*Objectives 1.1, 1.6, 1.7, 1.8, 1.9 and 1.10*)

State, giving reasons where appropriate, whether each of the following statements is true or false.

(a) According to the standard model for the formation of the Solar System, the planets were derived from the solar nebula, which contained 1–2% of the original gas and dust that went into the formation of the Sun and planets.

(b) Planets are bigger than planetesimals, which, in turn, are bigger than planetary embryos.

(c) Planetesimals would have coalesced into planetary embryos of mass about 10^{23} kg within the remarkably short time of about 10^5 years, but it would have taken 10 times longer for the planets to form fully.

(d) If the gas component of the solar nebula had been present throughout the Earth's formation, the planet would have acquired a hydrogen-rich primordial atmosphere by gravitational attraction.

(e) Volatiles have lower vaporization temperatures than do refractories.

(f) Iodine (I) is less volatile than suphur (S).

(g) Titanium (Ti) is more refractory that sulphur (S).

(h) Earlier accretion hypotheses envisaged the terrestrial planets growing by the attraction of planetesimals, but the standard model for the formation of the Solar System acknowledges that the planetesimals themselves would have coalesced to form larger impacting bodies.

(i) One rival to the standard model assumes that the solar nebula coagulated into individual proto-planets before cooling.

(j) The giant-impact hypothesis suffers from the problem that it cannot explain the orbital rotation characteristics of the Earth–Moon system.

(k) Under the pressure in the solar nebula, fluorine (F) would condense out before ziconium (Zr).

(l) Under the pressure in the solar nebula, $CaTiO_3$ would condense out before $MgSiO_3$.

(m) All material in the solar nebula would have been in vapour form at a temperature of 2 000 K.

(n) Heat travels in the Earth mainly by conduction.

(o) An Earth formed by the accretion of planetesimals would possess original heat.

(p) An Earth formed by accretion of planetesimals would possess initial heat.

(q) Heat due to accretion, heat due to self-compression and heat due to core formation are all derived from gravitational energy.

(r) A planetesimal striking the growing Earth would have had all its kinetic energy converted into heat.

(s) All long-lived radioactive isotopes in the Earth are substantial heat producers.

SAQ 8 (*Objectives 1.6 and 1.11*)

Calculate how many planetary embryos of mass about 10^{23} kg each it would have taken to form the rocky core of Jupiter, assuming the core to have a mass 15 times that of the whole Earth.

SAQ 9 (*Objective 1.7*)

Make a brief list of Solar-System phenomena that could be explained by giant impacts.

SAQ 10 (*Objective 1.9*)

Explain briefly how the homogeneous and heterogeneous accretion models differ, particularly in relation to the formation of the Earth's iron core.

SAQ 11 (*Objective 1.9*)

What are the two chief differences between the 'blob' model of core formation (1960s) and the rainout model (1990)?

SAQ 12 (*Objectives 1.10*)

List six possible sources of heat in the accreting Earth.

SAQ 13 (*Objectives 1.10 and 1.11*)

A body of mass 10^{17} kg strikes another of mass 6×10^{23} kg with a velocity of $1\,\mathrm{km\,s^{-1}}$ and becomes embedded in it. Assuming that all the impacting body's kinetic energy is converted into heat, which spreads rapidly and uniformly through a volume equivalent to one-sixth of the target's mass, what will the temperature rise in that mass be? (Take the specific heat capacity of planetary material to be $7.5 \times 10^{2}\,\mathrm{J\,kg^{-1}\,K^{-1}}$.)

SAQ 14 (*Objectives 1.10 and 1.11*)

In episode 1, two planetesimals of the same mass strike a planetary embryo with velocity v. In episode 2, a body of twice the mass of each of the planetesimals strikes the same planetary embryo with half the velocity of the planetesimals. Which episode imparts the greater energy to the planetary embryo?

SAQ 15 (*Objective 1.10*)

Explain how a giant impact at a late stage in the Earth's accretion would accentuate the Earth's thermal crisis.

1.4 ELEMENTS IN THE EARTH AND UNIVERSE

The Earth is a mass of chemical elements, many of which have been combined into compounds. We shall look at the compounds that constitute the Earth later in the Course, but for the time being we shall concentrate on elements, where they came from and how they were formed. The question of where they came from is easy to answer in the simple sense. The Earth acquired its elements from the solar nebula and, before that, from the interstellar cloud that existed before the Solar System formed. But knowing where the Earth's elements originated in the context of hypotheses of the formation of the Solar System is not the same as knowing how they formed in the first place.

According to the most generally accepted hypothesis, that of the **big bang**, the Universe began about 15 000 million years ago as a 'fireball' of very hot, dense matter. Right at the very beginning the temperature was so high that no atomic nuclei would have been stable. (**Atomic nuclei** consist of protons and neutrons, except for the hydrogen nucleus, which comprises just one proton.) Within a few seconds of the big bang, however, the temperature would have dropped to about 10^{10} K, at which point protons (hydrogen nuclei) and neutrons could exist; and during the next few minutes the temperature would still have been high enough to enable some of the hydrogen nuclei (protons) to fuse with each other and with neutrons to produce, first deuterium ('heavy hydrogen', containing one proton and one neutron) and then nuclei of the next-highest element, helium (two protons and two neutrons; see Figure 1.5 for the order of elements in the Periodic Table). But that's all. Because the new Universe cooled and expanded very rapidly, no helium would have been produced after the first few minutes; and because the temperature would not have been high enough once the helium had been generated, few, if any, nuclei of elements heavier than helium could have been produced at all. The immediate legacy of the big bang, therefore, was a vast cloud of just hydrogen and helium.

But we know that the Sun and planets contain elements heavier than helium — i.e. elements with higher **atomic number** (= the number of protons in the nucleus). So how did these heavier elements originate? The answer seems to be that most of them formed in the interiors of stars, where high temperatures are sustained over a long enough period of time to enable **fusion reactions** (i.e. the fusion of lighter elements into heavier ones) to continue beyond the generation of helium. A star begins to form when a local region of an interstellar cloud collapses under its own gravity (see Section 1.3). Compression then gradually raises the temperatures in the star's deep interior to the point (about 10^7 K) at which hydrogen can fuse into helium. As this fusion process, known as **hydrogen burning**, results in the emission of heat (i.e. it's what is known as an **exothermic** reaction), the star's interior remains very hot and is thus for a long time prevented from contracting further. Once the supply of hydrogen is used up, however, no more helium can be produced, the supply of energy from the fusion process is cut off, and the star begins to contract again. But, of course, the resulting compression generates more heat, and the temperature rises higher than ever before.

What happens next depends on the star's mass. If this is less than about half the Sun's mass, the star's temperature never becomes high enough to enable fusion to proceed beyond helium (atomic number = 2), and the star contracts into a 'dead', compact object known as a **white dwarf**. However, in more massive stars the temperature of the contracting core reaches about 10^8 K, at which point further nuclear fusion reactions can begin. Meanwhile the outer layers have expanded and cooled, resulting in a highly luminous variety of star known as a **red giant**. In the core, the first new fusion reactions to occur involve helium nuclei, and so are known as

helium burning. These generate elements up to carbon (atomic number = 6) and oxygen (atomic number = 8). For stars less than 3 or 4 solar masses this is the end of the line, and once the supply of helium is exhausted they contract into white dwarfs rich in carbon and oxygen. More massive stars can 'burn' carbon by a further series of fusion reactions, leading in stars of about 8 solar masses to the generation of elements up to iron (atomic number = 26) and nickel (atomic number = 28). Because the higher the temperature the heavier are the elements that can be produced by fusion, and because temperature in a star increases towards the centre, a massive star in which fusion reactions have proceeded as far as iron and nickel ends up with an 'onion' structure, with the heavier elements towards the centre (Figure 1.10).

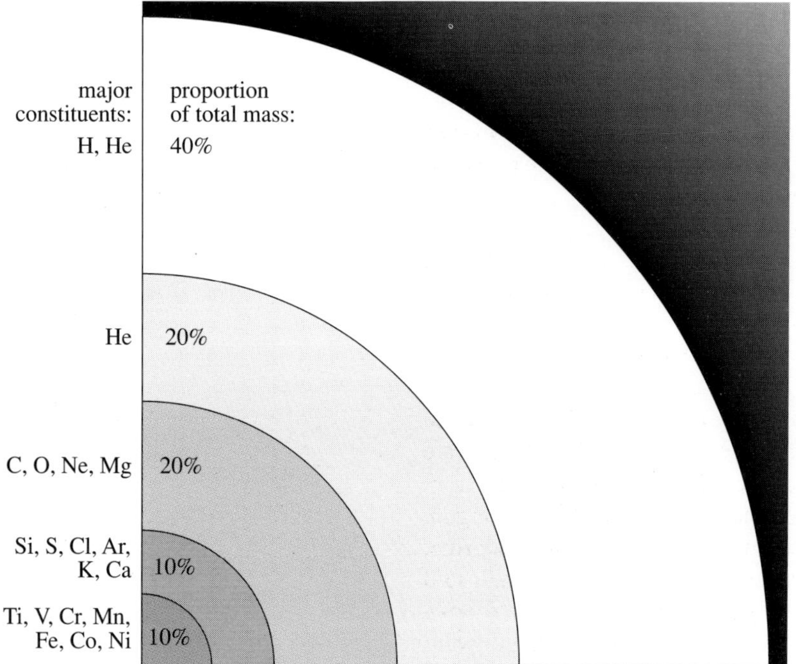

Figure 1.10 The 'onion', or 'shell', structure of a massive star just before its destruction in a supernova explosion (see later). The diagram is not to scale; the inner shells occupy very much less space than indicated, although the relative masses are shown numerically.

Nuclear fusion cannot produce elements heavier than iron and nickel because, for elements with atomic numbers greater than 26–28, the fusion reactions are not exothermic but **endothermic** — i.e. they do not emit energy but require the input of energy to enable them to proceed. In this sense, iron and nickel are the most stable of all elements. Nevertheless, stars do produce elements heavier than iron, although not by fusion but by a process known as **neutron capture**. Some nuclear reactions in stars release neutrons, which combine with heavy nuclei to form, via a series of reactions, even heavier nuclei. This process takes place slowly in stars and is thus known as the **s-process**. Elements up to bismuth (atomic number 83) are produced in this way.

All elements heavier than bismuth, as well as many that are lighter (and most of these heavier than iron), are also produced by neutron capture, but in circumstances very different from those mentioned in the last paragraph. By the time that a massive star has acquired a metallic core by fusion processes, it is nearing the end of its life. Fusion in the core may be able to proceed no further, but the relentless gravitational forces continue to compress the core and thus raise its temperature to at least a billion kelvins. As a result, some of the elements in the core start to disintegrate and ultimately the centre of the star suddenly collapses. The star's outer layers fall inwards at first, but are then thrown out again in what is known as a **supernova explosion**, in which most of the star's constituents are scattered through a region many light years across. Moreover, the explosion itself generates a flood of neutrons which are absorbed immediately by the existing heavy nuclei to produce even heavier ones in what is known as the **r-process**, rapid neutron capture. A supernova explosion thus distributes more heavy elements than were present in the star immediately before it exploded.

As the big bang produced only hydrogen and helium, all elements heavier than helium now within the Universe must have been distributed by supernova explosions that took place since the big bang occurred about 15 000 million years ago. This being so, one would expect that the older the star the lower would be the abundance of heavy elements within it (because the older the star the smaller would have been the number of supernova explosions taking place up to the time of its formation), and this indeed appears to be the case. It also follows that as time goes on, and as more supernova explosions take place, the abundance of heavier elements in the Universe will increase.

A supernova explosion is one of the most spectacular of all natural events, and one of the most energetic. Within its first 10 seconds it radiates 100 times more energy than will have been emitted by the Sun throughout its whole 10-billion-year life (the Sun is now about half-way through that life). Unfortunately, until recently, supernova explosions were difficult to study in detail because those observed are so far away. Telescopes reveal a dozen or so each year in distant galaxies, but the last one seen in our own Galaxy occurred in 1604, just before the invention of the telescope. On 23 February 1987, however, a supernova explosion (SN 1987A) was observed, which occurred 160 000 light years away (i.e. it actually took place 160 000 years ago) in the Large Magellanic Cloud, a galaxy close to ours, and within a day it was being monitored by instruments throughout the Southern Hemisphere (it was not visible in the north). Spectacular demonstration of the supernova's ability to generate heavy elements came in November 1987, when the spectra of light reaching the Earth from SN 1987A revealed iron, nickel, cobalt, argon, carbon, oxygen, neon, sodium, magnesium, silicon, sulphur, chlorine, potassium, calcium and possibly aluminium. Moreover, the indicated abundances of these elements were far higher than those likely to have been present in the original star, confirming that heavy elements were generated in the explosion itself.

It's interesting to ponder the fact that without supernova explosions the Solar System would consist entirely of hydrogen and helium. Indeed, it would not exist in its present form. The creation of elements (**nucleosynthesis**) heavier than helium would presumably still go on within certain stars, but the heavier elements thus generated would remain there. There would be no heavier-than-helium elements in the Sun, no rocky planets, no life, no civilization and, indeed nothing of interest whatsoever. You are only able to be here to study this Course by virtue of phenomena such as those shown in Plates 1.5 and 1.6.

1.4.1 The chemical constitution of the Earth

As the Earth's interior is largely inaccessible and is known to contain layers of different material (see Figure 1.3), accurate determination of the planet's overall (i.e. average) composition would seem to be an impossible task. And it is. Nevertheless, a reasonable hypothesis about the Earth's average composition may be proposed on the basis of data not from the Earth itself but from (a) measurement of the spectrum of light received from the Sun and other stars, and (b) the chemical analysis of certain types of meteorite. We explain here how this remarkable feat is achieved.

The sharp-edged visible surface of the Sun is a layer some 500 km thick (less than 0.1% of the solar radius), called the photosphere. This layer, at a temperature of about 6 000 K, radiates a continuous spectrum of light, often referred to as 'white light'. Before it reaches the Earth, however, this light has to pass through another layer, above the Sun's photosphere but continuous with it and in which the temperature falls to about 4 000 K in about 500 km. Atoms of elements in this second, cooler layer absorb some of the radiation from the photosphere, each element absorbing light of a particular, characteristic frequency or set of frequencies. When the Sun's white light is observed by instruments on the Earth, therefore, the spectrum is seen to be crossed by a series of dark lines corresponding to the absorbed frequencies. A diagram of these lines is shown in Figure 1.11 and part of the Sun's spectrum is reproduced as Plate 1.7. The dark lines are called **Fraunhofer lines** after J. Fraunhofer, who first studied them in detail (although he didn't discover them) during the early 19th century.

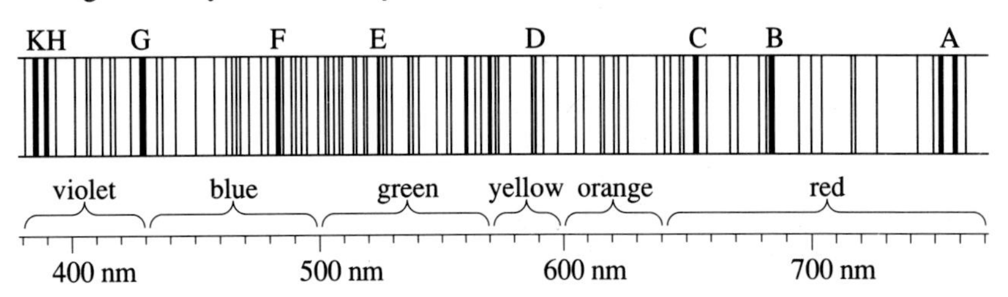

Figure 1.11 A diagram of the solar spectrum with the most prominent absorption lines labelled (as by Fraunhofer) as follows: A, B = oxygen, C, F = hydrogen; D = sodium; E, G = iron; H, K = calcium. The scale at the bottom shows wavelengths in nanometres (nm), where 1 nm = 10^{-9} m. The relationship between frequency (f) in hertz (Hz) and wavelength (λ) in metres is given by $f = c/\lambda$, where c is the speed of light $= 3 \times 10^8\,\text{m s}^{-1}$.

The frequencies of the Fraunhofer lines can be measured, and hence the elements causing them identified, by comparing the lines with the light frequencies emitted by elements heated in the laboratory; and the relative abundances of the elements in the solar atmosphere can be determined from the intensities of the Fraunhofer lines. The end result is knowledge of the *relative* abundances of the elements in the Sun (Figure 1.12). Note that the method does not give the *absolute* abundances of the elements (e.g. x kg of element X) but only the relative abundances (e.g. 1 atom of element X for each 10 atoms of element Y). As the relative amounts of the solar elements are known, knowledge of the absolute abundance of just one element in the Sun would enable all the other absolute abundances to be determined, but we do not have that knowledge.

S7

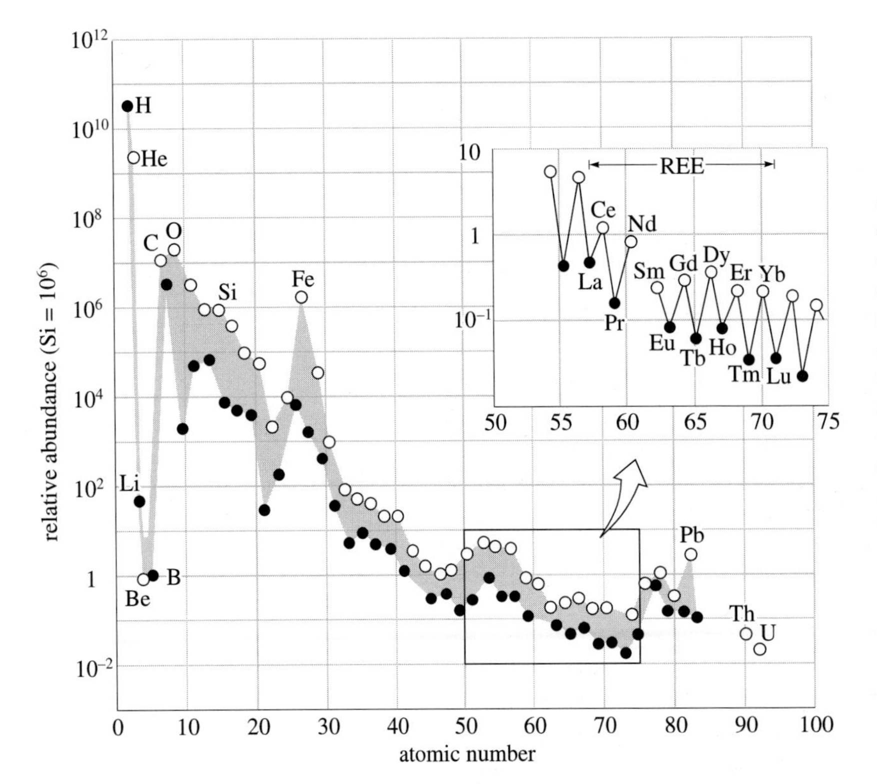

Figure 1.12 The relative abundances of the elements in the Sun. The abundances (vertical scale) are expressed relative to silicon (Si), the abundance of which is set arbitrarily at 10^6 atoms. The elements are expressed in terms of their atomic numbers (horizontal scale), and certain elements are labelled in the diagram. The elements with even atomic numbers are plotted as open circles, those with odd atomic numbers are closed circles. Note that the vertical scale is logarithmic. REE = rare earth elements (see Figure 1.5). The relative abundances of the elements have also been determined from certain stars, many of which give a result similar to that of the Sun.

You shouldn't worry if you don't fully understand the principles of absorption lines as enunciated in the above two paragraphs. What you should appreciate, however, is the general point that by examining the spectrum of light from the Sun it is possible to determine solar relative abundances and that the data thus obtained are as shown in Figure 1.12. You should also have spotted that in the discussion of Fraunhofer lines there is an inconsistency in the argument.

❑ Figure 1.12 purports to show the relative abundances of elements in Sun. Does it in fact do so?

■ Strictly, no. Figure 1.12 is based on absorption data (i.e. on the Fraunhofer lines produced by the absorption of radiation in the Sun's outer layer). It therefore provides information on elements in the Sun's outer layer rather than in its interior. However, in the case of the Sun, we can only suppose that the outer layer comprises material being thrown up from part of the interior, although in the last analysis it is impossible to be entirely sure that the proportions of elements in the outer layer and interior are exactly the same. On the other hand, as we shall soon see, the consistency of solar data with those from other sources encourages confidence that analysis of the Sun's absorption lines does in fact provide reliable information on the solar interior.

ITQ 11

Examine Figure 1.12 and then make a list of what you perceive to be its most important features.

You may by now be wondering just what relevance the solar data in Figure 1.12 have to the Earth. The Earth is, after all, a rocky planet with, on the face of it, little in common with a star in which the chief elements are hydrogen and helium. So why should the composition of the Sun tell us anything at all about the composition of the Earth? To appreciate the answer to this question, you should bear in mind that, as we explained in Section 1.4, the elements heavier than helium in the Sun were not produced there but in bigger stars and in supernova explosions, and were already in the interstellar cloud before the Sun formed. But the Earth formed from that same cloud and in the same region of the Universe. We might speculate, therefore, that the relative abundances of the heavier-than-helium elements in the Sun should be similar to those in the Earth. We might go even further and set up a formal hypothesis to that effect.

❑ Accepting this hypothesis as reasonable, would you nevertheless expect to find a perfect match between the abundances of elements in the Earth and Sun? If not, why not?

■ No. The obvious difference is that the Sun is fuelled by hydrogen burning whereas the Earth is not. Even if the comparison is limited to elements heavier than hydrogen and helium, however, the match is unlikely to be perfect. There are major differences between the histories of the Sun and the Earth, and these are likely to have influenced certain elements in different ways. This is especially true of volatile elements, which are particularly affected by temperature.

In the absence of any corroborating evidence, the hypothesis that the relative abundances of the elements in the Earth are similar to those in the Sun would be fairly weak. But fortunately, there is some corroborating evidence — from meteorites. As we saw in Section 1.2.1, most meteorites are fragments of asteroids. Since meteorites were first discovered they have been, and continue to be, studied in great detail, not only because they are interesting objects in their own right but because there has long been a suspicion that they — or at least some of them — are the most primitive materials in the Solar System. By 'primitive' here, we mean that meteorites and their parent bodies, the asteroids, are thought to be the closest known representatives of the original material of the solar nebula and to have remained comparatively unaltered by events occurring subsequent to their formation. Most meteorites, for example, have ages greater than those of any known rocks on Earth and comparable with that of the Earth itself, suggesting that they have been around in more or less their present form ever since the Solar System came into being.

It has to be admitted, however, that at first sight the sheer variety of meteorites would seem to belie any such universal significance. Briefly, there are

(1) **Stony meteorites**, which account for about 95% of all known meteorite falls and are subdivided into chondrites and achondrites:

(a) **Chondrites** are so called because most of them contain small, once-molten globules, or **chondrules**, of silicate. They consist of iron–magnesium silicates, dispersed grains of iron–nickel alloy and the iron-sulphide mineral **troilite** (FeS). Chondrites, or chondritic meteorites, are by far the most common type, accounting for about 85% of all known falls.

(b) **Achondrites** not only, as their name implies, lack chondrules, they also lack abundant metallic grains. They are composed mainly of silicate minerals rich in iron and magnesium and account for about 10% of all known meteorite falls.

(2) **Iron meteorites** are almost entirely (greater than 90%) metallic, the predominant component being an iron–nickel alloy containing 4–20% nickel. They also contain some troilite. Iron meteorites are those most often found because they contrast with most terrestrial rocks, but they only account for about 3.5% of all falls.

(3) **Stony–iron meteorites** are hybrids, containing a roughly 50:50 mixture of iron–nickel alloy and silicate phases. They form about 1.5% of all falls.

Meteorites may be primitive, but the fact that there are several quite distinct types of meteorite rather suggests that some might be more primitive than others. If that is so, which are the *most* primitive? In fact, this seems to be a fairly easy question to answer in terms of the above simple classification. The very least one could expect of a 'most primitive' asteroid is that, before disintegrating and sending its fragments on their way to Earth and elsewhere, it should not already have differentiated into crust, mantle and core — or, at least, mantle and core — as the Earth has done. In other words, it would seem reasonable to suggest that a differentiated asteroid represents a greater departure from primitivism that an undifferentiated one. In fact, there is evidence that some asteroids have departed from primitivism in this way.

❑ Can you deduce what that evidence is? (Hint: Compare the materials in the list of meteorite types above with the materials of the Earth's main layers as shown in Figure 1.3.)

■ The core of the Earth (see caption to Figure 1.3) — and, by implication, that of a large differentiated asteroid — consists largely of iron. The iron core of an asteroid is thus an obvious source of iron meteorites. Similarly, achondrites match what, by analogy with the Earth, one might expect the mantle of a differentiated asteroid to be (i.e. iron-rich and magnesium-rich silicates). Less obviously, perhaps, the stony–iron meteorites might have come from an iron/silicate boundary region such as the boundary between an asteroid's core and mantle. Because achondrites, iron meteorites and stony–iron meteorites are thought to have come from differentiated parent bodies, they are sometimes called **differentiated meteorites**, although you should realize that the meteorites themselves are not differentiated. The important point about all this is that because differentiated meteorites came from different parts of differentiated asteroids they cannot, by definition, be representative of the average composition of the solar nebula.

That leaves chondrites. As chondrites have no obvious source regions within differentiated asteroids, they are usually regarded as being from an undifferentiated body and hence more representative of the average composition of the solar nebula — i.e. they are the most primitive. However, as chondrites account for about 85% of all known falls, by excluding achondrites, iron meteorites and stony–iron meteorites from the primitivism stakes, we haven't actually achieved very much! As you may have suspected, or even feared, by now, there are chondrites and chondrites; not all are the same. On the contrary, they are quite varied in their chemical and mineralogical compositions; so the search for the most primitive requires further categorization.

There are three main chemical classes of chondrites, or chondritic meteorites, one of which has three subclasses. The three main classes are **ordinary chondrites**, so called because they are the most abundant type, **enstatite chondrites** (**E-chondrites**), so named because they have high abundances of the mineral enstatite ($MgSiO_3$), and **carbonaceous chondrites** (**C-chondrites**), which contain, in addition to silicate minerals, a tarry mixture of abiogenic organic compounds. One of the most important differences between these three types appears to be their temperature of formation, C-chondrites having formed at lower temperatures than E-chondrites, and the latter at lower temperatures than ordinary chondrites.

❑ Bearing in mind what we said about volatiles in Section 1.3, can you suggest how the three types of chondrite might differ?

■ The lower the temperature at which the chondrites formed, the higher should be their volatile content, because higher temperatures are likely to drive off more volatiles. That this is indeed the case is demonstrated rather elegantly in Figure 1.13. (Note, however, that volatile content can depend not only on formation temperature but on the temperatures to which the materials concerned have been subjected subsequent to formation.)

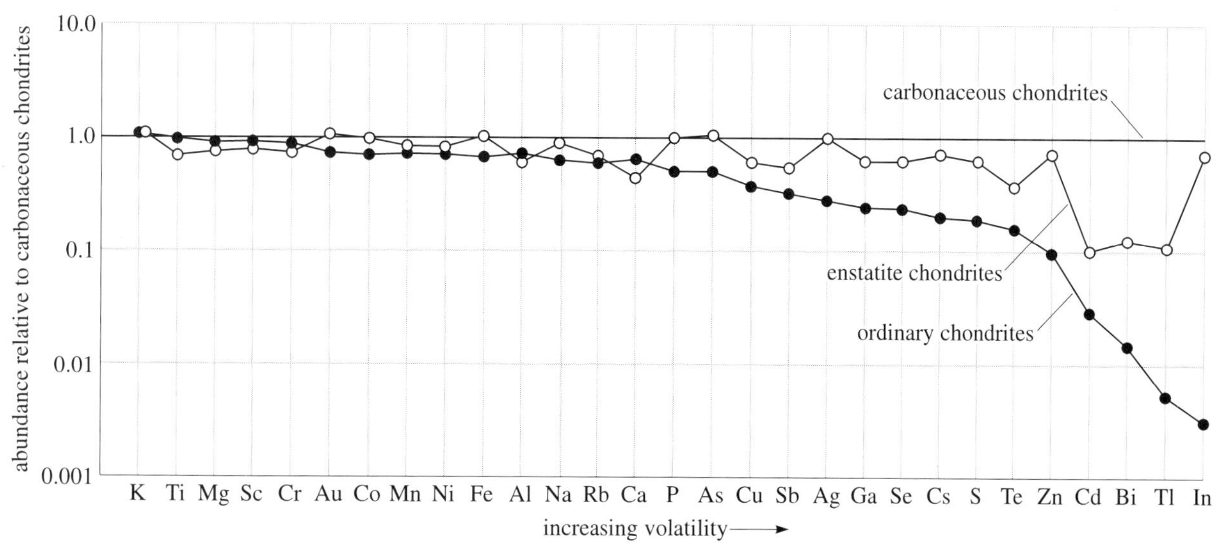

Figure 1.13 Different types of chondrite have different degrees of depletion of volatiles. In this diagram, elements are listed in order of increasing volatility to the right. The abundances of the elements for ordinary and enstatite chondrites are then plotted relative to their abundances in carbonaceous chondrites (set at 1.0 for each element). What the diagram shows is that ordinary chondrites are more depleted in volatiles than are enstatite chondrites, and that the latter are more depleted than carbonaceous chondrites.

Another important difference between the three types of chondrite is their degree of oxidation. In C-chondrites, most of the iron is oxidized, forming chiefly silicates; in ordinary chondrites, some of the iron is in a reduced state, as metal and iron sulphide grains; and in E-chondrites, most of the iron is reduced to metal or iron sulphide. The order of increasing oxidation is therefore: E → ordinary → C. On the basis of oxidation state, ordinary chondrites are also divided into the subclasses **H-chondrites, L-chondrites** and **LL-chondrites** in which the order of increasing oxidation is: H → L → LL.

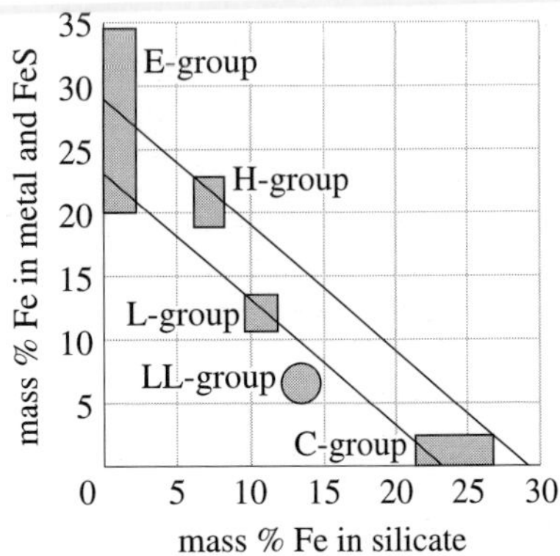

Figure 1.14 The degree of oxidation of the five classes of chondrites. The horizontal axis shows the mass per cent of iron in silicates (high oxidation state) and the vertical axis shows the mass per cent of iron as metal or in iron sulphide (low oxidation state). Materials lying on the same diagonal line have the same total iron content, which can be read off at either end of the line.

ITQ 12

Examine Figure 1.14, which represents the degrees of oxidation of the five classes of chondrite, and then arrange the five classes in order of (a) increasing degree of oxidation, and (b) increasing total iron content.

Chondrites may be classified in terms not only of their oxidation states but also of their mineralogical texture. Although chondrites are thought to be the type of meteorite least altered since their formation, they are not completely unaltered and some have been altered more than others. This alteration, or **metamorphism**, which is generally due to reheating, has changed the texture of the mineral. With increasing temperature, for example, the chondrules gradually become obliterated and new crystals grow. Note that we are not talking here of total melting, which would have separated the iron silicate phases completely (and hence removed any suggestion that the chondrites are 'primitive'), but only of heating, allowing alteration to take place in the solid state.

We do not propose to go into the textural classification in great detail. Suffice it to say that meteorite investigators recognize six classes (1–6 in order of increasing metamorphism), sometimes called **petrological types**, and that they then plot these against chemical class as shown in Figure 1.15. The matrix thus produced is not merely some esoteric artefact of interest only to meteorite specialists, however; it holds within it the answer to the question of interest to us, namely, that of which type of chondrite most closely resembles the early Solar System.

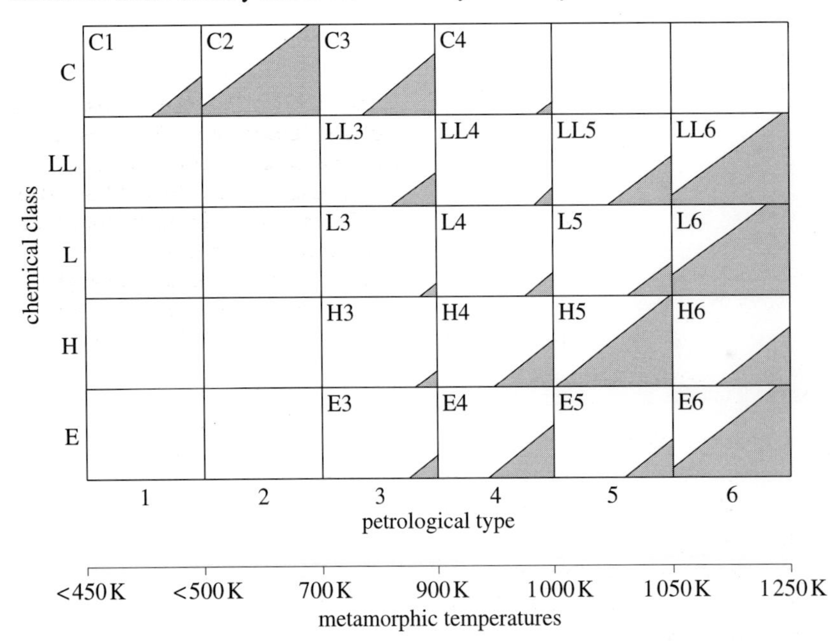

Figure 1.15 The classification of chondrites. Petrological types (1– 6 in order of increasing alteration) are plotted horizontally, and chemical classes are plotted vertically. The number of classes thus produced is 30 (C1, LL3, H4, E6, etc.). The shaded areas represent the proportion of each chemical class within each petrological class (e.g. about 50% of H-chondrites are of petrological type 5, there are almost no C-chondrites of petrological type 4, and so on). Empty boxes have no known chondrites. The scale at the bottom represents the maximum temperatures at which metamorphism took place.

❑ In order to answer this question, which end of the spectrum of petrological types is of most interest?

■ Clearly, in trying to identify which chondrites are the most primitive, we need to identify which have been least altered since they formed, and this means looking at the lower-numbered petrological types. As Figure 1.15 makes clear, most of the chondrites of these petrological types are carbonaceous chondrites; and so it is these that are regarded as the most primitive in the sense in which we have been using that word.

Armed with this knowledge, we can at last go back to the question of the overall chemical composition of the Earth. The average relative abundances of the elements in carbonaceous chondrites are now well known from laboratory analyses; and as we saw in the earlier part of this Section, the relative abundances of the elements in the Sun and stars are also known. So how do the two sets of data compare? Figure 1.16 shows how. When the two sets of relative abundances (for elements above helium) are plotted against each other, the elements fall more or less on a straight line at 45°, showing that to a first approximation the relative abundances of the elements heavier than helium in carbonaceous chondrites are the same as those in the Sun. Of course, the match is not perfect. Lithium (Li) and boron (B) lie well off the curve. These elements were discussed in the answer to ITQ 11, and we shall not pursue them further here. More instructive is the behaviour of nitrogen (N), oxygen (O) and carbon (C), which also lie off the curve. These light, volatile elements are consistently more abundant in the Sun than in C-chondrites, presumably having been depleted in the latter either during formation of the chondrites or later.

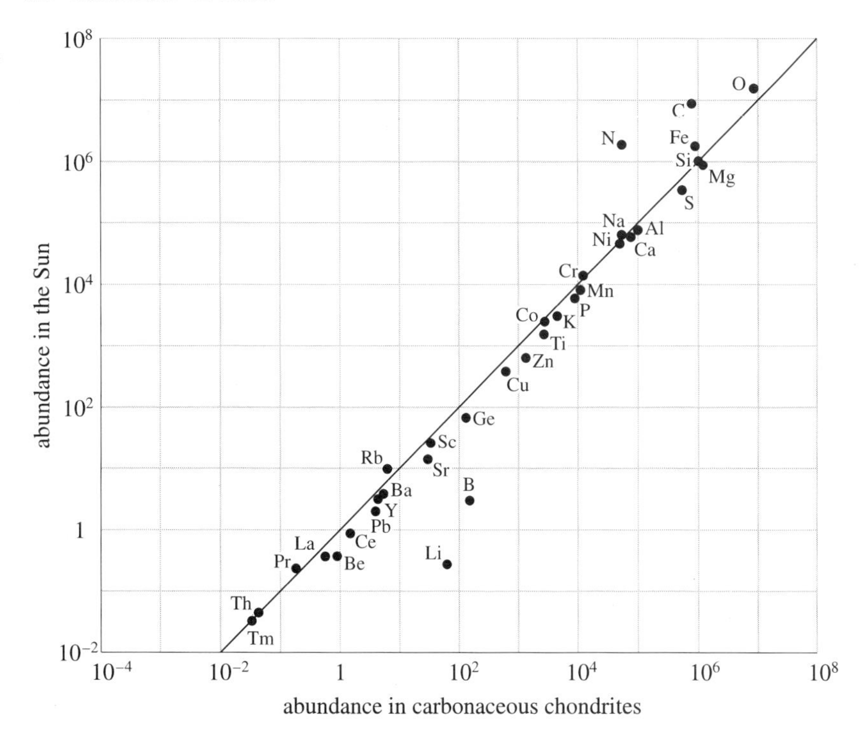

Figure 1.16 The relative abundances of the elements in the Sun plotted against those in carbonaceous chondrites, based on Si = 10^6 atoms as in Figure 1.12. An element with equal abundances in the Sun and carbonaceous chondrites would plot exactly on the 45° line shown. Note that both scales are logarithmic.

A few volatile elements apart, however, the match between elemental abundances in the Sun and carbonaceous chondrites is remarkable. This consistency considerably strengthens the hypothesis that the overall composition of the Earth is well represented by the relative abundances of the elements in C-chondrites. For this reason, the abundances concerned are sometimes referred to as the **cosmic abundances of the elements**, although this is a rather anthropocentric nomenclature given that it has not been tested much beyond the Solar System.

1.4.2 THE CHONDRITIC EARTH MODEL

An Earth taken to have the same overall composition as that of carbonaceous chondrites is called a **chondritic Earth**, and the hypothesis that the Earth is indeed chondritic is often referred to as the **chondritic Earth model** (CEM). These simple definitions need some qualification, however. As Figure 1.16 shows, the relative abundances of many elements in carbonaceous chondrites are the same as those in the Sun, and so for these elements a chondritic Earth is also a solar Earth (i.e. one having the same relative abundances as those of the heavier-than-helium elements in the Sun). But solar and chondritic relative abundances for some elements are *not* identical. As we saw in Section 1.4.1, N, C and O, for example, are depleted in C-chondrites relative to solar abundances; and we suggested that this was perhaps because these volatile elements were depleted in C-chondrites in an absolute sense (i.e. that there is 'too little' N, C and O in C-chondrites rather than that there is 'too much' N, C and O in the Sun — that the C-chondrites rather than the Sun deviate from 'true cosmic abundances').

We cannot be absolutely sure that it is the C-chondrites that are the deviants, but let's suppose for the moment that they are. Where does that leave the Earth? Is a 'chondritic Earth' to be taken to be one having the same relative abundances of N, C and O as do C-chondrites, in accordance with the strict definition of 'chondritic Earth' given above, or are we suggesting that it is one having solar relative abundances of N, C and O, presumed now to be closer to the true cosmic relative abundances of the elements? Earth scientists tend to be vague about this, but perhaps it doesn't matter too much. If we could measure the overall abundances of N, C and O in the Earth (we can't), there is no guarantee that they would be the same as those in either the Sun or C-chondrites anyway. That's the problem with volatile elements. Their behaviour may be predictable in the chemical sense, but we know neither the precise conditions under which the constituent bodies of the Solar System formed nor just what processes those bodies have been subjected to. The history of the behaviour of the volatile elements in the specific circumstances in which the bodies of the Solar System originated and evolved is probably unknowable and undeducible. Volatiles thus play an uneasy part in the CEM, which is perhaps best regarded as being mainly about the more refractory elements, which are much better behaved, as Figure 1.16 shows.

The important question is: Is the CEM valid? — i.e. is the Earth really chondritic overall, at least in respect of the more refractory elements? The CEM is difficult to test.

❑ Can you see why?

■ The Earth is neither physically nor chemically uniform; it comprises crust, mantle and core, each of which evidently has a different chemical composition. Moreover, the mantle and core are inaccessible to direct sampling. There are therefore insufficient analytical data from which an average composition of the Earth could be calculated.

In fact, it *is* possible to put the CEM to a test of sorts. It may not be possible to sample the mantle and core, but it is possible to make some deduction about their compositions and hence to work out an average composition of the Earth. We shall return to this later in Block 1. In the meantime, it is instructive to look at the Earth's crust, or at least the uppermost part of it, which can, of course, be sampled directly.

As the crust can be sampled, you might suppose that it would be an easy matter to determine its average composition. But it's not. The difficulty arises from the huge variety of rocks making up the upper crust. Each rock type can be chemically analysed very accurately; the problem is to estimate just how much of each rock type there is. Moreover, even if we could determine accurately the proportion of each rock type present in the accessible near-surface layer, that would be of little help in respect of the crust's deeper levels.

Table 1.3 provides a summary of 36 estimates of crustal abundances made between 1916 and 1976. (the table is limited to 20 elements, some of obvious importance in the Earth and some less common.)

Table 1.3 Estimates of abundances of selected elements in the Earth's crust. (ppm = parts per million)

Element	Lowest value (ppm)	Highest value (ppm)	Ratio of highest to lowest value
C	200	4902	24.5
O	452341	495200	1.09
Na	15208	28500	1.87
Mg	10191	33770	3.31
Al	74500	88649	1.19
Si	257500	315896	1.23
Ar	0.04	4	100
K	15773	32625	2.07
Ca	16438	62894	3.83
Mn	155	1549	10.0
Fe	30888	64668	2.09
Co	12	100	8.33
Ni	23	200	8.70
Cu	14	100	7.14
Ru	0.0001	0.05	500
Ag	0.02	0.1	5.00
Cd	0.1	5	50.0
Ba	179	1070	5.98
W	0.4	70	350
Au	0.001	0.005	5.00

❑ Examine Table 1.3. What is the most startling conclusion to be drawn from it?

■ The estimates of crustal abundances are highly variable. The ratio of highest-to-lowest values for the 20 elements listed ranges from 1.09 to 500; and if all natural elements were to be included, the range would be $1.09–2.5 \times 10^8$!

You may be thinking that, as Table 1.3 contains data from as far back as 1916, the picture the table gives could be deceptive. After all, haven't analytical techniques improved considerably over subsequent years? Yes, they have; but that has made little difference, for the problem is not the accuracy of the chemical analyses but that of estimating the proportions of the various rock types in the crust. The latter estimations may also have improved, but evidently not enough. There is little evidence that the more recently estimated abundances have been converging towards a 'true' value. The truth is that, even now, we have only a very poor knowledge of crustal abundances.

This may come as a surprise to you. Before reading the last few paragraphs you might have imagined that knowledge of crustal abundances would have amounted to one of the near-certain 'facts' about the Earth, especially as the crust (though, actually, not all of it) is accessible. But the Earth is so complex that accurate facts about the planet on the larger scale are hard to come by. This is an example of the general point we made in Section 1.2. The overall composition of the Earth's crust is just another case of the tenuousness of much of the data that Earth scientists have to use.

In order to make crustal-abundance data usable, geochemists adopt what they sometimes call 'consensus values', which are simply 'best guesses' lying somewhere — usually roughly mid-way — between the highest and lowest values in Table 1.3. Figure 1.17 shows what happens when consensus relative abundances are plotted against the relative abundances of elements in the Sun.

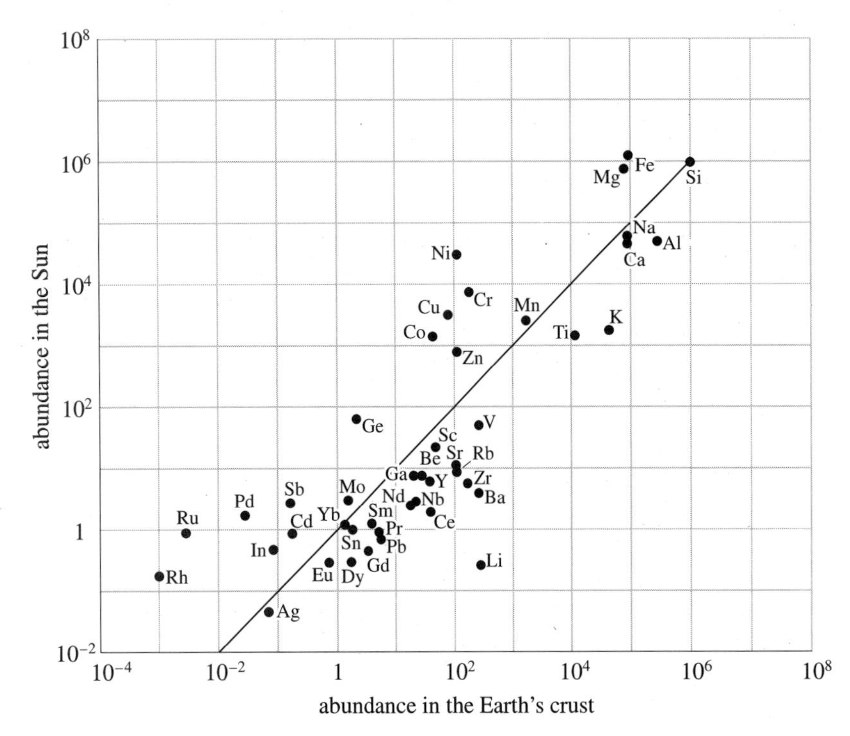

Figure 1.17 The (consensus) relative abundances of the elements in the Earth's crust plotted against those in the Sun, based on Si = 10^6 atoms as in Figure 1.12.

❑ What is the most conspicuous difference between Figures 1.16 and 1.17?

■ The data points in Figure 1.17 are clearly much more scattered than those in Figure 1.16.

This is only to be expected. We have already made the point that, because the Earth is differentiated, the crust is unlikely to be chemically representative of the whole planet. Indeed, there is no reason in principle why the relative abundances of elements in the crust should be anything like those of the Earth as a whole. In this sense, therefore, the surprise is not that the data points in Figure 1.17 should be so scattered but that they should be as clearly grouped around the 45° line as they are. The fact that they are so grouped is, in fact, an important piece of evidence in favour of a chondritic Earth.

❑ Can you formulate the reasoning behind this conclusion?

■ If the Earth were far removed from being chondritic, any one of its layers (crust, mantle, core) would be chondritic only by the merest chance. Even if the Earth were chondritic overall, none of its individual layers need be. The fact that the crust turns out to be not precisely chondritic but with clear chondritic tendencies (i.e. the points in Figure 1.17, though scattered, do lie around the 45° line) strongly suggests that, unless nature is playing a cruel trick, the crust is likely to be derived from a chondritic or near-chondritic planet. At the very least, the data expressed in Figure 1.17 could hardly be considered as being inconsistent with the CEM.

No less encouraging for the CEM is the behaviour of some of the specific elements in Figure 1.17.

ITQ 13

If the chondritic Earth model is valid, all the departures from the straight line in Figure 1.17 must be due to element enrichments or depletions in the crust. From Figure 1.17 and the information provided by Figure 1.3 and by meteorites (Section 1.4.1), decide (a) which of the elements nickel (Ni) and magnesium (Mg) is/are depleted or enriched in the crust, and in each case by how much, and (b) whether or not the enrichments and/or depletions are consistent with an initially chondritic Earth that became layered.

Elements to the left of the line in Figure 1.17 are depleted in the crust relative to the Sun (and hence to carbonaceous chondrites) and *may* be so because they have preferentially migrated into the mantle and/or core. Nickel and magnesium certainly seem to have migrated in this way. Elements to the right of the line in Figure 1.17, however, are enriched in the crust relative to the Sun and *may* be so because they have preferentially migrated towards the crust. If such migrations have taken place, it follows that the crust's departure from chondritic composition would not necessarily be inconsistent with an overall chondritic Earth. (If, on the other hand, nickel — to take one example — could be shown to be depleted in the mantle and core as well as the crust, it would follow that, in respect of nickel at least, the Earth could not possibly be chondritic overall.)

Support for the CEM is certainly not weakened by the knowledge that different elements have different geochemical affinities (which means that each element tends to enter into a particular type of chemical combination) and that these affinities are generally consistent with the sort of migrations described in the last paragraph. Some elements, for example, preferentially enter into combination with oxygen or as silicates, and are said to be **lithophile elements**.

❑ Where would lithophile elements be most likely to be found?

■ In the crust or mantle, both of which comprise largely silicates and oxides.

All the elements to the right of the line in Figure 1.17 are lithophile and enriched in the crust relative to chondritic abundances. Mg is also lithophile, however, although because it is enriched in the mantle rather than the crust, it lies above the line in Figure 1.17. Elements having a preferential tendency to form sulphides are called **chalcophile elements**, examples of which are Zn, Pb, Cd and Ag. Elements that prefer to exist as metals, rather than enter into combination as sulphides, silicates or oxides, are called **siderophile elements** and tend to be concentrated in the core. Examples are Ni, Pd and Rh.

Elements have different geochemical affinities because of their different electronic and other properties, although these properties are not the sole arbiter of whether an element is lithophile, chalcophile or siderophile. Opportunity also comes into it, making it possible for some elements to lie in two or even three of these groups. Iron is an important example. Iron is basically chalcophile, but there is far too little sulphur in the Earth to utilize all the iron available. Some is therefore siderophile (iron metal in the core) and some is lithophile (in silicates in the crust and mantle).

The general point we are making here is that, although the Earth's crust clearly deviates from a chondritic composition, it does so for many elements in a way that is understandable in terms of the known geochemical behaviour of the elements, bearing in mind that the Earth is differentiated into crust, mantle and core. There is therefore nothing we have said so far which is inconsistent with the Earth's being chondritic on average. The chondritic Earth model represents the best insight we have into the Earth's overall composition, even though, ironically, it was based on no terrestrial data at all. The CEM will appear again later in the Course; but in the meantime, we turn to the Earth's physical structure and the principles on which it is determined.

SUMMARY OF SECTION 1.4

The 'big bang' that gave rise to the Universe about 15 000 million years ago produced hydrogen and helium, but the new Universe cooled and expanded so rapidly that no heavier elements could be generated. The elements heavier than helium now within the Solar System were formed in stars (but not the Sun). Depending on their masses and temperatures, stars can generate elements as heavy as iron (atomic number = 26) and nickel (atomic number = 28) by fusion reactions, and as heavy as bismuth (atomic number = 83) by slow neutron capture. The more prolific process for the production of elements heavier than iron and nickel, however, is rapid neutron capture, which takes place when stars end their lives in supernova explosions. Such explosions not only generate heavy elements but scatter both them and the lighter elements from the original stars throughout part of the Universe, providing more interstellar material to be incorporated into new stars. Without supernova explosions, there might still be a Sun, but there would be no rocky planets.

By examining light from the Sun, and in particular the Fraunhofer lines resulting from absorption in the solar outer layer, it is possible to identify the elements in the Sun and to estimate their relative abundances. It is also possible to measure the abundances of elements in meteorites, which are apparently the most primitive material in the Solar System (the most representative of the original solar nebula). In fact, there are various types of meteorite — stony (chondrites and achondrites), iron and stony–iron — some of which are more primitive than others. It would appear that the chondrites are the most primitive, because, unlike other types of meteorite, they came from undifferentiated asteroids; and classification of the chondrites on the basis of chemistry and mineralogical texture reveals that carbonaceous chondrites are the most primitive of all.

When the relative abundances of heavier-than-helium elements in the Sun are compared with those of elements in carbonaceous chondrites, the two sets of data are seen to be almost identical, except for certain volatile elements. This agreement supports the view that the overall composition of the Earth, volatile elements apart, is the same as that of carbonaceous chondrites; and this is the basis of the chondritic Earth model (CEM), which cannot be tested directly because the Earth is divided into layers (crust, mantle and core) with different compositions and, as far as the mantle and core are concerned, is inaccessible to direct sampling. The crust can be sampled (although it is still difficult to estimate the true average abundances of the elements within it) and is, as expected, found to be non-chondritic. However, the closeness of the crust to being chondritic is such as to suggest that the Earth's surface layer is not likely to have been formed from an initially non-chondritic Earth. Moreover, the crustal deviations from a chondritic composition are often consistent with the known geochemical behaviour of elements (which can be lithophile, chalcophile or siderophile, or more than one of those) in the context of an Earth that has differentiated. The CEM is the best view we have of the Earth's overall composition.

Objectives for Section 1.4

When you have completed this Section, you should be able to:

1.1 Recognize and use definitions and applications of each of the terms printed in the text in bold.

1.12 Outline the nature of the various sources of elements in the Universe (the 'big bang', stellar interiors, supernova explosions), the processes by which the elements were formed (fusion, neutron capture, supernova explosions) and the range of elements generated by each source/process.

1.13 Describe the main types of meteorite and show how, by classifying meteorites, it is possible to determine which type is most representative of the initial solar nebula (i.e. the most primitive).

1.14 Demonstrate that, by correlating the relative abundances of the elements in the Sun with those in carbonaceous chondrites, it is possible to adduce strong circumstantial evidence in favour of the hypothesis that the Earth is chondritic on average, at least in respect of the more refractory elements — and that the evidence is strengthened even further by interpreting the composition of the Earth's crust in the light of the known geochemical behaviour of elements.

1.15 Perform simple calculations relating to the material covered in Objectives 1.12–1.14.

Apart from Objective 1.1, to which they all relate, the three ITQs in this Section test the Objectives as follows: ITQ 11, Objectives 1.12 and 1.14; ITQ 12, Objective 1.13, Objective 1.14.

You should now do the following SAQs, which test other aspects of the Objectives.

SAQs for Section 1.4

SAQ 16 (*Objectives 1.1, 1.12, 1.13 and 1.14*)

State, giving reasons where appropriate, whether each of the following statements is true or false:

(a) The only elements generated in the 'big bang' origin of the Universe were hydrogen and helium.

(b) Hydrogen burning and helium burning are endothermic reactions.

(c) Since the 'big bang' took place, the only source of new elements has been the interiors of stars.

(d) The core of a red giant ultimately becomes a dead, compact body called a white dwarf.

(e) Fusion reactions in stars can generate elements up to bismuth (atomic number = 83).

(f) Fusion reactions in the Sun generate only helium.

(g) Supernova explosions both generate new elements and scatter them (as well as elements from the original star) throughout part of the Universe.

(h) The proportion of heavy elements in the Universe is decreasing as more and more stars are destroyed in supernova explosions.

(i) Fraunhofer lines originate because of the absorption in the Sun's outer layer of light from its photosphere.

(j) The relative abundances of elements heavier than helium in the Sun generally decrease (but not smoothly) with increasing atomic number.

(k) Chondritic meteorites (chondrites) are the least common type of meteorite in terms of known falls.

(l) The order of increasing temperature of formation of chondrites is ordinary chondrites, carbonaceous chondrites, enstatite chondrites.

(m) Ordinary chondrites contain fewer volatiles than do enstatite chondrites or carbonaceous chondrites.

(n) The order of increasing degree of oxidation of chondrites is: enstatite, ordinary, carbonaceous.

(o) The relative abundances of C, N and O in the Sun are identical to those in carbonaceous chondrites.

(p) L-chondrites are more highly metamorphosed than are C-chondrites.

(q) Estimates of the average abundance of magnesium in the Earth's crust range from 10 191 ppm to 33 770 ppm, giving a highest-to-lowest ratio of 3.31.

(r) Both Ge and Zr are depleted in the Earth's crust relative to the Sun.

(s) A siderophile element is most likely to be concentrated in the Earth's crust.

SAQ 17 (*Objective 1.15*)

In the spectrum of light from a star, very strong Faunhofer lines are observed at frequencies of 4.580×10^{14} Hz, 6.186×10^{14} Hz and 6.977×10^{14} Hz. What element(s) does the star contain?

SAQ 18 (*Objective 1.15*)

Which of the elements C, Pb and Nd is/are enriched and which is/are depleted in the Sun with respect to silicon? In each case, determine by how much the element is enriched or depleted.

SAQ 19 (*Objective 1.13*)

Summarize the characteristic features of the principal meteorite types (a–d) by matching them with the appropriate characteristic (A–G).

(a) chondrites

(b) achondrites

(c) stony–irons

(d) irons

(A) consist largely of Fe–Ni alloy

(B) are largely Mg–Fe silicates

(C) contain approximately equal amounts of Fe–Ni alloy and Mg–Fe silicates

(D) contain chondrules

(E) contain little or no Fe–Ni alloy

(F) account for most finds but few falls

(G) account for most falls

SAQ 20 (*Objective 1.15*)

A carbonaceous chondrite having the same mass as an enstatite chondrite contains $100x$ atoms of Cd. How many atoms of Cd does the enstatite chondrite contain?

SAQ 21 (*Objective 1.15*)

Analyses of gold in chondrites indicate concentrations of about 1.5 ppm in the metal phase and about 0.005 ppm in the silicate phase. Assuming that the relative abundance of gold in the metal and silicate phases of chondrites is the same as in, respectively, the core and mantle of the Earth, what is (a) the mass of gold in the Earth's core and mantle, and (b) the proportion of the Earth's gold that is in the core? (Mass of mantle = 4×10^{24} kg; mass of core = 2×10^{24} kg.)

SAQ 22 (*Objective 1.1*)

Match the possible forms or occurrences of elements (a–f) with the geochemical classifications of the elements (A–C).

(a) form silicate minerals

(b) occur as sulphides

(c) found as native metals

(d) occur as oxides

(e) may be alloyed with metallic iron (e.g. in core)

(f) found in combination with sulphur as ZnS

(A) lithophile

(B) chalcophile

(C) siderophile

1.5 SEISMIC SOURCES AND SEISMIC WAVES

As the Earth's deep interior is inaccessible, the only way of determining its physical structure is to probe from the planet's surface or from above it (e.g. using satellites). The 'probes' concerned here are not, of course, artefacts that penetrate the deep Earth but techniques that make use of a variety of natural physical phenomena such as seismic waves, gravity, heat, magnetism and electric currents. The most powerful of these by far in terms of the amount and quality of information provided is **seismic waves**, the vibrations of the Earth generated by earthquakes, explosions and other shock-producing mechanisms. The study of the passage of seismic waves through the Earth is called **seismology**.

Because of its importance, we shall look at seismology in some detail, beginning briefly with the nature of seismic sources and waves. You will no doubt be aware, not least from the public media, that earthquakes also have an important human dimension in that they pose a severe threat to people and their settlements. Indeed, it has been estimated that during historic time over 14 million people have lost their lives in earthquakes or because of earthquake after-effects such as landslides, fires and tsunamis (large sea waves caused by submarine seismic or volcanic disturbances). This aspect of earthquakes is not part of this Course, however. Here we shall concentrate on earthquakes (and explosions) mainly as the sources of vibrations that enable us to probe the Earth's interior and hence in some measure determine 'how the Earth works'. There is, of course, some irony in the fact that the natural phenomenon that has killed more people than any other is also the phenomenon that enables us to determine most about the interior of the planet on which we live.

1.5.1 EARTHQUAKES AND SEISMIC WAVES

In general terms, an earthquake is a sudden release of strain energy in a comparatively localized region of the Earth's crust or upper mantle. Because of the movement of material in the Earth's interior, some rocks in the planet's uppermost layers are subjected to forces (stress), and these forces cause the rocks to distort (strain). When, for whatever reason, the strain is suddenly eliminated, the energy that was gradually accumulating as the strain built up is released almost instantaneously. Some of this energy is dissipated as heat, but most is transmitted away from the point of release as seismic waves. (We should mention here that the terms 'stress' and 'strain' have precise scientific meanings and mathematical definitions, but we shall return to those later.)

Seismic waves, and hence the earthquakes that gave rise to them, are detected and recorded using instruments known as **seismometers**, the principles of which are illustrated in Figure 1.18. Each of the instruments shown comprises a rigid frame attached to the Earth, a chart recorder connected to the frame, a heavy mass suspended from a spring or wire, a constraint (e.g. the hinge in Figure 1.18b) which restricts motion to one direction, and a pen fixed to the mass. When the Earth and frame are jolted by a seismic vibration, the suspended mass tends to remain stationary because of its great inertia. However, the frame moves with the Earth, and the relative motion between the frame and the suspended mass is traced on the chart by the pen. The movement of the mass is restricted so that the instrument records vibrations in one direction only, which means, in turn, that different instruments are needed to record horizontal and vertical motions of the ground. An earthquake recording station must therefore comprise three instruments, one such as that in Figure 1.18b for measuring vertical motions and two such as that in Figure 1.18a for measuring horizontal motions at right angles (e.g. north–

south and east–west). The full signal is then calculated using a simple mathematical procedure which combines the three component signals.

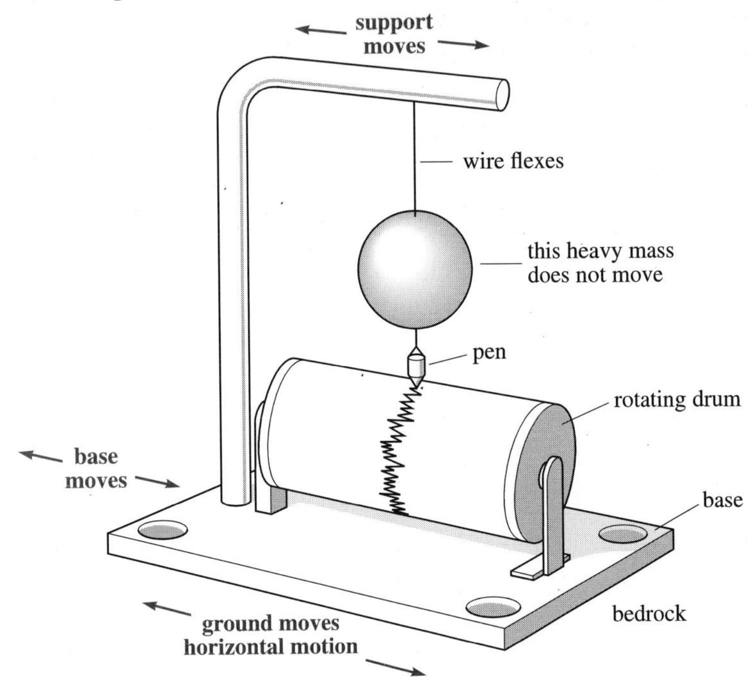

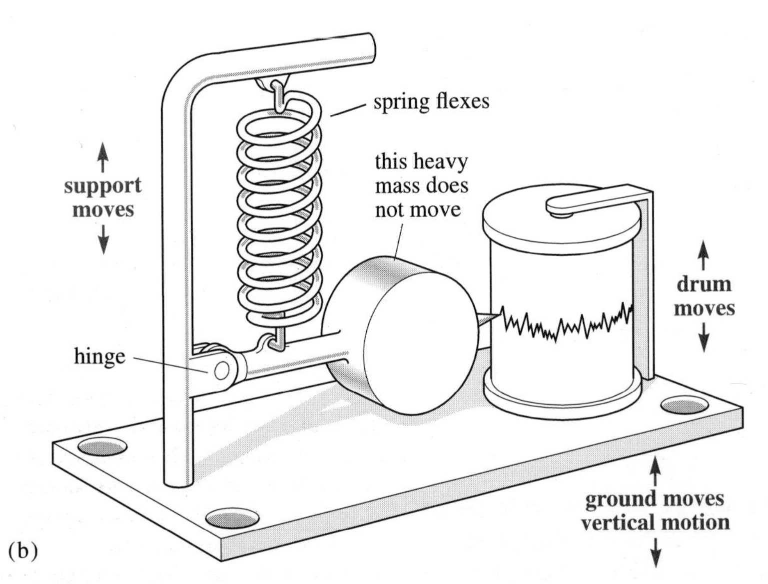

Figure 1.18 The principle of the seismometer. The instrument in (a) detects and records horizontal ground motions; that in (b) detects and records vertical motions.

Incidentally, you should not imagine that modern seismometers look anything like those in Figure 1.18. These days, seismometers are all grey boxes and digital readouts, as often as not connected directly to computers. Some instruments still have chart recorders, but others record the digital data onto magnetic tape or disc from which a chart recording can be recovered later. But however sophisticated a seismometer is, it is still based on the principles illustrated in Figure 1.18.

A seismometer detects and records seismic waves, of which there are four main types (Figure 1.19):

(1) **P-waves**, sometimes called longitudinal waves, involve the transmission of compressions and dilatations (expansions). As a P-wave passes through a material, the particles of the material vibrate backwards and forwards along the direction of wave travel.

(2) **S-waves**, or transverse waves, involve shear displacements. The motion of the material particles as the wave passes through is at right angles to the direction of wave travel and can be vertical (as in Figure 1.19b) or horizontal, or a combination of the two, depending on the nature of the earthquake.

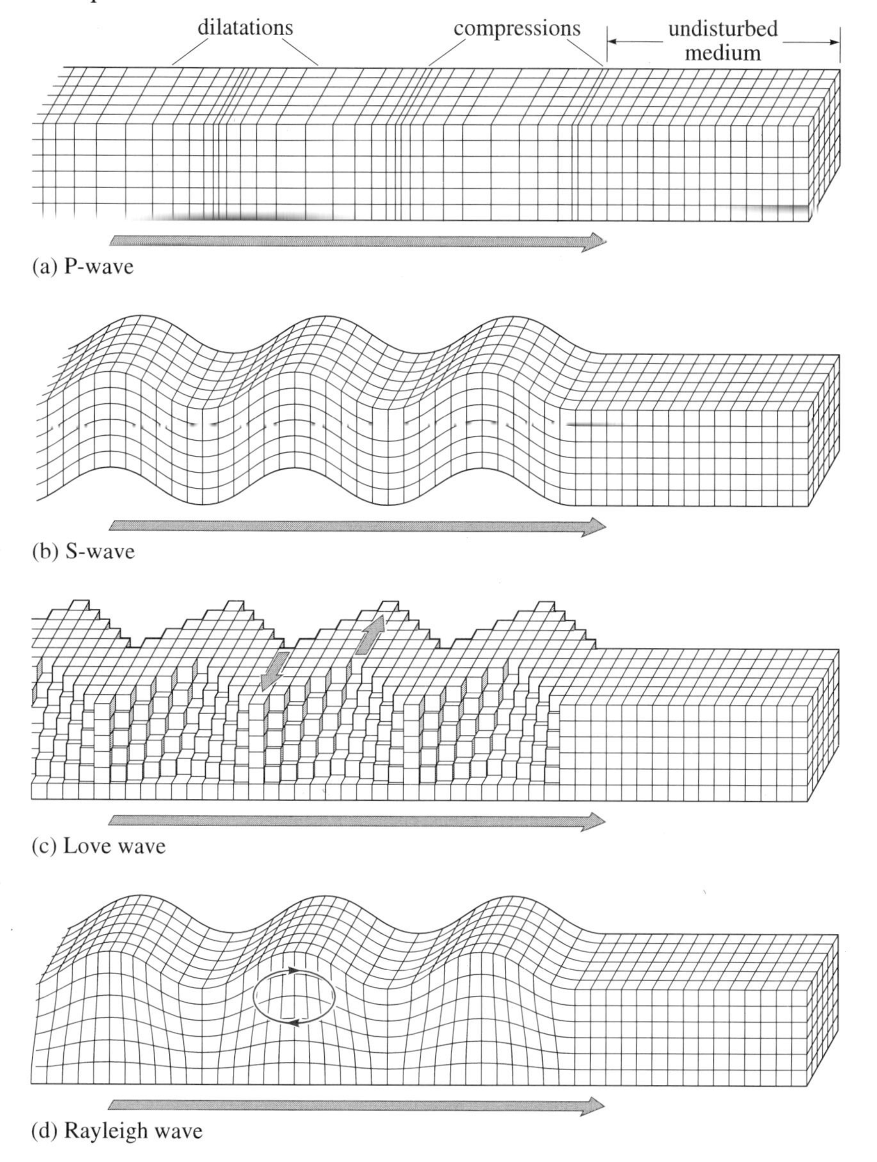

Figure 1.19 The four main types of seismic wave: (a) P-wave, (b) S-wave, (c) Love wave, and (d) Rayleigh wave. The broad arrows indicate the direction in which the waves are travelling.

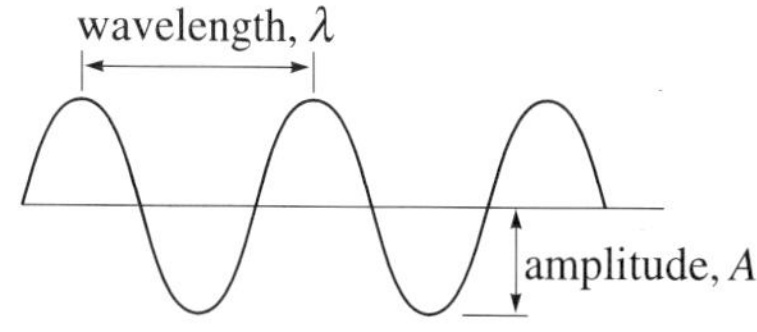

Figure 1.20 The general (perfect) wave. The wavelength (λ) of the wave is the distance between two adjacent peaks (or troughs). The amplitude (A) of the wave is half the height between a peak and a trough. In this diagram, the wave should be imagined as moving to the right. The time taken for adjacent peaks (or troughs) to move past a fixed point is called the wave period (T). Thus T is also the time it takes the wave to travel a distance λ. The number of peaks (or troughs) that pass a fixed point in one second is called the wave frequency (f). There is a simple relationship between λ, T, f and the velocity (v) of the wave, namely, $v = \lambda/T = f\lambda$. The units are: for λ, m; for T, s; for f, cycles per second = Hz; for v, m s^{-1}; and for A, m. Real waves are much more complicated than that shown here because they are usually packages of waves with different frequencies.

P-waves and S-waves are known as **body waves**, because they travel through the Earth's interior (i.e. through the body of the Earth). However, earthquakes also generate **surface waves**, so called because they are restricted to the vicinity of the Earth's surface (i.e. they travel near the ground surface). There are two main types of surface wave:

(3) **Love waves**, like S-waves, involve shear displacements, but they differ from S-waves in that the displacements are only horizontal and in that the **amplitude** of the waves decreases rapidly with increasing depth in the Earth. (The amplitude of a wave is its 'height' and is explained, along with other wave terms, in Figure 1.20.)

(4) **Rayleigh waves** are rather like ocean waves in that the particles of the medium through which the wave passes travel in vertical ellipses. As with Love waves, the amplitude of the disturbance decreases rapidly with increasing depth.

Seismic waves are often called **elastic waves**, although this is somewhat of a misnomer because the term 'elastic' really applies not to the waves but to the behaviour of the rocks through which they pass. When a seismic wave passes through a material, the material deforms momentarily as its particles vibrate but then returns to its previous configuration once the wave has passed; the wave is a disturbance that passes and leaves no trace. In other words, like a rubber band, the material behaves elastically. 'Elastic' here contrasts with 'plastic', which describes the behaviour of material which, when subjected to stress, remains permanently deformed (e.g. putty). Of course, the same material can behave differently when subjected to different forces. Rocks which behave elastically where seismic waves are concerned can also behave plastically when subjected to other kinds of influence (e.g. heat or pressure).

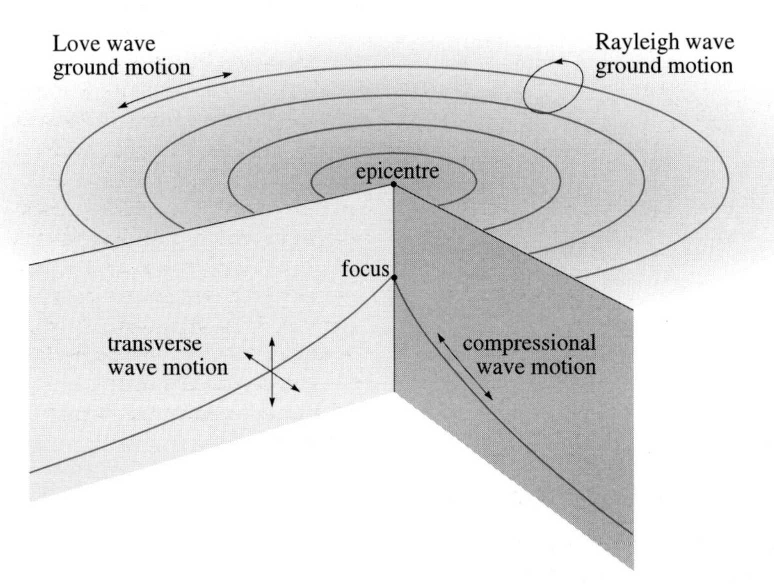

Figure 1.21 Earthquake waves. The shaded area at the bottom represents a cut-away view of the Earth's interior, revealing the focus vertically below the epicentre. The arrows indicate the directions of particle motion resulting from the four types of wave.

The waves from an earthquake source are transmitted outwards in all directions (see Figure 1.21). The source is usually a small volume of perhaps several cubic kilometres rather than a point, but for most practical purposes it can be regarded as a point, which is known as the **earthquake focus**, or hypocentre. The point on the Earth's surface vertically above the focus is called the **earthquake epicentre**. Because waves travel outward in all directions from the focus, they are potentially capable of being detected by seismometers anywhere, although whether or not a wave will actually reach a distant seismometer will depend partly on the size of the original earthquake and partly on how rapidly the wave attenuates (i.e. how rapidly its energy is dissipated) as it travels through or around the Earth. The bigger earthquakes generate waves that pass right through (or around) the Earth and emerge at the other side.

Within a given type of rock or mosaic of rocks, P-waves travel faster than S-waves, body waves travel faster than surface waves, and Love waves travel faster than Rayleigh waves. Thus as waves radiate outwards from an earthquake focus, the different types of wave separate out from one another in a predictable way.

❑ If a seismometer is used to detect waves from a distant earthquake, will the P-waves reach the seismometer before or after Love waves, assuming that the paths of the two types of wave are of roughly equal length and that the waves pass through more or less the same types of rock?

■ Before, because P-waves travel faster than Love waves.

Because the different types of wave travel at different speeds*, they arrive at the same seismometer at different times. This enables the different waves to be distinguished on ***seismograms***, the seismic traces recorded on the seismometer's chart. In the seismogram shown in Figure 1.22, the P-waves arrive first (because they travel the fastest), followed by S-waves and then surface waves.

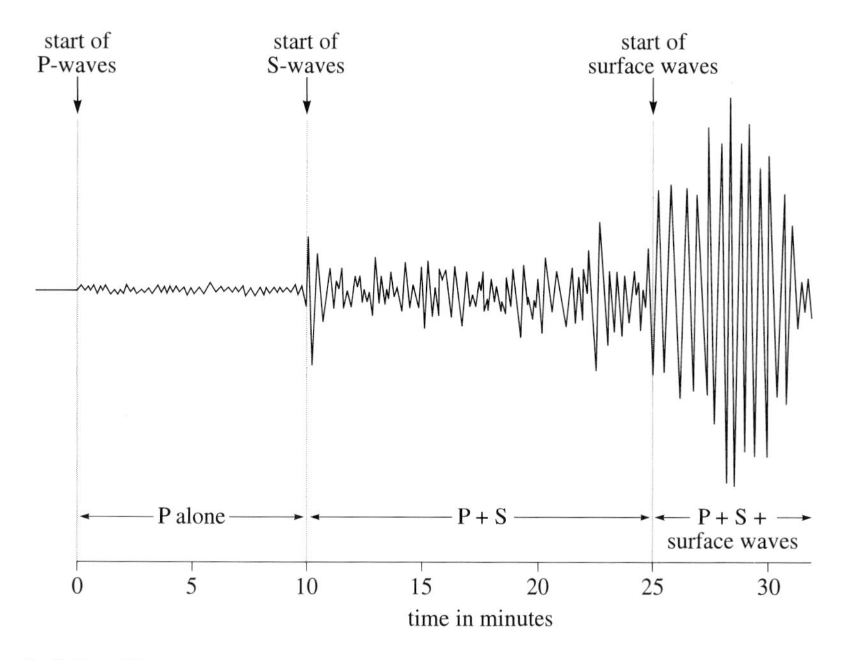

Figure 1.22 A simple seismogram, in this case produced by an earthquake that took place in Turkey in 1939 as recorded in Cambridge, Massachusetts, USA. The recording runs from left to right. The first arrivals (left) are P-waves. Then come the larger deflections made by S-waves (and the continuing P-waves) and finally the even larger deflections made by surface waves (plus the P-waves and S-waves). The Love and Rayleigh waves are not distinguished separately here (although it is possible to do this by computer analysis of the wave form); but as Love waves travel faster than Rayleigh waves, the very first surface waves to arrive must be Love waves. Note that on this particular seismogram the times are given in minutes.

1.5.2 EARTHQUAKE LOCATION AND DISTRIBUTION

Large amounts of information may be obtained from seismograms, not all of which are as simple as that shown in Figure 1.22; but here we shall concentrate on just two aspects of them, namely, what they tell us about the locations of the earthquakes that produced them, and (in the next Section) how they help in the design of a scale to specify the sizes of earthquakes.

Suppose that a seismometer at a point A on the Earth's surface records a seismogram such as that in Figure 1.22, the result of an earthquake that occurred close to the Earth's surface and an unknown point X. What we would like to know is how far away from A did the earthquake take place — in other words, what is the distance AX (which we shall call d). Velocity (strictly, speed) is, by definition, distance divided by time; so if v_P is the velocity of P-waves (in km s^{-1}) and t_P (in s) is the time it took the first P-waves to reach A from X (i.e. to travel the distance d), we have

$$v_P = \frac{d}{t_P} \qquad \text{(Equation 1.4)}$$

and similarly for the S-waves

$$v_S = \frac{d}{t_S}\,. \qquad \text{(Equation 1.5)}$$

Forget about v_P and v_S for a moment and think about t_P and t_S. These are the times it took the first P-waves and S-waves, respectively, to reach A from X. We know neither of these because we do not know when the earthquake occurred; indeed, we didn't even know that an earthquake had

* Seismic-wave speeds are quoted without reference to direction; in other words, they are indeed 'speeds' in the mathematical sense. However, incorrect though it may strictly be, seismic-wave speeds are invariably referred to as 'seismic-wave velocities', and we shall follow that convention here.

taken place until the seismometer detected it some time later. But the seismogram does tell us something about a relationship between t_P and t_S.

❑ Can you spot what the relationship is?

■ The seismogram in Figure 1.22 reveals that the S-waves reached the seismometer 10 minutes (600 s) after the P-waves. In other words, the seismogram gives us $t_S - t_P$.

So rearranging Equations 1.4 and 1.5, we have

$$t_P = \frac{d}{v_P} \quad \text{and} \quad t_S = \frac{d}{v_S}$$

and so

$$t_S - t_P = \frac{d}{v_S} - \frac{d}{v_P} = d\left(\frac{1}{v_S} - \frac{1}{v_P}\right)$$

and d (which is what we are after) is given by

$$d = \frac{t_S - t_P}{\left(\frac{1}{v_S} - \frac{1}{v_P}\right)} = \frac{t}{\left(\frac{1}{v_S} - \frac{1}{v_P}\right)} \qquad \text{(Equation 1.6)}$$

if, for simplicity, we replace $t_S - t_P$ by t.

The quantity t is given by the seismogram; so if only we knew v_P and v_S, we could determine d. Well, we do know v_P and v_S to a certain extent. Thousands of measurements made in the past show that in the upper continental crust, for example, the average seismic-wave velocities are $v_P = 5.60\,\text{km s}^{-1}$ and $v_S = 3.40\,\text{km s}^{-1}$. So putting these figures into Equation 1.6 gives

$$d = 8.65t \qquad \text{(Equation 1.7)}$$

This is a very simple result indeed, but, like many simple results, has its limitations. Equation 1.7 is based on the average values of v_P and v_S in the uppermost crust, but an earthquake may take place in a region where the actual values of v_P and v_S differ somewhat from these average values. Moreover, as the Earth is curved, if the earthquake (X) and the seismometer (A) are far apart, the waves travelling from X to A will pass through deeper parts of the Earth where v_P and v_S are higher than the values given above (for, as we shall see later in the Course, v_P and v_S increase with depth in the Earth). In practice, this means that Equation 1.7 really only applies when d is no more than a few hundred kilometres and that, for greater distances, other v_P and v_S values have to be used. Finally, the earthquake may not, as we have assumed, occur at or close to the surface but tens or even hundreds of kilometres down. If the earthquake focus is deep, the waves travelling from it to the seismometer will most certainly not go entirely through the uppermost crust, even if the seismometer is close to the epicentre. For all these reasons, Equation 1.7 must be regarded as an approximation only.

ITQ 14

A seismometer in California detects a near-surface earthquake that occurred within that State. The first P-waves arrive at the seismometer at 11 minutes and 12 seconds after noon and the first S-waves arrive at 11 minutes and 36 seconds after noon. Approximately how far from the seismometer did the earthquake take place?

❑ The answer to ITQ 14 gives the distance from the earthquake focus to the seismometer and hence, because the focus is close to the surface, the distance from the epicentre to the seismometer. Is this information sufficient to locate the earthquake?

■ No, because it doesn't give the direction of the earthquake relative to the seismometer. As it stands, the earthquake could have occurred anywhere along the circumference of a circle centred on the seismometer and having a radius equal to the answer to ITQ 14.

In fact, to locate an epicentre it is necessary to use a minimum of three seismometer stations well spaced apart. You should now do ITQ 15, which shows how an epicentre is determined.

ITQ 15

Figure 1.23 is a 'map' of an area in which a near-surface earthquake is known to have occurred because of recordings at seismic stations A, B and C. The P-wave and S-wave arrival times are as follows:

Seismic station	P-wave arrival times (GMT)			S-wave arrival times (GMT)		
	hours	minutes	seconds	hours	minutes	seconds
A	09	33	12.1	09	33	24.2
B	09	33	22.4	09	33	44.3
C	09	33	15.1	09	33	37.1

Determine the difference between the P-wave and S-wave arrival times for each station, convert these to distances, and then determine the position of the epicentre by drawing arcs of circles centred on A, B and C, with radii corresponding to the distances you have worked out.

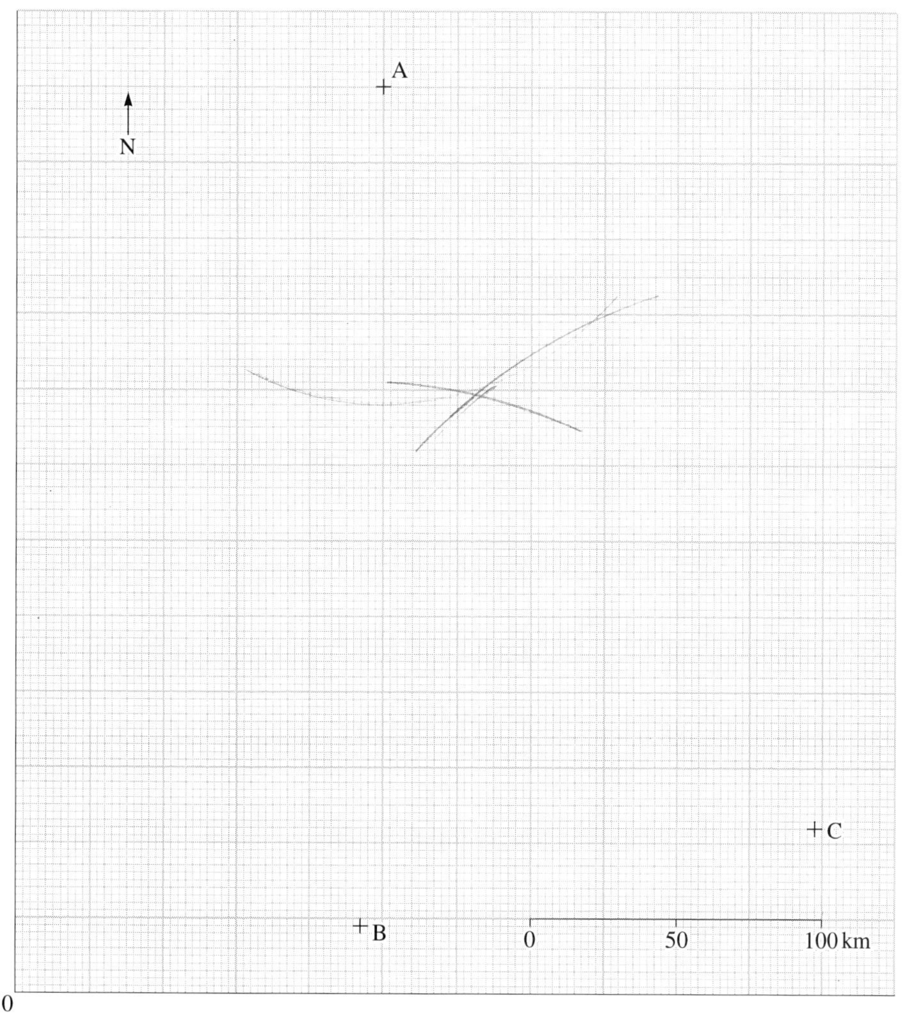

Figure 1.23 The positions of three seismic recording stations (for use with ITQ 15).

You will have seen from the answer to ITQ 15 that the arcs do not intersect exactly at a point. This is not uncommon and arises partly because Equation 1.7 is approximate, partly because experimental errors occur in the determination of t, and partly because earthquakes seldom occur conveniently at, or near, the Earth's surface. When the epicentre of a real earthquake is determined, seismograms from as many recording stations as possible (often over 60) are used in the hope that all errors and approximations will cancel out to give an accurate location.

There are now more than 1 000 continuously recording earthquake observatories operating around the world, which can detect and record even quite small earthquakes at very great distances. However, not all such observatories are lone stations. One of the more important developments in the 1960s was the establishment of **seismic arrays**, clusters of seismic stations (each with three or more seismometers), often arranged in geometric patterns. Figure 1.24 shows, as an example, the pattern of stations set up in Taiwan in 1980. Not only does such an array provide more accurate determinations of epicentres and other earthquake-source characteristics, it can provide directional information from analyses of the ways in which the seismic waves sweep across the array, especially when the digitized signals from each component station are immediately fed into a computer and thus become available for instant analysis.

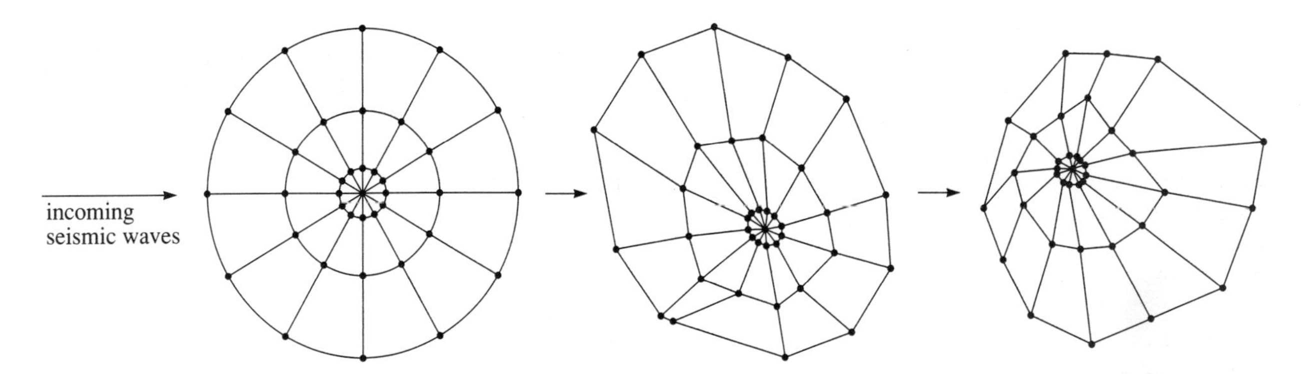

Figure 1.24 The arrangement of the seismic array in Taiwan. The array comprises 37 seismic stations arranged in three rings with radii of, respectively, 100 m, 1 km and 2 km, and with a station in the middle. When seismic waves impinge upon the array, each site is slightly displaced by the wave motion, producing distorted rings (seen here as plotted by computer graphics at two separate times — distortion highly magnified).

Tens of thousands of earthquake epicentres have now been determined; and when they are plotted on a world map, it immediately becomes clear that they are not randomly distributed across the Earth's surface but are mainly grouped into bands that weave around the globe. These are shown in Figure 1.25 and even more impressively on the Smithsonian Map. The banding of earthquake epicentres is highly significant in terms of the behaviour of the Earth's uppermost layers and will be discussed further in Block 2.

Nor are earthquakes randomly distributed with depth in the Earth. The **focal depth**, or **depth of focus**, of an earthquake (i.e. the depth at which the focus lies below the Earth's surface) may be determined using a technique similar to, but rather more refined than, that used to determine epicentres. All earthquake foci occur within the upper 720 km of the Earth. (The deepest shock ever recorded appears to have occurred at 720 km beneath the Flores Sea, East Indies, on 29 June 1934.) For analytical purposes, earthquakes are classified on the basis of depth of focus as **shallow** (0–70 km), **intermediate** (70–300 km) or **deep** (below 300 km).

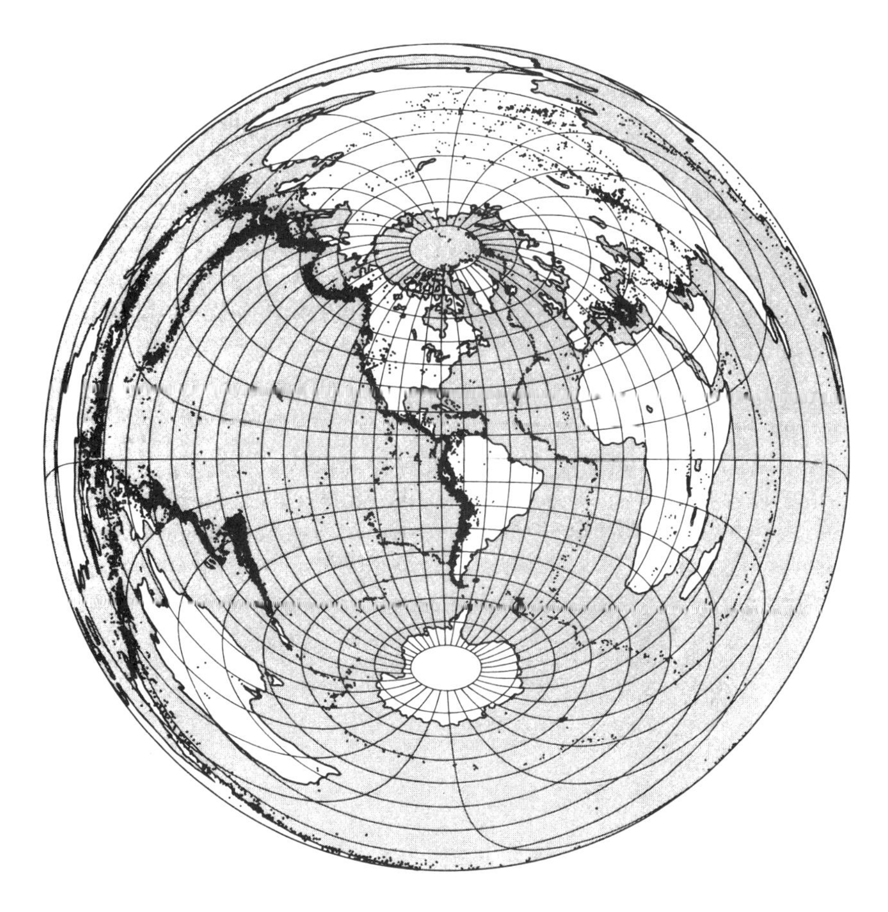

Figure 1.25 A view of global seismic activity, showing earthquakes of magnitude 4.5 or greater from 1963 to 1973. (Magnitude will be explained in Section 1.5.3.)

Seventy-five per cent of shallow earthquakes, 90% of intermediate earthquakes and almost all deep earthquakes occur around the margin of the Pacific, in the so-called circum-Pacific belt (see Figure 1.25). Most of the other large earthquakes occur in the Alpine–Himalayan mountain belt. Earthquakes are also concentrated along the ocean-ridge system (for reasons that will become clear in Block 2), although these are usually shallow-focus events and comparatively small (typical magnitudes 4.0–6.0; we shall discuss magnitudes in the next Section). In fact, the world's earthquakes are heavily biased in favour of shallow-focus events, as Figure 1.26 shows. Over 75% of the energy released by earthquakes derives from shallow-focus events and only about 3% from deep-focus events; so it is hardly surprising that shallow-focus earthquakes produce most of the damage at the Earth's surface. But to return to depth distribution, earthquakes below any given region are not necessarily limited to a particular depth nor even to a narrow depth range, and where earthquakes occur throughout a wide depth range they need not be distributed uniformly. Beneath Japan, to take but one example, earthquakes occur to depths in excess of 500 km, but there are significantly more foci in the 300–400 km depth range than in the 400–500 km or the 500–600 km ranges. In California, by contrast, most earthquakes are limited to the upper 20–25 km.

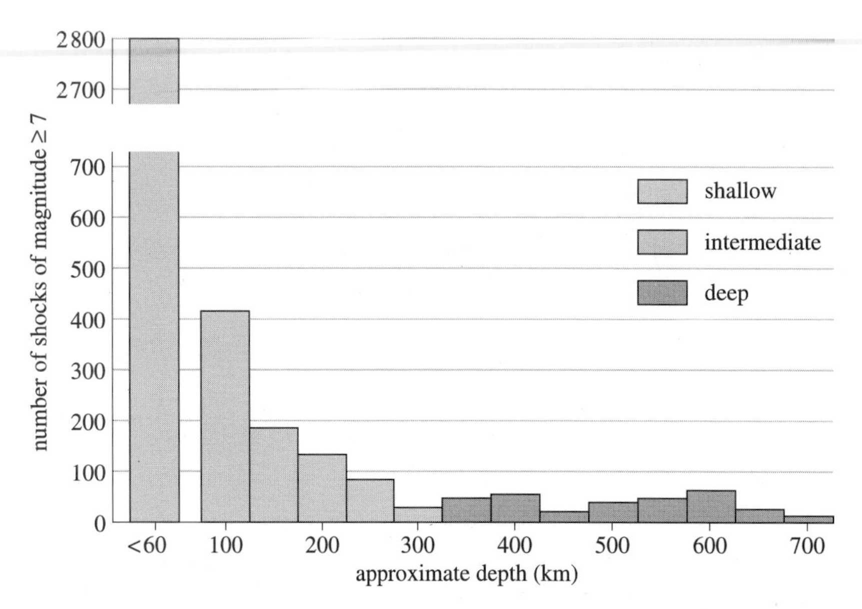

Figure 1.26 The variation in the number of earthquakes with depth for world events recorded between 1904 and 1945.

Although earthquakes are often unevenly distributed with depth, there is one particular type of environment in which they form a regular pattern which has a profound significance. Beneath the deep trenches that lie along certain continental edges, particularly around the margin of the Pacific Ocean, earthquake foci lie in a narrow band that usually dips at an angle of about 45° away from the ocean side and may even extend beneath the continent. Figure 1.27 illustrates a typical example — the shallow, intermediate and deep earthquake foci that lie beneath the Tonga trench in the southern Pacific. The Tonga trench lies above a subduction zone, a region in which oceanic lithosphere (crust and upper mantle) is plunging down beneath the adjacent continent; and the earthquake foci mark the position of the subducting mass. Earthquake patterns of this type were discovered long before subduction, and came to be called **Wadati–Benioff zones**. Such zones later came to be seen as important evidence for subduction; and as such they are part of the story of plate tectonics, which we shall be considering in more detail in Block 2.

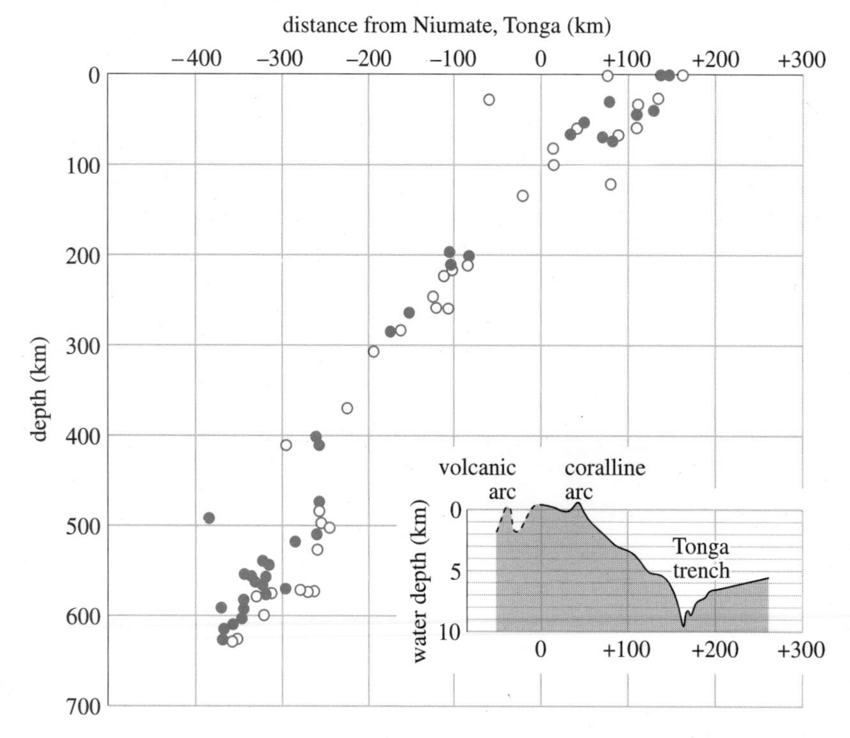

Figure 1.27 A vertical section perpendicular to the Tonga trench showing earthquake foci for the year 1965. The open circles represent foci projected from within 150 km to the north of the section; the filled circles represent foci projected from within 150 km to the south.

Finally, earthquakes, especially the larger ones, often occur as part of a series. Large-magnitude earthquakes are usually followed by a number of smaller **aftershocks**, occurring fairly close to the focus of the main shock (see Figure 1.28). Aftershocks are thought to be due to mechanical readjustments in the crust or mantle following the disruption caused by the main shock, and their frequency usually falls off gradually with time (often over several years for very large main shocks). Some large earthquakes are also preceded by smaller **foreshocks**; but as the characteristics of foreshocks are no different from those of other earthquakes, they cannot actually be identified as foreshocks until the main event has occurred.

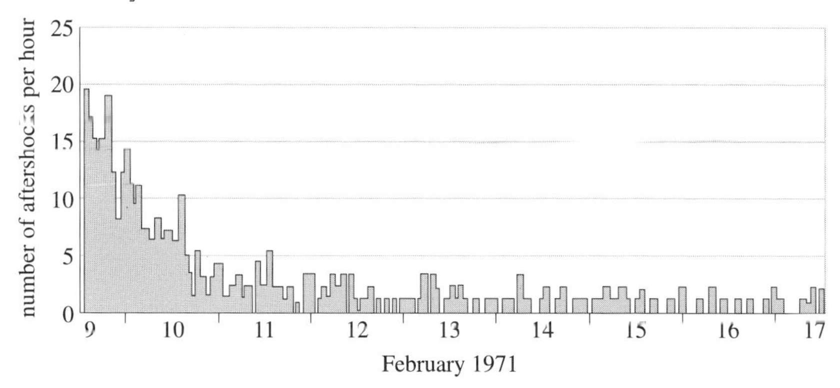

Figure 1.28 The hour-by-hour number of aftershocks of the San Fernando, California, earthquake of 1971.

1.5.3 EARTHQUAKE MAGNITUDE

It's quite obvious that earthquakes vary widely in size, from those producing huge amounts of damage to those so small that they cannot be felt but only detected by very sensitive seismometers. It's also clear that the damage and/or deaths that earthquakes cause are related not only to the sizes of the shocks. A large earthquake in California, for example, will produce much more damage than one of equal size in the middle of the Atlantic. Even on land, the effects of comparable earthquakes can be quite different. A seismic shock beneath a city generally causes more havoc than one of the same size in a rural region (simply because there are more people and buildings in a city); but equal-sized earthquakes occurring beneath two different cities can also have quite different consequences, depending partly on the types of construction involved, partly on the nature of the rock on which the cities stand (hard rock — igneous — is generally 'safer' than soft rock — sedimentary) and partly on the focal depth. In short, any attempt to devise a measurement of earthquake size based on the damage the event causes is doomed to failure.

Seismologists have therefore devised a much more objective measure, called **earthquake magnitude**, which is an absolute measure of the size of an earthquake based on the amount of energy released when the earthquake takes place. What they actually measure is the amplitude of the seismic waves (see Figure 1.20), which is related to the energy (the greater the energy, the greater the amplitude). The amplitudes of waves may be read from seismograms such as that shown in Figure 1.22, but not directly. Seismometers contain amplifying devices that make the amplitudes on the seismogram greater than those of the waves they represent; but as a seismometer's amplifying factor is known, it is a simple matter to determine the true amplitude of a wave from measurements of the seismogram.

❑ The use of wave amplitude to define earthquake magnitude immediately poses a practical problem. Can you think what that might be?

■ An earthquake may generate waves of a particular amplitude, but as the waves travel further and further from the earthquake focus they will gradually get weaker and spread out (i.e. attenuate), and thus suffer a decrease in amplitude, as their energy is dissipated. A seismometer closer to the earthquake focus will therefore record larger amplitudes than a seismometer further away.

It follows that any attempt to devise a magnitude scale must take the attenuation of waves with distance into account. (The alternative would be to insist that every seismometer must be at the same distance from the earthquake focus, but that is totally impossible!) To devise a magnitude scale, it is also necessary to allow for the fact that waves of different frequencies (or periods) have different amplitudes; so to standardize measurements, it is necessary to specify the frequency or period of the waves being used to define the magnitude. A general equation for magnitude (M) that takes all such factors into account is

$$M = \log\left(\frac{A}{T}\right) + af(\Delta,h) + b \qquad \text{(Equation 1.8)}$$

where

log is logarithm to the base 10,

A is the maximum amplitude of the wave at the recording site, measured in micrometres (10^{-6} m),

T is the period of the wave, measured in seconds (see Figure 1.20),

$f(\ )$ means 'function of' — i.e. a quantity that depends on the factors in the brackets,

Δ is the angular distance from the earthquake epicentre to the point at which the amplitude measurement is made, measured in degrees (this is called the **epicentral angle** and is shown in Figure 1.29),

h is the focal depth of the earthquake, measured in kilometres, and

a and b are constants.

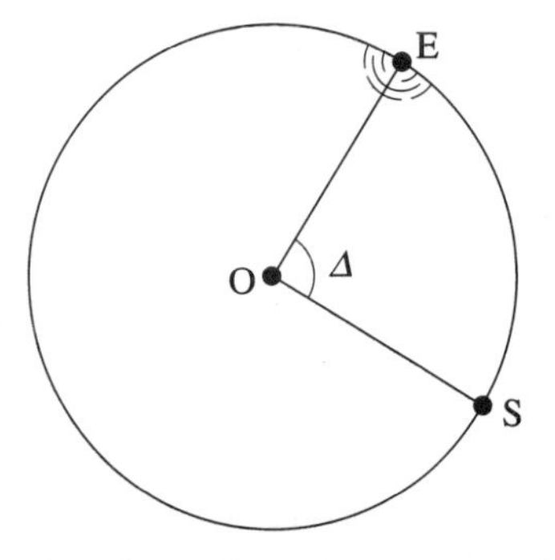

Figure 1.29 The epicentral angle (Δ, measured in degrees) is the angle subtended at the centre of the Earth (O) by that part of the Earth's circumference between the earthquake (E) and seismometer (S).

Don't worry; we know how you feel. Most of you are no doubt now asking: Where the devil has Equation 1.8 suddenly appeared from? What is it meant to *be*? What does it *mean*? Why does it contain those particular variables? Why are they arranged in that particular way? If it's any consolation to you, every member of the team producing this Course has asked himself/herself the very same questions on many occasions, for scientists are very fond of suddenly popping equations into the argument. To some extent, we all have to take equations such as 1.8 on faith, in the knowledge that past experience has shown them to work and that scientists have devised them on rational principles. Certainly, in a Course such as this we cannot always take time or space to delve into the development of every equation that appears. All the same, you should be able to see in general terms why Equation 1.8 is as it is.

Amplitude (A) appears in the top line because we have decided to *define* magnitude (M) in terms of amplitude (the larger the amplitude, the larger the earthquake). The period (T) appears because, as we have already mentioned, amplitude depends on period and so the relationship between the two must be built in. We have also mentioned that there is attenuation of the amplitude with distance; so there has to be a factor that allows for this distance effect (distance here being specified as an angle, Δ), which is also influenced by the depth (h) at which the earthquake takes place. The precise numerical relationships between the variables cannot be known from theory; so the constants a and b have to be included and then determined by observation. Finally, why the logarithm? This is just a matter of convenience. Because the range of energies released by earthquakes is very large, it is handier to contract the magnitude scale by making it logarithmic. There are several earthquake scales, and they are *all* logarithmic. This means that an earthquake of magnitude 7 is ten times larger (i.e. its amplitude is ten times greater) than one of magnitude 6, 100 times larger than one of magnitude 5, 1 000 times larger than on of magnitude 4, and so on.

The first magnitude scale was devised by the seismologist C. F. Richter in 1935 to compare the sizes of shallow earthquakes in California. Since then the concept of magnitude has been extended to earthquakes around the world and can involve the amplitudes of either body waves or surface waves. However, all magnitude scales based on Equation 1.8 are called 'Richter scales'; there is no such thing as *the* Richter scale.

One magnitude scale in widespread use is that for **surface-wave magnitude**, designated M_S, which is based on the maximum amplitude (A) of the horizontal component of surface waves (usually Rayleigh waves) having a period of 17–23 s. This is particularly suitable for shallow-focus earthquakes of focal depth less than 50 km whose waves are recorded at epicentral angles (Δ) in excess of 20°. Many observations have been used to determine that under these circumstances Equation 1.8 can be written as

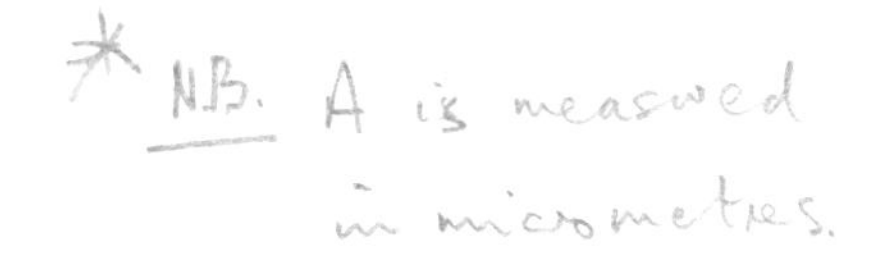

$$M_S = \log\left(\frac{A}{T}\right) + 1.66 \log \Delta + 3.3. \qquad \text{(Equation 1.9)}$$

Because earthquakes deeper than 50 km are not so efficient at generating surface waves, for such deeper earthquakes a correction of up to 0.4 (dependent on depth) must be added to the value of M_S determined by Equation 1.9. Moreover, at epicentral angles of less than 20° a correction must also be made for differences in wave absorption, scattering, spreading and dispersion. Different scientists have estimated this latter correction to be anywhere from 0.1 to 0.6. Although you will not be asked to apply such corrections in exercises, we mention them here because they illustrate a most important point, which is that, although it is easy to determine surface-wave magnitude, the result is not necessarily a very exact description of an earthquake. Because so many factors are involved, different groups of seismologists can easily disagree by up to 0.3 in the M_S of a particular earthquake under favourable conditions; and under less favourable conditions (e.g. inaccurate seismometers, corrections not applied properly, and so on), the disagreement in M_S can be much bigger. You may well have noticed such discrepancies in newspaper reports of large earthquakes, where different reporters have gone to different observatories to ask what the shock magnitude was. Usually, observatories later pool their data so that after several weeks or so they agree on an average value of magnitude, which then gets recorded in the scientific data books, but that is too late for the newspapers.

ITQ 16

An earthquake occurs at a depth of 10 km beneath California. At an epicentral angle of 30°, the Rayleigh waves having a period of 20 s are found to have a maximum amplitude of 2×10^{-4} m. Calculate M_S for this earthquake.

Because the deeper-focus earthquakes are not very effective in generating surface waves, their magnitudes are often determined using body waves (usually P-waves). This results in a **body-wave magnitude**, designated m_b, which is determined using body waves having a period of 12 s. In this case, observations suggest that the form of Equation 1.8 to use is

$$m_b = \log\left(\frac{A}{T}\right) + 0.01\Delta + 5.9. \qquad \text{(Equation 1.10)}$$

Note that this equation contains Δ, not $\log \Delta$.

ITQ 17

The earthquake in ITQ 16 also generated P-waves, those of period 12 s having an amplitude of 6×10^{-5} m. Calculate m_b for the earthquake.

❑ What do you notice by comparing the answers to ITQs 16 and 17?

■ That the two magnitude values (M_S and m_b) are different, even though the earthquake is the same.

There is, indeed, a slight difference between the surface-wave magnitude and body-wave magnitude scales. On the other hand, there is a relationship of sorts between M_S and m_b. What we mean by 'of sorts' will become clear when you have done ITQ 18.

ITQ 18

M_S and m_b were measured separately for 12 earthquakes, with the following results:

Earthquake	M_S	m_b
1	1.3	3.3
2	1.9	4.0
3	2.3	4.8
4	3.0	4.9
5	3.5	5.5
6	4.0	4.0
7	4.5	5.3
8	5.4	5.9
9	5.5	6.5
10	6.2	5.9
11	7.0	7.5
12	7.9	7.0

Plot these data on a graph with M_S up the vertical axis and m_b along the horizontal axis. What picture does the result give?

For worldwide M_S and m_b data, the equation of the best straight line that can be drawn through the data points is

$$m_b = 2.94 + 0.55M_S. \qquad \text{(Equation 1.11)}$$

This, then, is the equation to be used to calculate m_b when only M_S is known, or M_S when only m_b is known. Remember, however, that it can generally only give approximate answers because there is no strict one-to-one relationship between M_S and m_b.

The M_S and m_b scales are theoretically open-ended, which means to say that they have no upper limit. In practice, however, the Earth appears not to be able to produce earthquakes with M_S greater than about 9. The greatest known earthquake in Japan, for example, was that of 1933 which had an M_S of 8.9. By contrast, the San Francisco earthquake of 1906 had M_S = 8.3, the Tokyo earthquake of 1923 had M_S = 8.2, the Alaska earthquake of 1964 had M_S = 8.6, the Tangshan (China) earthquake of 1976 had M_S = 7.6, and the Mexican earthquake of 1985 had M_S = 7.9. Large earthquakes such as these are, of course, those that make the news, but they are actually quite rare. As Table 1.4 shows, on average there is only one earthquake of magnitude 8.0+ per year, and the smaller the earthquake the more of them there are each year.

Table 1.4 The worldwide mean annual frequency of earthquakes, based on events recorded from 1918 to 1945. (There are millions of earthquakes each year of magnitude below 2.)

Magnitude range	Number per year
≥ 8.0	1
7.0–7.9	18
6.0–6.9	108
5.0–5.9	800
4.0–4.9	6 200
3.0–3.9	49 000
2.0–2.9	300 000

Finally, you will recall that at the beginning of this Section we said that earthquake magnitude is a measure of the energy released by the earthquake. What has to be measured is the seismic-wave amplitude because, unlike energy, it can be determined from the seismogram, but there is a known relationship between the amplitude and energy of a wave, and the wave energy can be related to the energy (E) released by the earthquake. The expression for energy that emerges is

$$\log E = 4.8 + 1.5M_S \qquad \text{(Equation 1.12)}$$

where E is measured in joules (J).

ITQ 19

Determine the energy released by the 1906 San Francisco earthquake.

ITQ 20

With each increase in one unit of a magnitude scale (e.g. from 5.0 to 6.0), the corresponding seismic-wave amplitude increases by a factor of 10. Determine by what factor the energy released by the earthquake increases.

ITQ 21

With reference to Table 1.4, determine whether the one annual earthquake with magnitude greater than 8.0 releases more or less energy than those in the magnitude 2.0–2.9 range. (Assume for the purposes of calculation that the average magnitudes are, respectively, 8.5 and 2.5.)

We mentioned in Section 1.5.1 that as a wave travels outwards from an earthquake source it gradually attenuates; indeed, ultimately it dies out completely. What, then, happens to the energy contained in the original wave? The answer is that it is gradually dissipated as heat. As a wave progresses, it vibrates the particles of the material through which it is passing; and as the particles rub against each other they extract energy from the wave to overcome the friction between them.

1.5.4 Earthquake causes and mechanisms

Many shallow earthquakes are quite clearly associated with faults visible at the Earth's surface or with known subsurface faults. Indeed, it is no less clear that it is violent slips of such faults that cause the earthquakes in the first place. The traditional explanation of fault-generated earthquakes is the **elastic rebound theory** proposed by H. F. Reid in the aftermath of the great San Francisco earthquake of 1906. The fault from which an earthquake derives is regarded as a common boundary between two blocks of the Earth's crust. If crustal forces operate in such a way as to try to move the blocks relative to each other, no movement will take place along the fault in the first instance because of friction on the fault surface. Initially, therefore, the blocks remain locked. As the crustal forces persist, however, strain builds up in the blocks in the vicinity of the fault until, ultimately, the restraints are overcome and the two blocks suddenly slip with respect to each other, producing the earthquake. This sudden movement removes or reduces the pre-earthquake strain, although if the crustal forces continue to operate, strain will rebuild to generate a second shock later.

The principle of the elastic rebound theory is illustrated in Figure 1.30. In (a) the fault is initially in an unstrained state with the imaginary line AOB drawn perpendicularly across it. As the A block tries to move relative to the B block in the directions indicated by the arrows (under the action of crustal forces), the material at A moves to A′ and that at B moves to B′; but the fault remains locked at O as shown in (b). Diagram (c) shows the situation obtaining immediately before the slip occurs, by which time A has reached A″ and B has reached B″, although the fault is still locked at O. By (d) the earthquake has occurred by fault slip; and the whole region is again in an unstrained state, albeit with a displacement of the original imaginary line across the fault. The fault has slipped by distance XY. In the case of the 1906 San Francisco earthquake, for example, XY was about 6.3 m near the epicentre. The lengths of the faults along which slippage occurs vary from a few metres for very small earthquakes to more than 1 000 km for the largest. (Given that faults can slip for such large distances, you may wonder how an earthquake can possibly be located at a particular point. Once a fault begins to slip, it appears to slip instantaneously along its whole length. In fact, though very rapid, the movement is not instantaneous; it begins at a particular point, the point of rupture, and propagates along the fault at high speed. The earthquake focus lies at the point of the original rupture.)

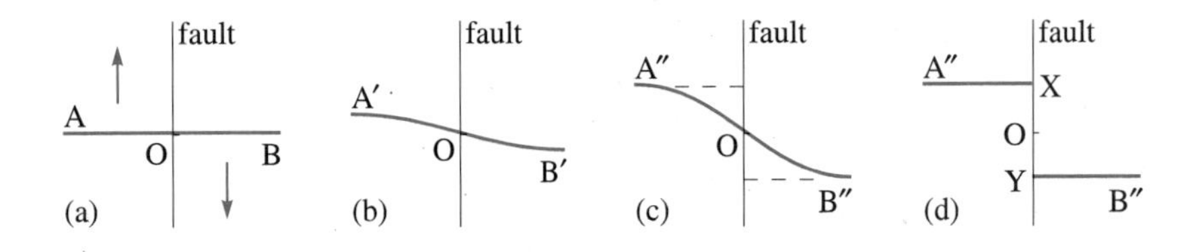

Figure 1.30 The elastic rebound theory of shallow earthquakes (see text for full explanation).

Although the elastic rebound theory appears to offer a reasonable representation of earthquakes associated with fault movements, it is by no means clear to what extent it applies in other circumstances. As you will see in Block 2, many earthquakes occur in regions where the nature of the faulting, if any, is difficult to discern. Moreover, there are problems even in continental regions where faults are obvious. The elastic rebound theory relies on the principle that there comes a point at which crustal forces have so great an effect that they overcome the friction locking the fault. At depths of more than a few tens of kilometres, however, the pressure between crustal blocks, and hence the frictional resistance along faults, becomes so great that no conceivable crustal forces could possibly prevail. In short, in theory at least, deep, intermediate and the deeper shallow earthquakes cannot be due to fault movement. The causes of earthquakes with foci deeper than a few tens of kilometres are still barely understood.

When an earthquake does occur, for whatever reason, the first wave from it to reach a seismic detector will be the P-wave, because P-waves travel faster than other types. When the records of a particular earthquake as obtained from many detectors are examined, however, the *first* P-wave disturbance to reach some stations is seen to be a compression (ground movement away from the earthquake focus) and at other stations is seen to be a dilatation (ground movement towards the focus). You should be able to see why this is so by looking at the P-wave illustrated in Figure 1.19a. Imagine a whole string of seismometers lying across the top of the wave. At any given moment, some instruments will be in zones of compression and some will be in zones of dilatation. Moreover, as the particles of the medium through which the wave is travelling are moving backwards and forwards along the direction of wave travel, a zone under compression at one moment will be under dilatation the next, and vice versa. It follows that when the P-wave first reaches any particular seismometer, it could do so as either a compression or a dilatation.

But the matter is not decided randomly. Whether the first P-wave arrives at a seismometer as a compression or dilatation depends on where the seismometer is located with respect to the **fault plane** (the fault surface). Figure 1.31a depicts a fault plane as viewed from above — i.e. what you see is the top edge of the fault; and the directions of the relative movements of the blocks on either side of the fault are indicated by the arrows. Also shown is the top edge of a plane at right-angles to the fault plane, known as the **auxiliary plane**.

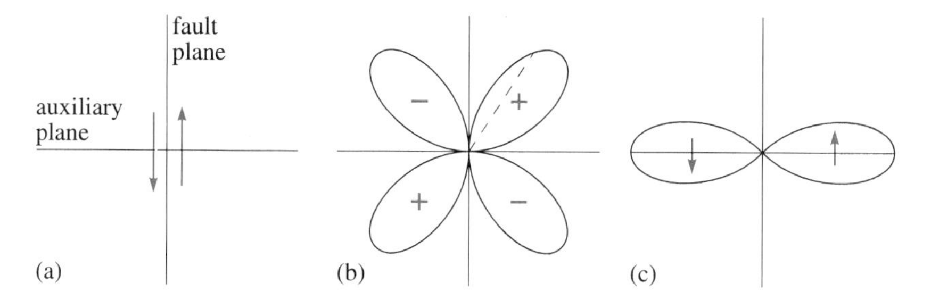

Figure 1.31 Plan view of (a) a single-couple (Type I) earthquake source, (b) the P-wave first-motion radiation pattern generated by it, and (c) the S-wave first-motion radiation pattern. Compression is indicated by + and dilatation by −. The arrows on the S-wave pattern indicate the first-shake direction. For each pattern, any line (e.g. the dashed line) drawn from the centre to a point on the perimeter of a lobe represents the amplitude of the ground motion.

When the fault slips, it generates P-waves that arrive first as compressions (+) in what in the perspective of Figure 1.31 are the top right and bottom left quadrants. In the top left and bottom right quadrants by contrast, the first P-waves arrive as dilatations (−). We must repeat that we are here talking only about **first motions**, the first disturbances of the ground as P-waves travel away from the earthquake focus. In fact, what a seismometer at any given point will 'see' is a series of alternating compressions and dilatations as the wave passes through; at issue here is merely which comes first. And which does come first can be determined from the seismogram by looking at whether the first P-wave to arrive sends the seismometer needle first up or down (up for compression, down for dilatation).

❑ Does the first P-wave in Figure 1.22 arrive as a compression or a dilatation?

■ The seismometer needle clearly first moved upwards, indicating a compressional first arrival. (Note: 'Up' indicates compression and 'down' indicates dilatation by convention. However, some seismometers could be wired 'the other way round', in which cases the reverse would apply in practice. It doesn't matter as long as the operator knows which way the instrument is connected.)

Although within a given quadrant the first-arrival P-wave is always the same, whether compressional or dilatational, the amplitude of the ground motion is not the same in all directions. It is a maximum at 45° to the fault plane and a minimum (zero, in fact) along the fault plane and at right-angles to it. The pattern of first arrivals, the **radiation pattern**, can thus be depicted as four 'lobes' (two compressional, two dilatational) in which a line (e.g. the dashed line in Figure 1.31b) from the centre to any point on the perimeter of a lobe represents the amplitude of the compression or dilatation — i.e. the length of the line (maximum at 45° and zero at 0° and 90° from the fault) is proportional to amplitude.

P-wave radiation patterns provide a way of determining the directions of fault and auxiliary planes. Suppose, for example, that an earthquake occurs in a region fairly heavily covered with seismometers but that the fault giving rise to the earthquake is hidden (i.e. not visible at the surface). By examination of the records of the various seismometers, it is possible to determine (i) where the P-wave first motions were compressional and where they are dilatational, and (ii) where the ground-motion amplitude was greatest and least. In other words, it is possible to construct a diagram such as Figure 1.31b, which immediately provides the directions of the fault and auxiliary planes. However, the answer is ambiguous in that, although the two planes can be defined, it's not possible to say which is which. At this point, it becomes necessary to consider the radiation pattern produced by the first S-waves. For the simple fault source shown in Figure 1.31a, the S-wave pattern possesses two lobes (Figure 1.31c). No S-waves are transmitted along the direction of fault propagation, but the S-wave amplitude reaches a maximum at right-angles to the fault. Although the S-wave radiation pattern is symmetrical, because there are only two lobes the relationship between the pattern and the fault plane is unambiguous, and so the fault plane may be determined uniquely. The fault source shown in Figure 1.31a is known as a **single-couple source**, or Type I source.

Unfortunately, it turns out that very few earthquakes give rise to two-lobe S-wave radiation patterns as shown in Figure 1.31c; most produce four lobes, as shown in Figure 1.32c. The reason for this is that fault planes are seldom, if ever, the perfectly plane surfaces depicted in diagrams. If they were, motion along the fault would be in one direction as implied by Figure 1.31a, and the fault would act as a single-couple source. In practice, however, fault planes are rough surfaces with projecting rocks, and their top edges (e.g. where the fault intersects the surface of the crust) are slightly wavy, rather than straight, lines. This means that there is almost always some component of fault motion at right angles to the obvious direction of the main fault movement. In effect, therefore, the fault source is a **double-couple source**, or Type II source, of the type shown in Figure 1.32a. This doesn't change the P-wave radiation pattern, but it does produce a four-lobe S-wave radiation pattern. The ambiguity inherent in the P-wave pattern is thus reflected in the S-wave pattern; and so the latter cannot, in practice, be used to distinguish between the fault and auxiliary planes. To make the distinction, it becomes necessary to use other information, such as knowledge of the local geology, to decide which of the two planes is most likely to be the fault plane.

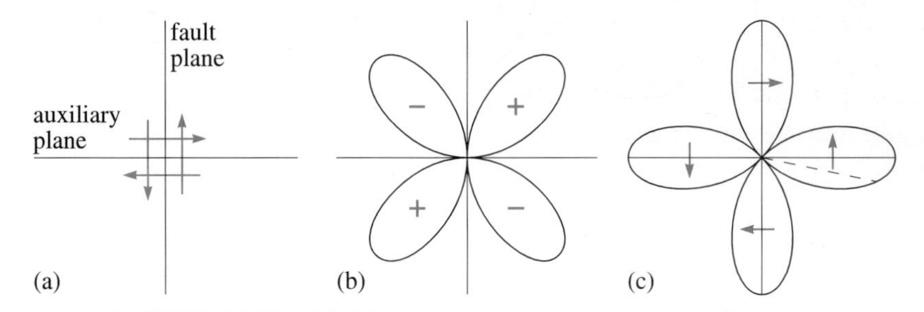

Figure 1.32 Plan view of (a) a double-couple (Type II) earthquake source, (b) the P-wave first-motion radiation pattern generated by it, and (c) the S-wave first-motion radiation pattern. Conventions are as for Figure 1.31.

You should have realized by now that the above discussion of first motions is rather a simplification, not least because it considers only two dimensions (i.e. the plan view of Figures 1.31 and 1.32). In practice, not only are the radiation pattern lobes three-dimensional, but the Earth has a spherical, rather than a flat, surface. This makes the geometry of radiation patterns much more complicated: but the *principles* remain the same as those discussed above. The determination of fault planes and auxiliary planes in the real, three-dimensional, spherical Earth are known as **fault-plane solutions**. We do not propose to go here into the complicated geometry and mental contortions needed to understand fault-plane solutions. You should appreciate, however, that they are extremely important because they enable Earth scientists to determine the directions in which rocks deep in the Earth are moving.

1.5.5 NON-EARTHQUAKE SEISMIC SOURCES

Earthquakes generate seismic waves which, as we shall see in the next Section, can be used to determine the Earth's internal structure. However, earthquakes are not the only phenomenon to produce seismic waves.

❑ Earlier in this Course, we mentioned another source of seismic waves, albeit one that is of no practical use in determining the Earth's structure. Can you recall what it was?

■ In Section 1.3.3, we noted that giant impacts generate seismic waves, one consequence of which was to distribute some of the heat of impact deep inside the Earth.

In fact, any 'bang' will generate seismic waves. You can produce them yourself by laying a metal plate on the ground and hitting it with a sledge-hammer. Geologists who wish to probe the Earth to explore for coal and minerals adopt the same principle, although as Figure 1.33 shows, they get a bigger bang. Explosions can also be used, including nuclear explosions. Of course, few of these artificial sources are large enough to send waves right through the Earth, as the larger earthquakes do. Manually hitting a plate will generate waves that penetrate only a few tens of metres at most, but sources such as that in Figure 1.33 can produce waves that penetrate tens of kilometres. Explosions (e.g. of TNT) can be made large enough to send waves down to the upper mantle; and waves from nuclear explosions, like those from earthquakes, can pass right through the Earth.

Figure 1.33 A seismic source: the device at the back end of this lorry thumps the ground, generating seismic waves.

There are definite advantages in using non-earthquake sources of seismic waves, namely, (1) that the time and position of the source are accurately known, (2) that waves may be produced at will, rather than being dependent on unpredictable natural events, and (3) that experiments can be planned in relation to geological structure and in regions where no earthquakes occur. The main disadvantage is that, with the exception of underground nuclear explosions, over which seismologists do not have control anyway, the magnitudes of non-earthquake sources cannot approach earthquake magnitudes, and so investigations are limited to comparatively shallow depths in the crust and upper mantle. On the other hand, non-earthquake sources are ideal for determining small-scale fine structure.

There is no difference in principle between the elastic waves produced by non-earthquake sources and those produced by earthquakes. Thus P-waves produced by an explosion will travel through a given medium with the same velocity as earthquake-generated P-waves. However, non-earthquake sources are not very efficient at producing either S-waves or surface waves, and so they are regarded primarily as sources of P-waves.

SUMMARY OF SECTION 1.5

The most productive series of techniques for determining the structure and physical state of the Earth's interior are those embodied in the science of seismology, the study of the behaviour of seismic waves (from earthquakes, explosions or other shock-producing mechanisms) as they pass through the Earth. Earthquakes are sudden releases of strain energy in the Earth's crust or upper mantle; and the waves from such releases and from non-earthquake sources may be detected and recorded by seismometers.

There are four main types of seismic wave: P-waves and S-waves (body waves, which can pass through the Earth's interior) and Love waves and Rayleigh waves (surface waves, which are restricted to the vicinity of the Earth's free surface). Seismic waves, which are 'elastic waves', are transmitted in all directions from the focus of a shock and gradually attenuate with distance as their energy is dissipated in the form of heat. The four types of wave travel with different velocities through a given rock type, and so can usually be identified individually on seismograms, the seismic traces recorded on a seismometer's chart. The four waves in order of decreasing velocity are: P-waves, S-waves, Love waves and Rayleigh waves.

By measuring the times of arrival at a seismometer of the P-waves and S-waves from an earthquake, it is possible to determine how far from the seismometer the earthquake took place. If the seismic waves are recorded at three well-spaced seismometer stations, it is also possible to determine the position of the earthquake's epicentre. Similar techniques may be used to determine the earthquake's depth of focus. Seismic arrays can provide more accurate determinations of earthquake-source characteristics. Neither earthquake epicentres nor earthquake foci are randomly distributed; most epicentres fall along narrow bands on the Earth's surface and, beneath trench systems, foci lie on dipping bands. Both of these distributions provide important evidence for plate-tectonic processes (and will be examined in more detail in Block 2).

Earthquake magnitude is an absolute measure of the size of an earthquake and is ultimately related to the amount of energy released when the earthquake takes place. In practice, however, it is measured in terms of the amplitude of the waves at the recording site; and because waves attenuate with distance from the earthquake focus, a formula for magnitude must be devised that takes into account the attenuation as well as such other factors as the depth of the earthquake's focus and the period of the waves. In fact, several scales, all known as Richter scales, can be devised, based on either surface waves or body waves. It is also possible to derive a formula for determining the energy released by an earthquake in terms of the earthquake's magnitude.

The simplest hypothesis for the cause of earthquakes, especially shallow ones, is that they are due to sudden slippage of faults along which strain has built up — a phenomenon incorporated into H. F. Reid's elastic rebound theory. However, it is difficult to see how the elastic rebound theory could apply to earthquakes at depths of more than a few tens of kilometres where, in theory, pressures should be too large to allow fault slippage. But whatever the cause of earthquakes, the first P-wave and S-wave disturbances to reach a seismometer can be in the form of either a compression or a dilatation, depending on the position of the seismometer with respect to the location of the earthquake. By observing where the first wave arrivals are, respectively, compressions and dilatations, it is possible to determine the directions of the fault plane and auxiliary plane. However, it is seldom possible to distinguish between the two on the basis of seismic evidence alone; and, in practice, other evidence (e.g. local geological conditions) is required.

Non-earthquake sources also generate seismic waves, but they are less efficient than earthquakes at producing S-waves. On the other hand, there are definite advantages in using non-earthquake sources, chiefly in that the timing and location of the sources are under the seismologist's control.

OBJECTIVES FOR SECTION 1.5

When you have completed this Section, you should be able to:

1.1 Recognize and use definitions and applications of each of the terms printed in the text in bold.

1.16 Describe in general terms the principle of the seismometer and recognize the need for three instruments at each seismic station.

1.17 Summarize the characteristics of the four main types of seismic wave (P-waves, S-waves, Love waves and Rayleigh waves) and of the general (perfect) wave.

1.18 Describe the chief characteristics of earthquakes, their location and distribution.

1.19 Explain how earthquake magnitude scales may be devised, used and extended to determine the energy released by an earthquake.

1.20 Outline the elastic rebound theory for the origin of shallow, fault-related earthquakes and note its limitations.

1.21 Show how the radiation patterns produced by the first P-waves and S-waves (first motions) to arrive at a seismometer may be used to determine the directions of an earthquake's fault and auxiliary planes.

1.22 List the advantages and disadvantages of using non-earthquake sources of seismic waves.

1.23 Perform simple calculations relating to the material covered in Objectives 1.16–1.22.

Apart from Objective 1.1, to which they all relate, the seven ITQs in this Section test the Objectives as follows: ITQs 14–15, Objectives 1.18 and 1.23; ITQs 16–21, Objectives 1.19 and 1.23.

You should now do the following SAQs, which test other aspects of the Objectives.

SAQS FOR SECTION 1.5

SAQ 23 (*Objectives 1.1, 1.16, 1.17, 1.18, 1.19, 1.21 and 1.22*)

State, giving reasons where appropriate, whether each of the following statements is true or false.

(a) An earthquake is, in general terms, a sudden release of stress in a local region of the Earth's crust or upper mantle.

(b) An earthquake recording station will comprise at least three seismometers, one for measuring vertical motions and two for measuring horizontal motions in directions at right angles.

(c) When a P-wave travels in a given direction, the particles of the material through which the wave is travelling vibrate backwards and forwards along the direction of wave travel.

(d) In Rayleigh waves, the particles of the medium through which the wave is travelling move in horizontal ellipses.

(e) S-waves are body waves.

(f) Love waves are body waves.

(g) Surface waves attenuate rapidly with depth in the Earth.

(h) Surface waves are elastic waves.

(i) The earthquake focus is a point (actually a small volume) at which an earthquake originates.

(j) All earthquake epicentres lie on the surface of the Earth.

(k) The four main types of seismic wave in order of decreasing velocity within a given type of rock are: S-waves, P-waves, Love waves and Rayleigh waves.

(l) More earthquakes occur around the margin of the Pacific Ocean than in any other region of the world.

(m) Wadati–Benioff zones are characteristic of oceanic trench environments.

(n) According to the elastic rebound theory, rocks on either side of a fault will revert to their original positions after an earthquake has occurred.

(o) Non-earthquake seismic sources are efficient generators of both P-waves and S-waves.

SAQ 24 (*Objectives 1.17 and 1.23*)

Determine the wavelength and frequency of a perfect wave that has a velocity of $6\,\mathrm{km\,s^{-1}}$ and a period of 20 s.

SAQ 25 (*Objectives 1.17 and 1.23*)

The unit of frequency is the hertz (Hz), a derived unit named after the physicist Hertz. What is the unit of frequency in terms of the base units kg and/or m and/or s?

SAQ 26 (*Objectives 1.18 and 1.23*)

A seismometer in Rome detects a near-surface earthquake that occurred in southern Italy. The first P-waves arrive at the seismometer 43.9 s before the first S-waves. Approximately how far from the seismometer did the earthquake take place?

SAQ 27 (*Objectives 1.18 and 1.23*)

Three seismic recording stations (A, B and C) in an array lie at the corners of a right-angled triangle. A is 16.5 km due north of B, and C is 10.5 km due west of B. P-waves and S-waves are received from a shallow earthquake at the following times (GMT):

Station	P-waves			S-waves		
	hours	minutes	seconds	hours	minutes	seconds
A	12	30	55.4	12	30	58.9
B	12	30	54.5	12	30	57.4
C	12	30	56.3	12	31	00.4

Determine (a) the position of the earthquake epicentre relative to station B and (b) the time of the event.

SAQ 28 (*Objectives 1.19 and 1.23*)

An earthquake occurs at a depth of 20 km. A seismometer at an epicentral angle of 41.7° from the earthquake records Rayleigh waves having a period of 20 s and a maximum amplitude of 2×10^{-4} m. Determine the magnitude M_S for the earthquake.

SAQ 29 (*Objectives 1.19 and 1.23*)

The earthquake in SAQ 28 also generates P-waves, those of period 12 s having an amplitude of 6.0×10^{-5} m at the seismometer. Determine the magnitude m_b for the earthquake.

SAQ 30 (*Objectives 1.19 and 1.23*)

Determine two values for the energy released by the earthquake in SAQ 28. Which of the two do you think is likely to be the more valid, and why?

SAQ 31 (*Objectives 1.19 and 1.23*)

Which would release the more energy, a single earthquake of $M_S = 8$ or a million earthquakes each of $M_S = 4$?

SAQ 32 (*Objectives 1.17, 1.18, 1.19, 1.20 and 1.21*)

Match each of the items (1)–(10) with *one* of the items (a)–(j).

(1) An earthquake

(2) Rayleigh waves

(3) Love waves

(4) Intermediate-focus earthquakes

(5) Nearly all deep-focus earthquakes

(6) Surface-wave magnitude

(7) The elastic rebound theory

(8) An earthquake epicentre

(9) Radiation patterns

(10) A double-couple source

(a) occur in the depth range 70–300 km.

(b) involves ambiguity in both the P-wave and S-wave radiation patterns.

(c) is a sudden release of strain energy.

(d) have particle motions restricted to the vertical plane containing the direction of wave propagation.

(e) applies to shallow, fault-related earthquakes.

(f) are, by definition, produced by first motions.

(g) occur around the margin of the Pacific.

(h) have particle motions restricted to the horizontal plane perpendicular to the direction of wave propagation.

(i) is the point on the Earth's surface vertically above the earthquake focus.

(j) is an absolute measure of earthquake size related to the seismic energy released.

1.6 SEISMOLOGY

An earthquake taking place near the Earth's surface emits waves in all directions. Some of the paths taken by those waves are shown in Figure 1.34, where they are assigned code letters or combinations of letters. PKIKP, for example, represents a P-wave from the earthquake that passes down through the mantle, outer core and inner core, then up again through the outer core and mantle. ScS is an S-wave that goes down through the mantle, is reflected at the surface of the outer core and returns through the mantle as an S-wave. SKP, on the other hand, is an S-wave that passes down through the mantle, enters the outer core as a P-wave (the outer core is fluid, and S-waves cannot pass through fluids) and returns through the mantle as a P-wave*. SKS is the same, except that it returns through the mantle as an S-wave. *You do not need to remember these wave designations.* What you should be able to do, however, is to spot a number of different characteristics of wave behaviour in the Earth.

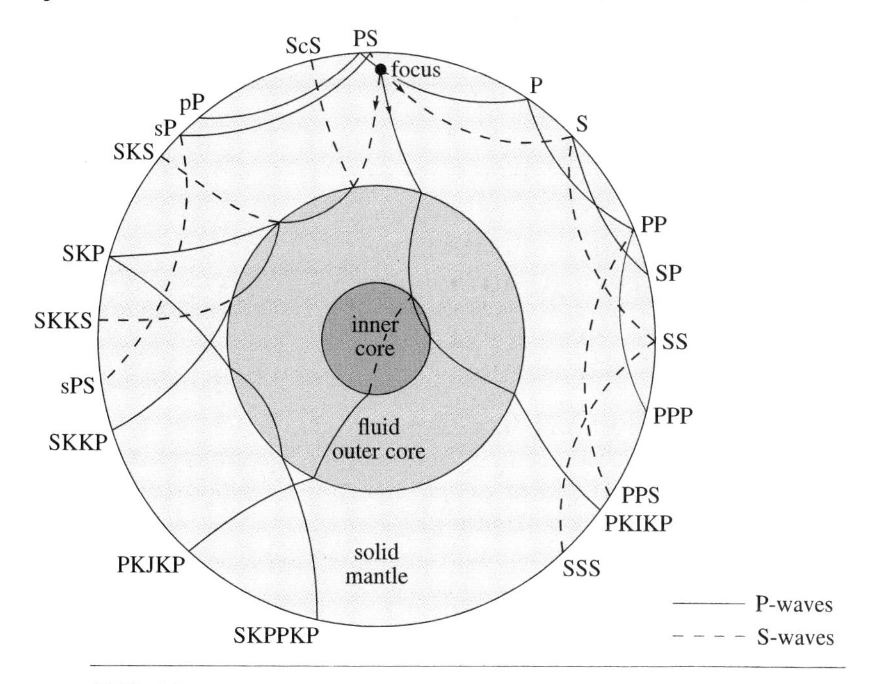

Figure 1.34 A selection of P-wave and S-wave paths through the Earth. The waves are emitted from the earthquake focus near the top of the diagram (see text for discussion of wave paths).

ITQ 22

Examine the wave paths in Figure 1.34 and then make a list of as many characteristics of wave behaviour as you can.

The picture of the various wave paths through the Earth is very confusing, especially bearing in mind that Figure 1.34 shows only a few representative examples. Moreover, the geometry and mathematics of such wave paths are complicated, because the Earth is a spherical body. The chief *principles* of wave behaviour are actually quite simple, however. What we propose to do, therefore, is to explain these principles in the context of simple, horizontal layers. We shall return later to the question of what the waves passing through the Earth tell us about the planet's interior.

1.6.1 SEISMIC-WAVE VELOCITIES

The velocity of a seismic wave passing through a medium depends on the properties of the medium. But which properties? In fact, there are two of relevance — elastic characteristics and density. You will recall that in

* For reasons that are beyond the scope of this Course, when P-waves or S-waves are reflected or refracted at boundaries in the Earth they can generate waves of the other type.

Section 1.5.1 we made the point that material reacts elastically to seismic waves, meaning that the waves disturb the material but, having passed through, leave it as it was originally. There must therefore be some property of material that tends to resist the deformation caused by the passing wave. We call this the **elastic modulus**, although as we shall see in a moment there is more than one elastic modulus.

If we apply a stress to a material, the material distorts, or strains, as long as the stress is maintained (remember that we are dealing here only with elastic behaviour, not plastic deformation). The elastic modulus is then defined in general as

$$\text{elastic modulus} = \frac{\text{stress}}{\text{strain}}.$$

If the material (e.g. rock) deforms easily, that means that its strain (deformation) is large in relation to the stress applied, and the elastic modulus is small. If, on the other hand, a material strongly resists when a stress is applied, the resulting strain will be small and the elastic modulus large.

❑ Which do you think will transmit seismic waves the more rapidly, easily deformed material (small elastic modulus) or material more resistant to deformation (large elastic modulus)?

■ The more easily deformable, or compressible, a material is, the longer it takes a wave to be transmitted through it and hence the lower is the wave velocity. The railway-wagon analogy is useful here: the more flexible the couplings, the longer it takes a disturbance to travel from one end of a line of wagons to the other. In short, the larger the elastic modulus, the higher is the wave velocity.

❑ How do you think density will affect wave velocity?

■ Density is defined as mass/volume. Thus the greater the density, the greater the mass in a given volume, and hence the greater the inertia. It will therefore take longer for a wave to be transmitted through more dense than through less dense material. In other words, the greater the density, the lower is the wave velocity. The railway-wagon analogy may be invoked here too. A disturbance will take longer to travel along a line of full wagons (higher density) than along a line of empty ones (lower density).

So putting the two properties (density and elastic modulus) together, we can say that

$$\text{wave velocity} \propto \frac{\text{elastic modulus}}{\text{density}}.$$

In fact, it can be shown experimentally that

$$\text{wave velocity} = \sqrt{\frac{\text{elastic modulus}}{\text{density}}}. \qquad \text{(Equation 1.13)}$$

Equation 1.13 means that an increase in wave velocity can be due to: (a) an increase in elastic modulus, (b) a decrease in density, or (c) an increase in both, but with elastic modulus increasing by more than density.

Different kinds of seismic wave will subject material to different forms of stress and, in turn, the material will undergo different types of elastic deformation, or strain. However, stress is always defined as the force applied to unit area, which means that if a force F is applied to an area A, the stress is given by F/A (unit: N m^{-2}).

Consider, first, a P-wave, in which the particles of the material through which the wave is passing are forced to move backwards and forwards along the direction of wave travel (see Figure 1.19a). In this case, the stress exerted on the material by the wave acts head-on, as shown by the arrow in Figure 1.35a. The effect of the stress is then to change the volume of a cube of the material (original volume, V) to $V - \Delta V$ (i.e. by a small amount ΔV). The strain is defined by

$$\text{strain} = \frac{\text{change in volume}}{\text{original volume}} = \frac{\Delta V}{V}$$

and stress = F/A, so we can write

$$\text{elastic modulus} = \frac{\text{stress}}{\text{strain}} = \frac{F/A}{\Delta V/V}$$

and this is called the **bulk modulus**, K. So

$$K = \frac{F/A}{\Delta V/V}\,. \qquad \text{(Equation 1.14)}$$

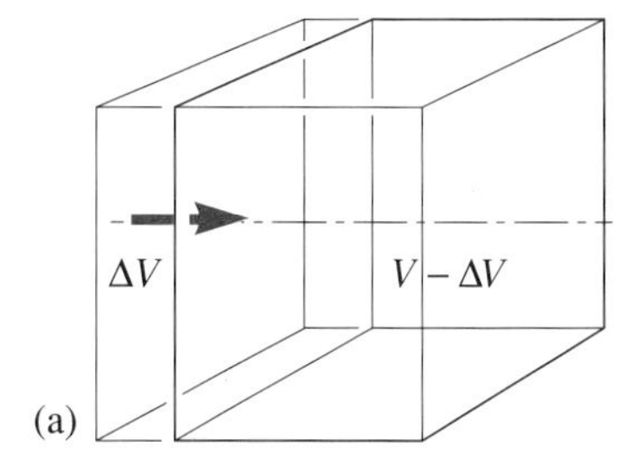

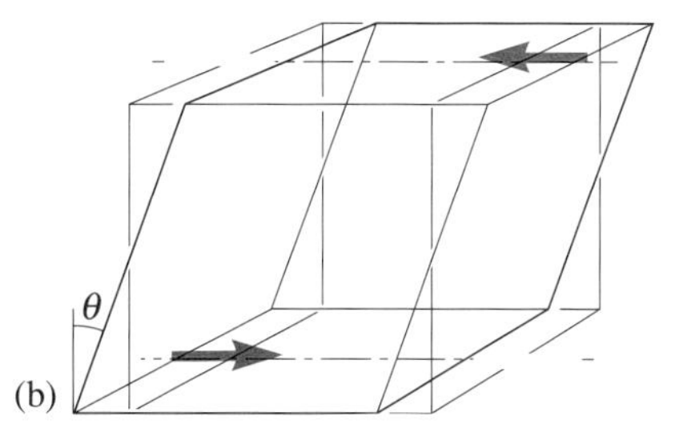

Figure 1.35 The behaviour of cubes of material under stress. That in (a) is subjected to longitudinal compressive stress as during the passage of a P-wave, and that in (b) is subjected to shear stress as during the passage of an S-wave. In each case, the heavy lines indicate the shape and dimensions of the deformed cube, and the fine lines the shape of the undeformed cube. V, ΔV and θ are defined in the text. Note that the arrows represent the direction of *particle* movement. In (a), this is the same as the direction of *wave* motion because the wave is a P-wave. In (b) however (S-wave), the particle movement is at right-angles to the wave motion; so the wave is directed into the paper.

For an S-wave the position is rather different. An S-wave is a shear wave, in which the particles of the material through which the wave is passing are forced to move at right-angles to the direction of wave-travel (see Figure 1.19b). In this case, the wave stress is exerted along the faces of the material, as shown in Figure 1.35b, and the strain is defined as the tangent of the angle (θ) by which a cube of the material is sheared. The modulus in this situation is called the **shear modulus**, or **rigidity modulus** (μ), given by

S8

$$\mu = \frac{F/A}{\tan\theta}\,. \qquad \text{(Equation 1.15)}$$

In the light of this discussion, it might appear that the elastic modulus to use in Equation 1.13 would be K for P-wave velocity and μ for S-wave velocity. In fact, this is not quite the case. Both theory and experiment show that if v_P is the P-wave velocity (units: $\mathrm{m\,s^{-1}}$), v_S is the S-wave velocity ($\mathrm{m\,s^{-1}}$) and ρ is the density of the material through which the waves are passing ($\mathrm{kg\,m^{-3}}$)

$$v_P = \sqrt{\frac{K + 4\mu/3}{\rho}} \qquad \text{(Equation 1.16)}$$

and

$$v_S = \sqrt{\frac{\mu}{\rho}} \qquad \text{(Equation 1.17)}$$

where the units of both moduli are $\mathrm{N\,m^{-2}}$. Thus S-wave velocity does indeed depend only on the modulus μ. P-wave velocity, by contrast, depends on both K and μ. The reason for this is that the longitudinal compressions and dilatations caused by P-waves also set up shearing motion in the surrounding rocks, and that affects the P-wave velocity. Things are not always as simple as they seem! Incidentally, Equation 1.16 is sometimes written as

$$v_P = \sqrt{\frac{\psi}{\rho}} \qquad \text{(Equation 1.18)}$$

where $\psi = K + 4\mu/3$ and is known as the **axial modulus**.

Equations 1.16 and 1.17 can be seen to be consistent with certain observations that we have already come across. For example, as K is always positive (by definition), $K + 4\mu/3$ must always be greater than μ. So, in any given material, the velocity of P-waves must always be higher than the velocity of S-waves. In addition, for a fluid $\mu = 0$ (a fluid cannot be sheared) and so $v_S = 0$; that is, S-waves cannot pass through a fluid. Finally, you should note that elastic moduli are not independent of density. However, as density increases, the elastic moduli increase relatively more. Thus, in general, the greater the density of a material, the higher the wave velocity within it, whereas, on the face of it, Equations 1.16 and 1.17 would appear to indicate precisely the opposite.

ITQ 23

Experiments carried out in a large granite quarry to determine the P-wave and S-wave velocities of granite gave the following results:

Distance from source	P-wave travel time	S-wave travel time
750 m	0.136 s	0.250 s
1 000 m	0.182 s	0.330 s

(a) Calculate the average values of v_P and v_S from these observations; and (b) determine the bulk and rigidity moduli of granite, assuming the density of granite to be $2\,700\,\mathrm{kg\,m^{-3}}$.

1.6.2 SEISMIC REFRACTION

The study of the way that seismic waves are refracted in the Earth provides information on the Earth's internal structure. The basis of the seismic refraction method is to initiate seismic waves at one point and then determine how long it takes for the waves to reach a series of observation points. In travelling from the source to the detectors, the waves may be refracted (or reflected) at any velocity boundaries, or **discontinuities**, which happen to lie in their path. A discontinuity in this context is a boundary between two layers in which the seismic-wave velocities are quite different. This could arise because the two layers are made of different materials, although there are also other possible reasons. The information that the seismic refraction method is designed to obtain is (i) the positions of discontinuities, and (ii) the velocities of the waves in the layers separated by the discontinuities. The ultimate aim, of course, is to find out what the various layers consist of, but, as we shall see later, this is not a question that seismological studies alone can answer.

The use of the terms 'refracted' and 'reflected' here is deliberate, for elastic waves impinging on physical boundaries in the Earth behave in the same way as light waves impinging on, for example, mirrors and lenses (the discontinuities in those cases being the boundaries between air and glass). The simple laws of optics may therefore be applied to elastic waves in the Earth.

Consider, for example, a seismic source (S) emitting elastic waves at the surface of a horizontal two-layer system in which the velocity of the waves is v_1 in the upper layer and v_2 in the lower layer, where $v_2 > v_1$ (Figure 1.36). The source will give off waves in all directions; but consider just one of the wave paths, namely, that which passes down through the upper layer and strikes the boundary between the two layers at an **angle of incidence**, i (see Figure 1.36 for the definition of i). As long as i is not greater than a certain critical angle, the wave will pass through the boundary and emerge into the lower layer at an **angle of refraction**, r, as shown in Figure 1.36a. Because $v_2 > v_1$, the angle r will be larger than the angle i, the exact relationship between i and r being given by **Snell's law**:

$$\frac{\sin i}{\sin r} = \frac{v_1}{v_2}. \qquad \text{(Equation 1.19)}$$

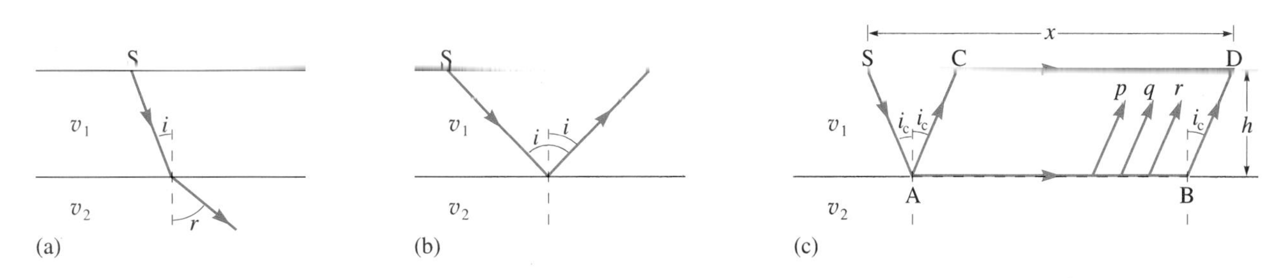

Figure 1.36 The travel paths of elastic waves from a seismic source S situated at the surface of a two-layer system in which the seismic-wave velocities in the upper and lower layers are, respectively, v_1 and v_2 (where $v_2 > v_1$). In (a), the wave strikes the boundary between the layers at an angle i, which is smaller than the critical angle, and is refracted into the lower layer at an angle r, which is less than 90°. (Note that angles of incidence and refraction are defined as the angles between the wave paths and the *perpendicular* to the boundary.) In (b), the angle of incidence is greater than the critical angle and so no refraction takes place at all — the wave is totally reflected back to the surface. In (c), the angle of incidence is equal to the critical angle and so the wave travels along the path AB in the lower layer. Secondary waves emitted along AB (p, q, r, and so on) are refracted back into the upper layer and may be detected at a point such as D. A direct wave that has travelled entirely in the upper layer is also received at D. No refracted wave may emerge over the range SC, the critical distance.

At one particular value of i, however, r will be 90°, in which case the refracted wave travels just inside the lower layer and parallel to the boundary, as shown by AB in Figure 1.36c. For this situation, $r = 90°$ and Equation 1.19 becomes:

$$\frac{v_1}{v_2} = \frac{\sin i}{\sin 90°} = \sin i_c, \quad \text{because } \sin 90° = 1. \qquad \text{(Equation 1.20)}$$

This particular value of i is known as the **critical angle**, i_c. If i is larger than i_c, the wave will not be refracted through the boundary at all but will be reflected back through the upper layer at the same angle, as shown in Figure 1.36b.

It is the case shown in Figure 1.36c that is of particular importance in seismology. Because the source (S) emits waves in all directions, there is bound to be one wave path that strikes the boundary at the critical angle; and so in the two-layer system we are considering, there is bound to be a wave travelling along the path AB in Figure 1.36c. Now, one very important property of waves is that all points on a wave-front emit **secondary waves** in all directions. Thus as the wave travels along the path AB, it will continuously emit secondary waves, some of which will be refracted back upwards into the upper layer (p, q, r, and so on). Each of these newly refracted waves will obey Snell's law; but because the wave is now passing from a layer with high velocity (v_2) to one with lower velocity (v_1), and because the new angle of incidence at points such as B is now 90°, Equation 1.19 becomes:

$$\frac{v_2}{v_1} = \frac{\sin i}{\sin r} = \frac{\sin 90°}{\sin r} = \frac{1}{\sin r}$$

or

$$\frac{v_1}{v_2} = \sin r. \qquad \text{(Equation 1.21)}$$

But we already know that

$$\frac{v_1}{v_2} = \sin i_c. \qquad \text{(Equation 1.20)}$$

Thus from Equations 1.20 and 1.21 it becomes clear that the angle of refraction and the critical angle of incidence have the same value; in other words, the waves such as p, q and r in Figure 1.36c will re-emerge into the upper layer at the critical angle. The first wave to behave in this way will be one of the secondary waves from point A, and this will strike the upper surface of the upper layer at C.

❑ What does this tell you about refracted waves arriving at the surface of the upper layer between S and C?

■ As A is the first point from which a refracted secondary wave can emerge, and as that secondary wave arrives at the surface at C, there can be no refracted waves arriving between S and C.

The distance SC is known as the **critical distance**. Refracted waves will, however, be received at the surface at every point beyond C, and a detector (D) placed at any one of these points will pick up a refracted wave (although, of course, there is bound to come a point at which the wave becomes too weak to detect, and so in practice there is a limit to the distance CD). Moreover, a detector placed at D will not only pick up the **refracted wave** SABD, it will also pick up the **direct wave** from S which has travelled along the path SD entirely along the top of the upper layer.

❑ Will a detector placed between S and C receive any wave at all?

■ Yes — it cannot receive a refracted wave, but it will receive a direct wave from S. In other words, there is no critical distance for the receipt of the direct wave. A detector placed beyond C (i.e. outside the range SC) will, of course, pick up both direct and refracted waves.

ITQ 24

In a simple two-layer system such as that shown in Figure 1.36, the seismic-wave velocity in the upper layer is $4.0\,\mathrm{km\,s^{-1}}$ and that in the lower layer is $8.0\,\mathrm{km\,s^{-1}}$. Determine the critical angle at the boundary between the two layers for a wave emitted from a source at the surface of the upper layer.

ITQ 25

In the system described in ITQ 24, a wave from the source at the surface strikes the boundary between the two layers at an angle of incidence of 20°. Determine the angle at which the wave is refracted into the lower layer.

ITQ 26

In the system described in ITQ 24, a detector is placed on the surface at a distance of 15 km from the source (S). Will the detector receive (a) the direct wave only, (b) a refracted wave only, or (c) both the direct wave and a refracted wave? The thickness of the upper layer is 10 km. (Note: To do this ITQ you will need to use very simple trigonometry.)

Now let's return to Figure 1.36c.

❑ Which wave do you think will reach D first, the direct wave SD or the refracted wave SABD?

■ On the face of it, the answer is obvious; the direct wave will reach D first, because it has less distance to go. But the matter is not that simple. Don't forget that, although the distance SABD is greater than the distance SD, the seismic-wave velocity in the lower layer is higher than that in the upper layer. If it is significantly higher, the refracted wave might well 'catch up' on the direct wave while travelling over the distance AB in the lower layer. In practice, whether the direct or refracted wave arrives first depends partly on just where the detector (D) is placed and partly on exactly how much higher the wave velocity is in the lower layer.

It is easy to calculate the time taken for the direct wave to reach D. Because velocity = distance/time, the time (t_1) for the direct wave to travel from S to D is simply:

$$t_1 = \frac{\text{distance}}{\text{velocity}} = \frac{x}{v_1} \qquad \text{(Equation 1.22)}$$

where x is the distance SD (see Figure 1.36c). Calculating the time taken for the refracted wave to reach D is somewhat more complicated because the refracted wave path has three separate segments (SA, AB and BD) in one of which (AB) the velocity is v_2 and in two of which (SA and BD) the velocity in v_1. In other words:

travel time SABD, t_2 = time over SA + time over AB + time over BD.

In fact, if the thickness of the upper layer is h:

$$t_2 = \frac{x}{v_2} + \frac{2h\sqrt{v_2^2 - v_1^2}}{v_1 v_2}. \qquad \text{(Equation 1.23)}$$

This equation is derived using simple trigonometry, but the derivation is time-consuming and not very instructive; so there is little point in going into detail here. However, we are now equipped to answer the question of whether the direct wave or the refracted wave reaches D first. This will depend on the relative values of t_1 and t_2 — in other words, on the quantities v_1, v_2, h and x. In qualitative terms, it is quite easy to see what will happen. If D is close to C, the direct wave will reach D first because the refracted wave, which in this case travels mostly in the upper layer, has the greater distance to cover. But as D is moved further away from C, there comes a point at which D receives the refracted wave first. This is because, as D is moved further from C, the higher-velocity part of the refracted wave's path, AB, becomes longer. There thus comes a point at which the time saved by the refracted wave's higher velocity in the lower layer outweighs the disadvantage of having to travel a longer distance — and the refracted wave 'overtakes' the direct wave. The position of D at which this happens is, of course, the point at which the refracted wave and the direct wave arrive at the same time. This is the point (x_d) at which $t_1 = t_2$, and so from Equations 1.22 and 1.23:

$$\frac{x}{v_1} = \frac{x}{v_2} + \frac{2h\sqrt{v_2^2 - v_1^2}}{v_1 v_2}$$

which gives (by routine mathematical steps which you don't need to understand):

$$x = 2h\sqrt{\frac{v_2 + v_1}{v_2 - v_1}} = x_d. \qquad \text{(Equation 1.24)}$$

It is thus clear that the point at which $x = x_d$ (at which the first wave to arrive at D changes from being the direct wave to being the refracted wave) depends on the values of h, v_1 and v_2.

ITQ 27

In the system described in ITQs 24–26, how far from the source (S) does the first arrival cease to be the direct wave and become a refracted wave?

Equation 1.24 also shows that, if we can determine v_1, v_2 and x_d, we can then determine h, the thickness of the upper layer. This is, of course, equivalent to determining the position of the discontinuity between the upper and lower layers.

Determination of v_1

Equation 1.22 represents a straight line of the form $y = mx$. Thus if t_1 is plotted against x, the result will be a straight line that passes through the origin and has a gradient of $1/v_1$. This is the line OP in Figure 1.37. This line expresses the variation of t_1 with x; in other words, it shows how the time taken for the direct wave to reach D in Figure 1.36c varies with the distance SD (x). v_1 may be obtained simply by measuring the gradient of OP, which is $1/v_1$.

Determination of v_2

Equation 1.23 represents a straight line of the form $y = mx + c$. Thus if t_2 is plotted against x, the result will be a straight line that has gradient $1/v_2$ but does not pass through the origin. This is the line QR in Figure 1.37. This line expresses the variation of t_2 with x; it shows how the time taken for the refracted wave to reach D in Figure 1.36c varies with the distance SD (x). Note that this line does not reach back to the time axis because there is a critical distance (x_c) over which D cannot receive the refracted wave. This is the distance SC in Figure 1.36c. v_2 may be obtained by measuring the gradient of QR, which is $1/v_2$.

Determination of h

v_1 and v_2 have now been determined; but we still need x_d, which is, you will recall from Equation 1.24, the distance at which both the direct and refracted waves are received at the same time. x_d may be obtained from Figure 1.37, because it is the point at which the lines OP and QR cross (i.e. point X in Figure 1.37). With x_d, v_1 and v_2 now known, h may be calculated from Equation 1.24, which, when rewritten, becomes:

$$h = \frac{x_d}{2}\sqrt{\frac{v_2 - v_1}{v_2 + v_1}}. \qquad \text{(Equation 1.25)}$$

Alternatively, h may be calculated by projecting the line QR back to the time axis and noting the time (t_0) at which the intercept occurs. Along the time axis in Figure 1.37, x is everywhere equal to zero. So putting $x = 0$ into Equation 1.23 to obtain the value of t_2 ($= t_0$) when $x = 0$ gives

$$t_0 = \frac{2h\sqrt{v_2^2 - v_1^2}}{v_1 v_2} \quad \text{or} \quad h = \frac{t_0 v_1 v_2}{2\sqrt{v_2^2 - v_1^2}}. \qquad \text{(Equation 1.26)}$$

As v_1 and v_2 are known and t_0 is obtained from the intercept, h may be determined.

The time–distance graph

Figure 1.37 is known as a time–distance graph and may be constructed by measuring the time it takes for the waves from S to reach D as D is moved further away from S (i.e. as x is increased). The line OXR represents the **first arrival** at D — i.e. the time it takes the first wave to arrive at D, whether it be the refracted wave or the direct wave.

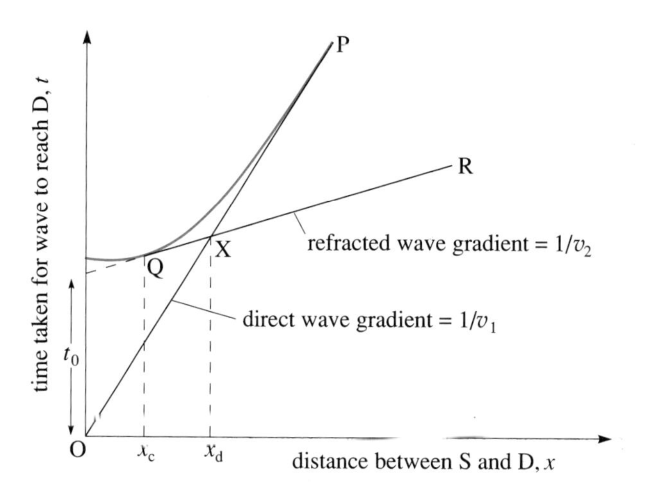

Figure 1.37 The time–distance graph for the situation shown in Figure 1.36c. OP represents the direct wave arriving at D, and QR represents the refracted wave. The first wave to arrive at D is represented by the line OXR. For $x < x_d$, this is the direct wave; but for $x > x_d$ the refracted wave becomes the first arrival. No refracted wave arrives at D at all when $x < x_c$. (The curved line is the time–distance curve for waves reflected at the boundary between the upper and lower layers. Don't worry about it during the discussion of refraction; we'll return to it later.)

❑ Which wave is the first arrival along OX (i.e. up to the distance x_d)?

■ OX is part of the line OP, which represents the direct wave. Along OX, therefore, the first arrival is the direct wave.

Beyond x_d (i.e. along XR), by contrast, the first arrival is the refracted wave. Conversely, up to distance x_d (but not over the range $x = 0$ to $x = x_c$) the refracted wave appears at D as the **second arrival**; and beyond x_d the second arrival is the direct wave.

ITQ 28

In a seismic experiment, the times (t) of the first arrivals at various distances (x) from the source were as follows:

x (km)	t (s)
5.0	1.6
10.0	3.3
15.0	4.8
20.0	5.5
30.0	6.8
45.0	9.0

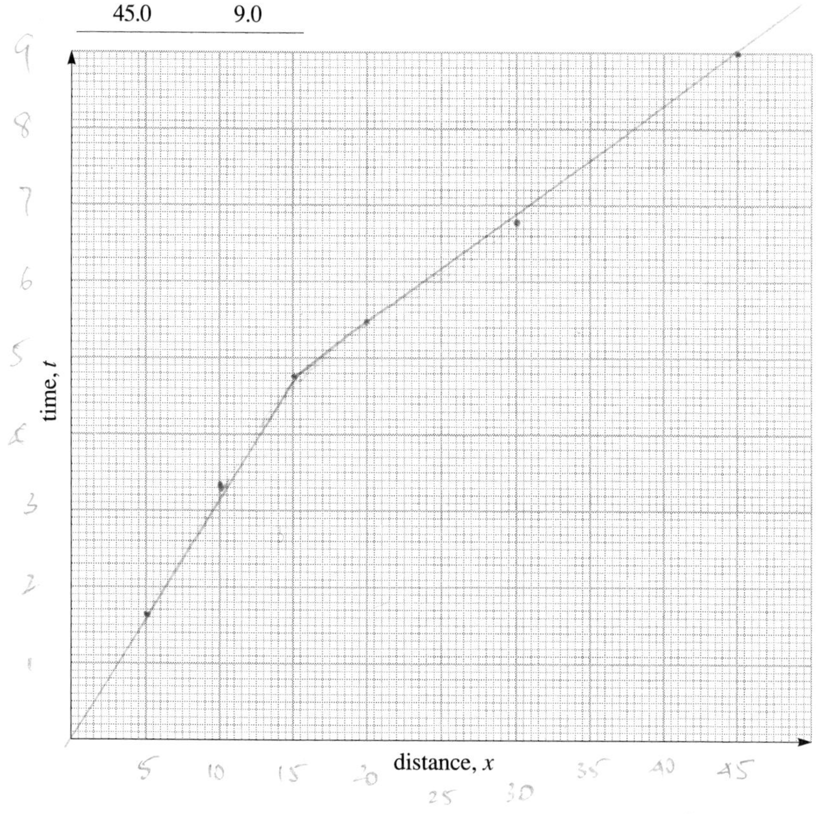

time, t

distance, x

Figure 1.38 Graph of time versus distance for the seismic data in ITQ 28.

Plot these data on Figure 1.38, and then (a) determine the seismic velocities of the layers involved, and (b) calculate the thickness of the upper layer by two different methods.

Other aspects of seismic refraction

(1) The model we considered above was the simplest one possible; it comprised two horizontal layers within each of which the seismic velocity was constant throughout, and the boundaries were plane surfaces. You may not think so, but the mathematics was correspondingly simple. In practice, there are often more than two layers present, in which case further refractions take place at the lower boundaries. For example, if there is a third layer in which the wave velocity is v_3 (shown in Figure 1.39a), one of the waves from S (which strikes the first boundary at an angle of less than i_c and is thus refracted into the second layer at $r < 90°$) will strike the second boundary at the critical angle (i'_c) for that boundary and will be refracted there at 90°. This wave will ultimately return to the surface to give another arrival at D and at a certain value of x it will become a first arrival, as long as $v_3 > v_2 > v_1$. The time–distance graph for the *first arrivals only* will then look something like that in Figure 1.39b. v_3 may then be calculated from the gradient of the third segment. However, the equation for the calculation of k, the thickness of the second layer down, is:

$$k = \frac{x_e(v_3 - v_2) - \frac{2h}{v_1}\left[v_2\sqrt{v_3^2 - v_1^2} - v_3\sqrt{v_2^2 - v_1^2}\right]}{2\sqrt{v_3^2 - v_2^2}} \qquad \text{(Equation 1.27)}$$

where x_e is the distance between source and detector at which the second refracted wave from the layer 2 – layer 3 interface becomes the first arrival.

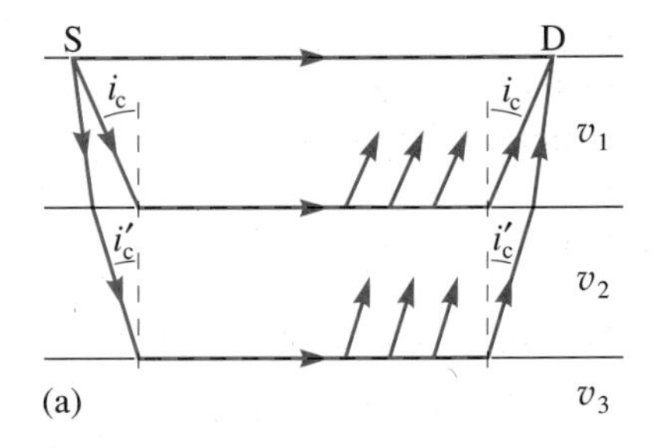

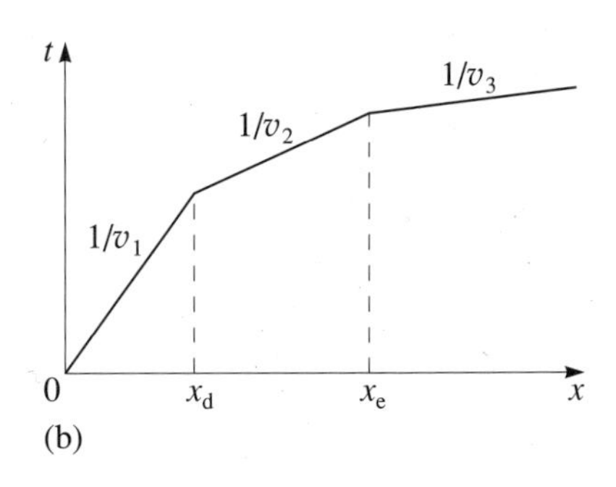

Figure 1.39 (a) Refractions in a three-layer system in which $v_3 > v_2 > v_1$, showing the paths of three waves from S recorded at D — a direct wave and a wave refracted from each of two interfaces. (b) The time–distance graph for first arrivals only. In the range 0–x_d the first arrival is the direct wave in the upper layer; in the range x_d–x_e the first arrival is the refracted wave from the first boundary; and beyond x_e the first arrival is the refracted wave from the second boundary.

You most certainly do not need to remember Equation 1.27, and we shall not ask you to use it. We reproduce it here, however, to make an important point, which is that even with just the second-simplest model — three horizontal layers within each of which the seismic velocity is constant throughout — the mathematics has already become very tedious. In a real experiment, of course, there may be more than three layers, the layers may not be horizontal, the boundaries between them may not be plane surfaces, the boundaries may be rugged, and the velocity within each layer may vary with depth or even laterally. In actual practice, therefore, the mathematics is not just tedious but horrendously complicated. *Yet the basic principles are exactly the same as those introduced above in the context of the two-layer model.* If you have understood those principles, you have mastered the essence of seismology.

(2) It's important to appreciate that the principles enunciated above, though sound, do not necessarily provide all the information that one would wish to have about Earth structure. You will recall that in the two-layer and three-layer models the seismic velocity in each layer was always higher than that in the layer above. But in the real Earth this condition does not always apply. If, for example, there is a layer in which the seismic velocity is lower than that in the layer above, it will not appear on the time–distance graph as a first arrival and hence will not be detected. It will remain a **hidden layer**. A layer may also remain hidden if its seismic velocity is higher than that of the layer above but if the velocity contrast is not very great, or if the layer is very thin. There are now advanced seismological techniques for detecting hidden layers, but they are way beyond the scope of this Course.

(3) Another complication that can arise is that the detector may pick up not only refracted waves and the direct wave but waves that have been

reflected at the various boundaries. In the simple two-layer system of Figure 1.36, the waves reflected from the first boundary are relatively weak in the SC range because most of the seismic energy is 'carried' by waves that are refracted into the second layer. The weak reflected waves within the SC range are known as **subcritical reflections** because they strike the boundary between the two layers at less than the critical angle, i_c. The strength, or amplitude, of reflected waves increases rapidly for distances greater than SC, where such reflections are known as **supercritical reflections**. Although they always take longer to arrive than do the direct or refracted waves, supercritical reflections may be the strongest waves to arrive at source–detector distances greater than SC. The time–distance curve for reflected waves is the curve in Figure 1.37. We shall look at seismic reflection in more detail in Section 1.6.3.

(4) Because the real Earth is more complicated than the simple models described above, it is quite insufficient in practice to have one seismic source and one detector, or even one set of detectors, as shown in Figure 1.40a. For example, if the boundary between the upper and second layers is inclined with respect to the Earth's surface (i.e. if it dips), even the set of detectors in Figure 1.40a will almost certainly fail to give an accurate value of h. This problem may be overcome by using source–detector arrangements that 'look' at the dipping boundary in two directions, thus enabling the boundary's angle of dip relative to the horizontal to be determined. More generally, to obtain the maximum information, various source–detector layouts can be used, depending on the circumstances. Some important layouts are shown in Figure 1.40b–e.

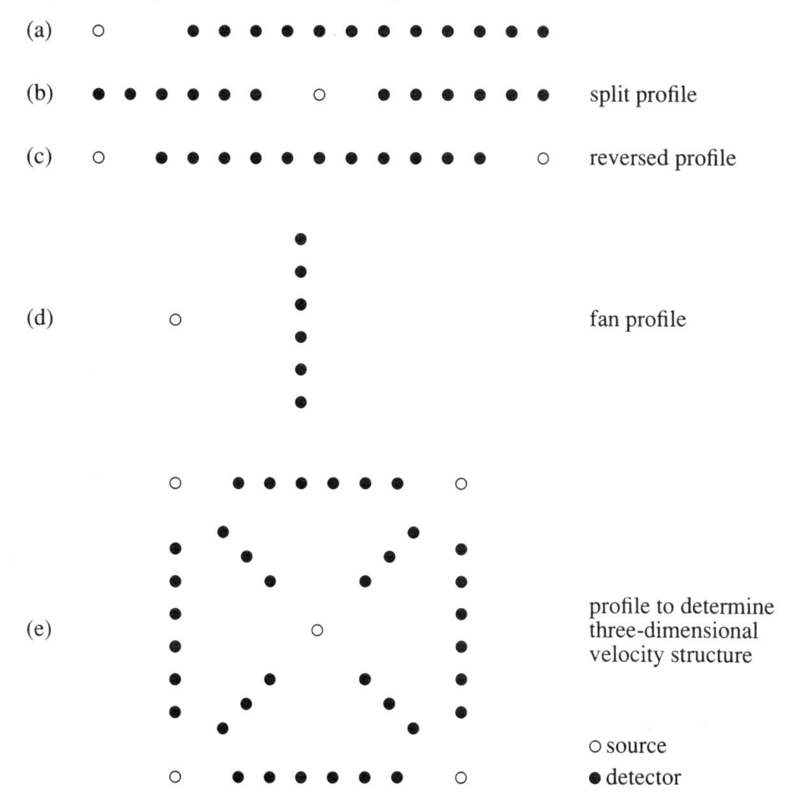

Figure 1.40 Possible arrangements (profiles) of seismic sources and detectors for use in seismic refraction experiments. Note that for refraction experiments at sea, the source and detector locations are usually reversed — see text.

(5) Seismic refraction surveys on land are often difficult and expensive, because it takes tonnes of explosive to produce a wave capable of penetrating the upper mantle, and detectors have to be placed out to distances of several hundred kilometres at spacings of less than 5 km. Moreover, seismometers are expensive; so they often have to be placed in position while one shot is detonated and then laboriously moved and set up again for the next shot elsewhere. Things are easier at sea, however. Marine seismometers are even more expensive, but explosions are comparatively cheap. It is therefore usual in marine surveys to reverse the

source and detector locations shown in Figure 1.40. What happens is that the ship sails on, firing depth charges or (if less penetration is required) an air gun as it goes. Moreover, because the oceanic crust is much thinner than the continental crust, refraction lines need only be a few tens of kilometres long to record waves refracted in the upper mantle.

(6) Time–distance graphs are often displayed not in the form shown in Figure 1.37 but as **reduced time–distance graphs**, in which instead of plotting time (t) against distance (x), the quantity plotted against x is:

$$t - \frac{x}{v_1}. \qquad \text{(Equation 1.28)}$$

v_1 is the seismic velocity of the direct wave in the uppermost layer, although other velocities are sometimes chosen for convenience in particular cases. The chief reason for having reduced time–distance graphs is simply to be able to present (in reports, for example) graphs with manageable scales. One consequence of plotting the reduced time–distance graph concerns the curve for the direct wave (assuming now that the **reduction velocity**, v_1, being used is indeed the velocity of the direct wave). From Equation 1.22, for the direct wave:

$$t_1 = \frac{x}{v_1}$$

and so

$$t_1 - \frac{x}{v_1} = 0.$$

In other words, for the direct wave the expression in Equation 1.28 is equal to zero. What this means is that the direct P-wave through the Earth's upper crust should appear on a reduced time–distance graph as a straight, horizontal line passing through the origin (i.e. where $t - x/v_1 = 0$). Figure 1.41 illustrates this direct wave, there denoted P_g, in a region where the upper crust has a seismic velocity of 6 km s^{-1} (i.e. the reduction velocity used is 6 km s^{-1}). The other curves are explained in the caption of Figure 1.41.

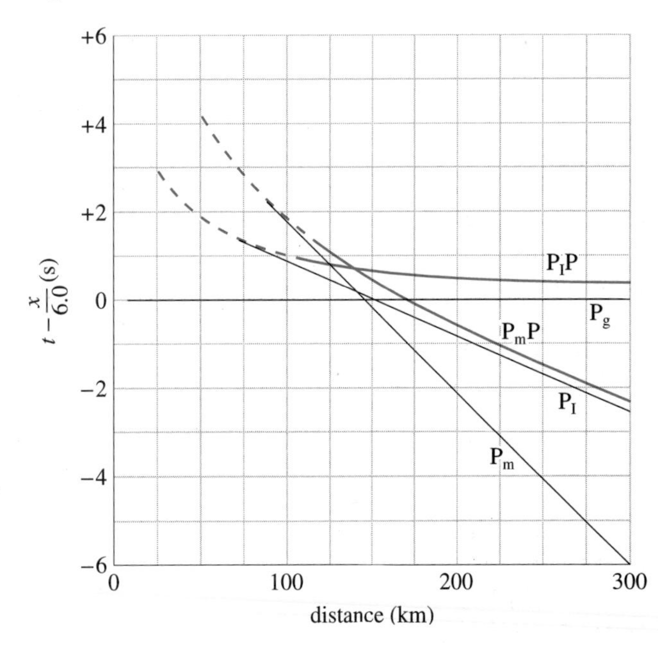

Figure 1.41 The reduced time–distance graph for a region in which the Earth's crust has two layers and the upper layer has a seismic-wave velocity of 6.0 km s^{-1}. The direct wave through the upper layer (labelled P_g) appears as a straight line passing through the origin. P_m is the wave refracted at the surface of the mantle and P_I is the wave refracted at the boundary between the two crustal layers. P_mP and P_IP are waves reflected at, respectively, the crust–mantle boundary and the mid-crustal boundary. The dashed lines represent subcritical reflections.

(7) Although the procedure culminating in the time–distance graph (Figure 1.37) or the reduced time–distance graph (Figure 1.41) well expresses the principles of seismic refraction, you should be aware that seismic data are not always presented in exactly such forms today. The

past decade or two have seen great advances in data processing techniques, computer power, digital analysis, and so on, with the result that data are now often displayed in forms more conducive to rapid processing. Seismic refraction data, for example, are now often plotted as a **record section**, as shown in Figure 1.42. Here the record from each seismometer is plotted at its appropriate epicentral distance. In fact, Figure 1.42 is a *reduced record section* in the sense that the vertical scale has been adjusted according to the principle explained in (6) above. An advantage of producing record sections or **reduced record sections** is that the travel-time and amplitude data are displayed together. The different waves can then be more easily correlated from seismic trace to seismic trace, thereby aiding the detection of, for example, hidden layers that would not show up on conventional time–distance graphs. You might have spotted, nevertheless, that Figure 1.42 is really just another form of time–distance plot because, of course, the high amplitudes on each trace are from the waves refracted at discontinuities.

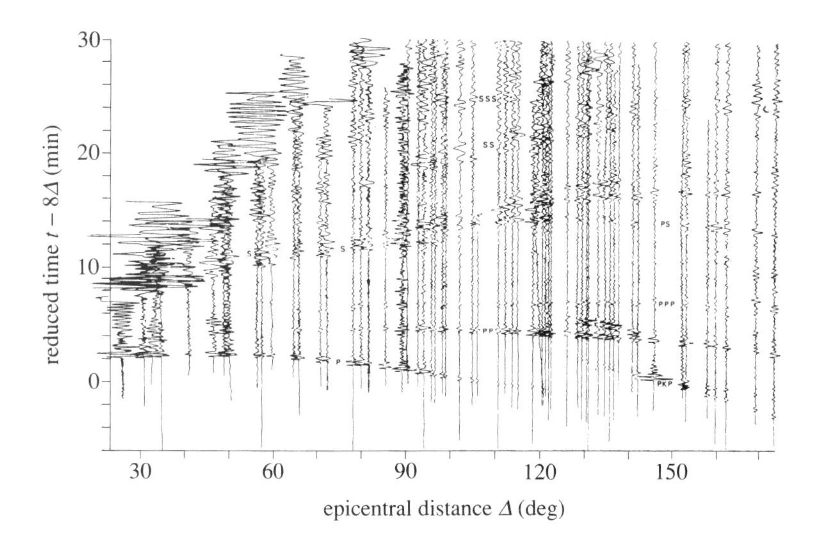

Figure 1.42 A reduced record section, in this case from an earthquake that occurred near Sumatra. The reduction velocity chosen is 13.9 km s^{-1} (because the first arrivals came from the deep mantle), which means that waves with this velocity appear horizontal. For ease of comparison, all the traces have the same amplitude scale. Identifying and recognizing the waves of an earthquake is a highly skilled task which we do not expect you to be able to carry out. You should, however, be able to answer the question in the text.

❑ Suppose that in Figure 1.42 the numbers on the horizontal scale (epicentral distance) had been omitted. Would you have been able to tell whether the short epicentral distances were on the right or left of the scale? If so, how?

■ Yes, because the amplitude of a seismic wave decreases with increasing distance from the source. In Figure 1.42, it is quite clear that the amplitudes of the waves generally decrease to the right. The seismic traces on the left were therefore obtained from the seismometers closest to the earthquake (low epicentral angles).

You should now look at the first part of video VB 01, 'Seismic Methods', which shows how a seismic refraction experiment is carried out. (This part of the video lasts about 35 minutes.)

1.6.3 SEISMIC REFLECTION

Up to the 1970s, academic Earth scientists had obtained most of their seismic information about the Earth's interior using the refraction method, with either earthquake or non-earthquake sources. The oil companies, on the other hand, had long made considerable use of **seismic reflection** to map subsurface sedimentary structures in the search for oil. There came a point, however, when the academics realized that the reflection technique had great potential for determining the fine physical structure of the crust (mainly) and upper mantle, since which time the method has been applied with some enthusiasm.

The principle of the seismic reflection method is extremely simple. P-waves from an energy source at the Earth's surface are reflected from subsurface discontinuities and, when they return to the surface, are recorded by detectors. Figure 1.43 shows how the method works for a single horizontal layer of constant thickness h and in which the P-wave velocity (v) is constant. Suppose that the source (S)–detector (D) distance is x. The path taken by the wave (SAD), reflected from the base of the layer, can be divided into two equal components, SA and AD (i.e. SA = AD). So:

$$\text{SAD} = 2 \times \text{SA} \qquad \text{(Equation 1.29)}$$

S9

and, by Pythagoras's theorem, for triangle SAO:

$$(\text{SA})^2 = (\text{OA})^2 + (\text{SO})^2$$
$$= h^2 + \left(\frac{x}{2}\right)^2.$$

Therefore,

$$\text{SA} = \sqrt{h^2 + \frac{x^2}{4}}$$

and so, from Equation 1.29 by substitution:

$$\text{SAD} = 2\sqrt{h^2 + \frac{x^2}{4}}$$

from which,

$$\text{travel time } (t) = \frac{\text{distance SAD}}{v} = \frac{2\sqrt{h^2 + x^2/4}}{v}. \qquad \text{(Equation 1.30)}$$

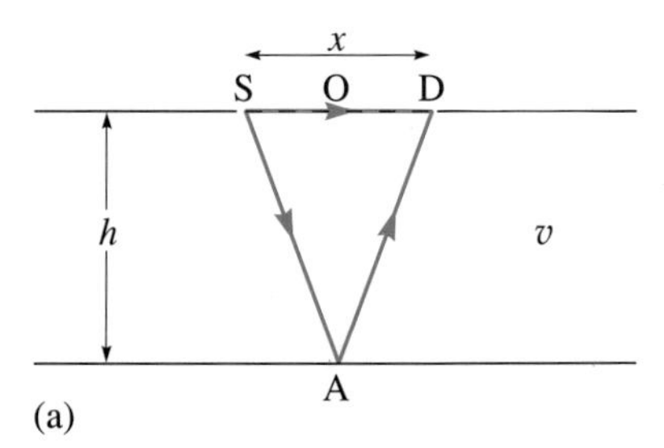

In this case, the time–distance graph is not a straight line, as it was in the refraction method, but a curve as shown in Figure 1.43b. This is because t is no longer proportional to x but to the square root of x^2 plus another term (h^2). The curve is, in fact, a hyperbola.

❑ Can you see what will happen to Equation 1.30 when x is very much bigger than h (i.e. $x \gg h$)?

■ If $x \gg h$, $x^2 \gg h^2$ and so the h^2 term can be ignored. Equation 1.30 then reduces to $t = x/v$, and this *is* a straight line (of the form $y = mx$). This is confirmed by Figure 1.43b, from which you can see that the hyperbola becomes, in effect, a straight line at large values of x.

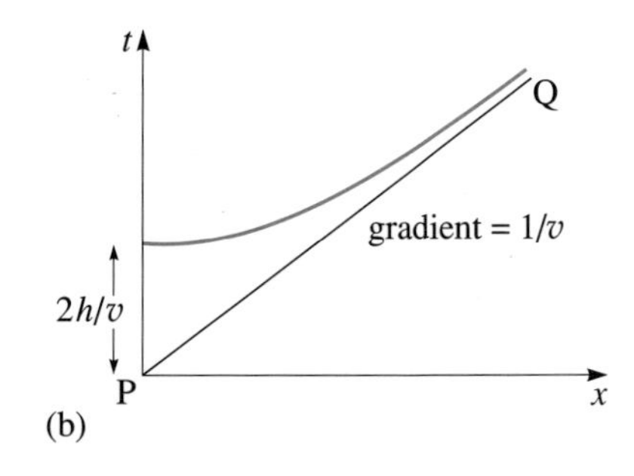

The gradient of the line is $1/v$; and so the seismic velocity in the layer may be obtained by drawing the line PQ through the origin in Figure 1.43b and measuring its gradient. The thickness of the layer (h) may then be obtained by determining where the hyperbola cuts the vertical axis, for along the vertical axis (where $x = 0$), Equation 1.30 becomes

$$t = \frac{2h}{v}$$

and so

$$h = \frac{tv}{2}. \qquad \text{(Equation 1.31)}$$

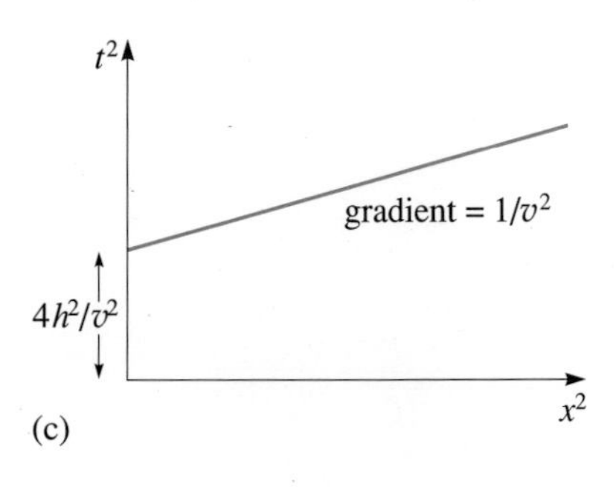

Figure 1.43 The principle of the seismic reflection method, illustrated for a single horizontal layer of thickness h in which the seismic velocity (v) is constant. A wave path through the layer is shown in (a); (b) is the time–distance (t–x) plot; and (c) is the t^2–x^2 plot.

The fact is, however, that it's easier to work with straight lines than it is with curves. Instead of plotting a time–distance (t–x) graph, it is more satisfactory to plot t^2 against x^2, which is a straight line (Figure 1.43c).

❑ Can you see why?

■ If both sides of Equation 1.30 are squared, we have

$$t^2 = \frac{x^2}{v^2} + \frac{4h^2}{v^2}$$

and this is a straight line of the form $y = mx + c$, where the two variable quantities are now t^2 and x^2.

The gradient of the straight line is $1/v^2$ (so measuring the gradient gives v) and the intercept on the vertical axis (where $x = 0$) is $4h^2/v^2$, which gives h (since v is known).

Another, perhaps more important, reason for using the straight-line method is that in many reflection experiments, the incident wave is directed almost vertically downwards. In other words, x is always small and so the hyperbola of the type shown in Figure 1.43b cannot be very well defined. This is less of a problem with a straight line, which can easily be defined over a short range of x (and hence x^2).

ITQ 29

In a seismic reflection experiment, the following data were obtained for first arrivals from a discontinuity in the Earth's crust:

Source–detector distance (km)	Arrival time (s)
6.00	2.5
8.94	3.0
11.49	3.5

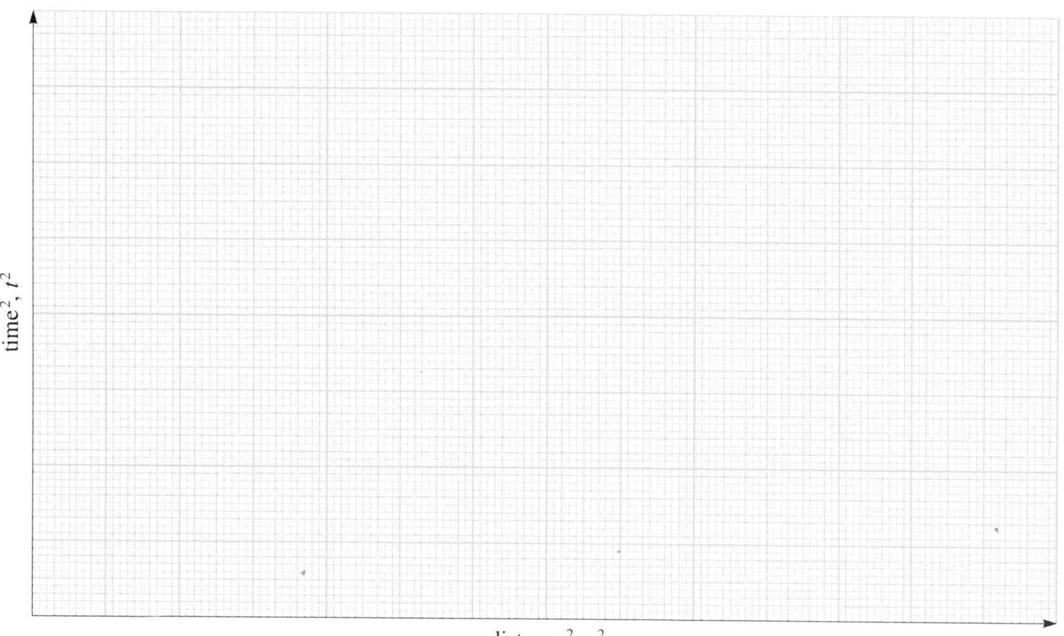

Figure 1.44 Graph paper for use with ITQ 29.

Use the graph paper provided (Figure 1.44) to determine (a) the seismic velocity above the discontinuity, and (b) the depth of the discontinuity.

As in the case of refraction experiments, it is customary to plot reflection data as a record section. The principle of the reflection record section is illustrated in Figure 1.45, which shows a two-layer system in which the base of the upper layer is sloping. The seismic trace of ground motion from one detector will contain two jumps in amplitude corresponding to the waves reflected from the two strong reflectors, namely, the bases of the two layers (Figure 1.45b). When the traces from several detectors are plotted together at their respective distances, the record section shown diagrammatically in Figure 1.45c results.

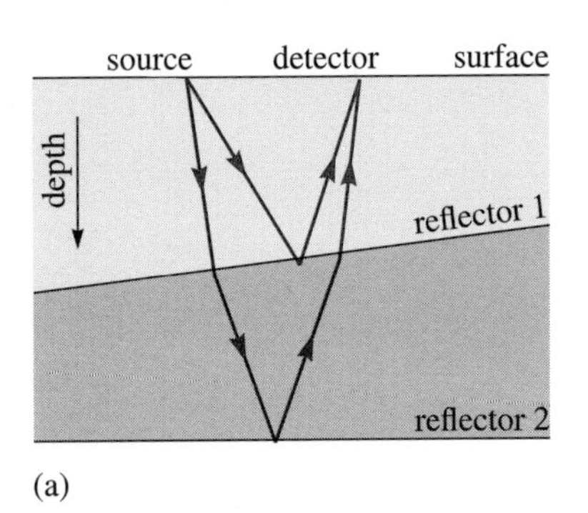

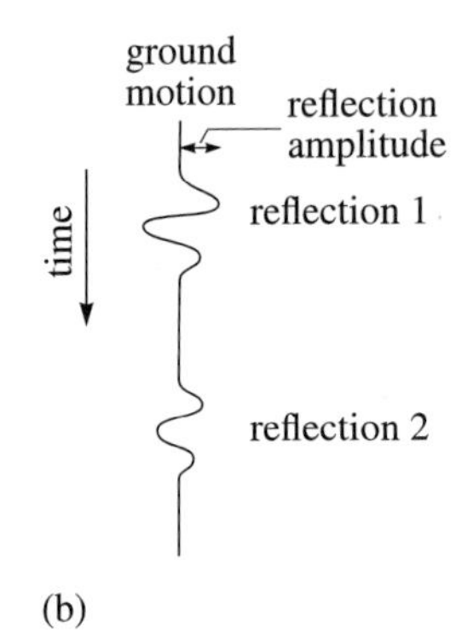

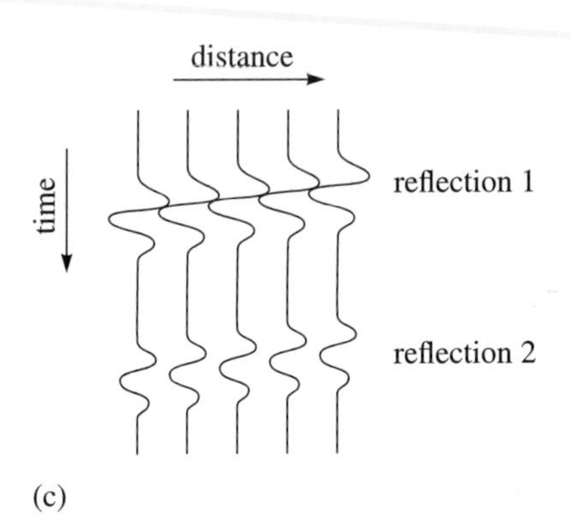

Figure 1.45 The construction of a reflection record section. (a) shows waves from a seismic source reflected from two boundaries, one of which is sloping (note that for a wave to be reflected from the second boundary, it first has to be refracted from the first layer into the second). (b) shows the form of the seismic trace produced by the two reflections (note that the vertical axis here is time rather than distance). (c) shows several traces placed side-by-side at the appropriate source–detector distances (note how the reflections can be followed from trace to trace).

When there are many seismic shots and many detectors, the sort of detailed record section shown in Figure 1.46 can be built up. The horizontal axis represents distance and the vertical axis represents depth, although it is customary to label it in terms of the time taken for the wave to travel down to the appropriate depth and back to the surface (the **two-way travel time**). What the record section represents, therefore, is, in effect, a cross-section through part of the Earth's crust/upper mantle. The enhanced amplitudes due to reflections show up as darker zones against the lighter background and mark the positions of the reflectors — the darker the lines, the more intense being the reflector and hence the greater the contrast in physical properties across that boundary.

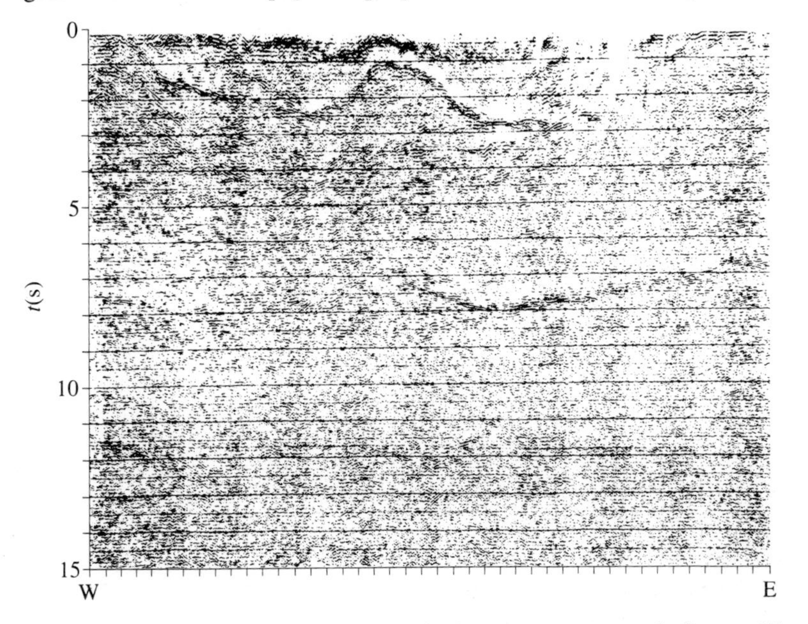

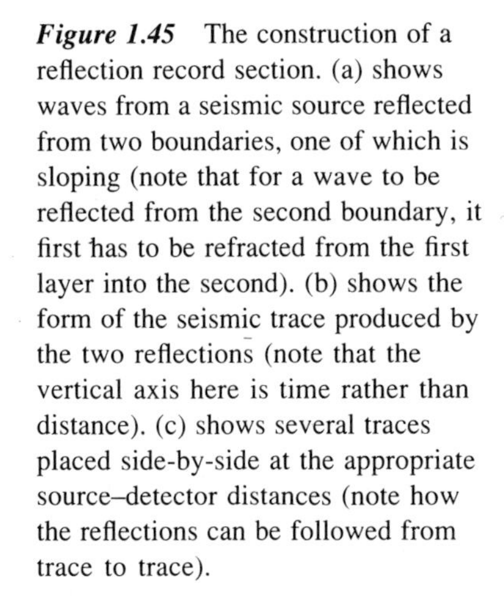

Figure 1.46 A reflection record section — in this case across the Rio Grande rift, near Socorro, New Mexico, USA. The horizontal scale is distance, each division representing 100 m. The vertical scale is two-way travel time (s). Particularly strong reflectors show up as dark bands against the lighter background.

The depth represented by any particular two-way travel time will, of course, depend on the velocity of seismic waves in the region concerned. But you can get a rough idea by assuming a reasonable P-wave velocity in the Earth's crust.

❑ If the crustal P-wave velocity is taken to be $\sim 6\,\text{km}\,\text{s}^{-1}$, what depth is represented by a 10 s two-way travel time?

■ If a P-wave travels at about $6\,\text{km}\,\text{s}^{-1}$, it travels 60 km in 10 s. However, if the two-way travel time is 10 s, the one-way travel time is 5 s; so the depth is about 30 km.

Reflection record sections are often converted into interpretative line drawings, which make the positions of the prominent reflectors rather clearer, especially to those who are not skilled in 'reading' the original sections. As an example, Figure 1.47 shows the line-drawing interpretation of the record section in Figure 1.46. Of course, you should appreciate that although seismic reflection techniques can delineate prominent reflectors in the Earth's crust and upper mantle, they offer few clues about what the reflectors represent in physical and/or chemical terms.

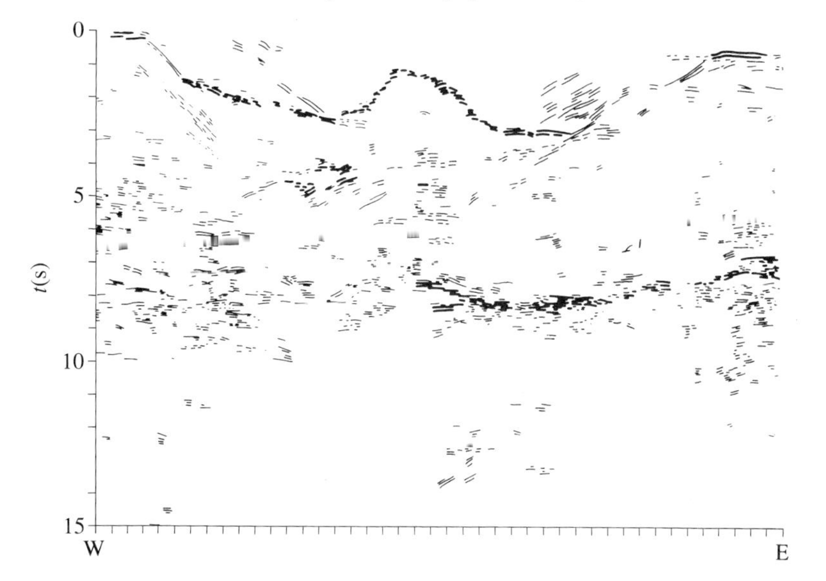

Figure 1.47 A line-drawing interpretation of the record section in Figure 1.46. The lines shown here represent the scientists' interpretation of the positions of the most prominent reflectors in the record section.

You should now look at the second part of video VB 01, 'Seismic Methods', which deals with seismic reflection surveys. (This part of the video lasts about 26 minutes.)

SUMMARY OF SECTION 1.6

The velocity of a seismic wave passing through a material depends on the material's elastic properties and density. In fact, velocity is given by the square root of elastic modulus divided by density (Equation 1.13), which indicates that velocity increases with increasing elastic modulus and with decreasing density. However, the relationship is not as straightforward as this would suggest, because elastic modulus also varies with density.

There are several different elastic moduli for any given material, the appropriate one(s) to use in particular circumstances depending on the type of seismic wave involved. For S-waves, for example, the appropriate modulus is the shear modulus, or rigidity modulus, μ; and as $\mu = 0$ for a fluid, S-wave velocity in a fluid is zero (i.e. S-waves cannot travel in fluids). For P-waves, by contrast, both the shear modulus and the bulk modulus are involved — and in such a way as to ensure that the velocity of P-waves is always higher than the velocity of S-waves.

Seismic waves striking boundaries between different materials obey the laws of refraction well known from optics. By initiating seismic waves at a point on the surface of the Earth and observing how long it takes them to reach a series of recording stations, also on the Earth's surface, it is possible to deduce the depths of the discontinuities at which the waves are refracted and the seismic velocities in the layers bounded by the discontinuities, although the mathematics gets very complicated if the layers are not horizontal and few. Moreover, if the velocity in a layer is not greater than that in the layer above, if the velocity is greater but not very much greater or if the velocity is very much greater but the layer is very thin, the layer in question may escape detection (hidden layer).

Seismic waves also obey the optical laws of reflection, providing an alternative method of detecting subsurface discontinuities and layer velocities. However, while both the refraction and reflection methods can reveal discontinuities, they cannot by themselves indicate what the discontinuities represent in physical or chemical terms.

OBJECTIVES FOR SECTION 1.6

When you have completed this Section, you should be able to:

1.1 Recognize and use definitions and applications of each of the terms printed in the text in bold.

1.24 Summarize the factors on which the seismic velocity within a material depends, and perform simple calculations relating to them.

1.25 Describe the principles of, and perform simple calculations based on, the refraction and reflection of seismic waves at an interface.

1.26 Plot and interpret time–distance curves for direct and refracted waves in horizontal two-layer and three-layer systems.

1.27 Plot and interpret time–distance curves for waves reflected at a subsurface interface.

Apart from Objective 1.1, to which they all relate, the eight ITQs in this Section test the Objectives as follows: ITQ 22, Objective 1.17; ITQ 23, Objective 1.24; ITQs 24–27, Objective 1.25; ITQ 28, Objective 1.26; ITQ 29, Objective 1.27.

You should now do the following SAQs, which test other aspects of the Objectives.

SAQS FOR SECTION 1.6

SAQ 33 (*Objectives 1.1, 1.24 and 1.25*)

State, giving reasons where appropriate, whether each of the following statements is true or false.

(a) The smaller the elastic modulus of a material, the more easily the material deforms.

(b) The smaller the elastic modulus of a material, the higher is the velocity of a seismic wave passing through it.

(c) The angle of incidence of a wave striking an interface is the angle between the wave path and the perpendicular to the interface.

(d) The critical distance is the distance from the seismic source at which refracted waves are first detected.

(e) The critical distance for the receipt of the direct wave from the seismic source is half that for the receipt of the refracted wave.

(f) Reflected waves may be received within the critical distance.

SAQ 34 (*Objective 1.24*)

The S-wave velocity in a material with a shear (rigidity) modulus of $3.5 \times 10^{10}\,\mathrm{N\,m^{-2}}$ is $3.0\,\mathrm{km\,s^{-1}}$. What is the density of the material?

SAQ 35 (*Objective 1.24*)

If the P-wave velocity in the material in SAQ 34 is $6.0\,\mathrm{km\,s^{-1}}$, what is the bulk modulus of the material?

SAQ 36 (*Objective 1.24*)

Determine the axial modulus of the material in SAQ 34.

SAQ 37 (*Objective 1.25*)

A seismic wave travelling at a velocity of $4.0\,\mathrm{km\,s^{-1}}$ in one material strikes the interface with another material at an angle of incidence of 30°. The wave is then refracted into the second material, in which it travels with a velocity of $6.0\,\mathrm{km\,s^{-1}}$. What is the angle of refraction into the second material?

SAQ 38 (*Objective 1.25*)

Suppose the wave in SAQ 37 had struck the interface between the layer at an angle of incidence of 45°. Would it still have been refracted into the second layer or would it have been reflected back into the first layer?

SAQ 39 (*Objectives 1.24 and 1.25*)

In the situation shown in Figure 1.36c, the critical distance is found to be 200 m. If the thickness of, and P-wave velocity in, the upper layer are, respectively, 100 m and 4.24 km s^{-1}, what is

(a) the P-wave velocity in the second layer?

(b) the density of the second layer if the bulk modulus is 6×10^{10} N m^{-2} and the shear modulus is 3×10^{10} N m^{-2}?

SAQ 40 (*Objective 1.26*)

In a horizontal two-layer system, a detector placed at a distance of 150 km from a seismic source at the Earth's surface receives the direct wave and the refracted wave from the source simultaneously. If the P-wave velocity in the upper layer is 4.0 km s^{-1} and that in the second layer is 6.0 km s^{-1}, what is the thickness of the upper layer?

SAQ 41 (*Objectives 1.25 and 1.26*)

In a seismic refraction experiment, a single explosion was generated and the waves from the explosion were picked up by 15 detectors. The detectors were arranged in a straight line with, at the same altitude as, and on the same side of, the explosion site. The results obtained were:

explosion–detector distance (km)	P-wave first-arrival times (s)
7	3.6
16	8.0
29	14.4
48	24.0
75	33.6
88	36.8
102	40.4
115	43.6
131	47.6
160	52.0
177	54.0
190	55.6
201	56.8
220	59.0
240	61.4

(a) On Figure 1.48 (*overleaf*), draw an accurate time–distance graph for these results.

(b) Use the graph to determine the number of distinct layers revealed by the data.

(c) Use the graph to determine the P-wave velocity in the uppermost and lowermost of the layers, assuming that all boundaries between layers are horizontal.

(d) Use the graph to determine the thickness of the uppermost layer

(e) Use the graph to determine the time of the *second arrival* at (i) the detector 48 km from the explosion and (ii) the detector 88 km from the explosion.

(f) Use the graph to determine the critical angle for the transmission of waves from the uppermost layer into the second layer.

Figure 1.48 Graph paper for SAQ 41.

1.7 The seismic view of the Earth

In Section 1.6, we looked at the theory of refraction and reflection at plane surfaces bounding horizontal layers within each of which the seismic velocity was assumed to be constant. The plane-boundary simplification may be an acceptable approximation for very local experiments; but on the scale of the Earth as a whole the boundaries are, of course, curved, and in the major crustal and mantle layers the seismic velocities are far from constant. Nevertheless, the waves passing through the Earth behave according to exactly the same *principles* as those developed for waves in Section 1.6. In this Section, we shall look at what the application of those principles can tell us about the internal structure of the Earth.

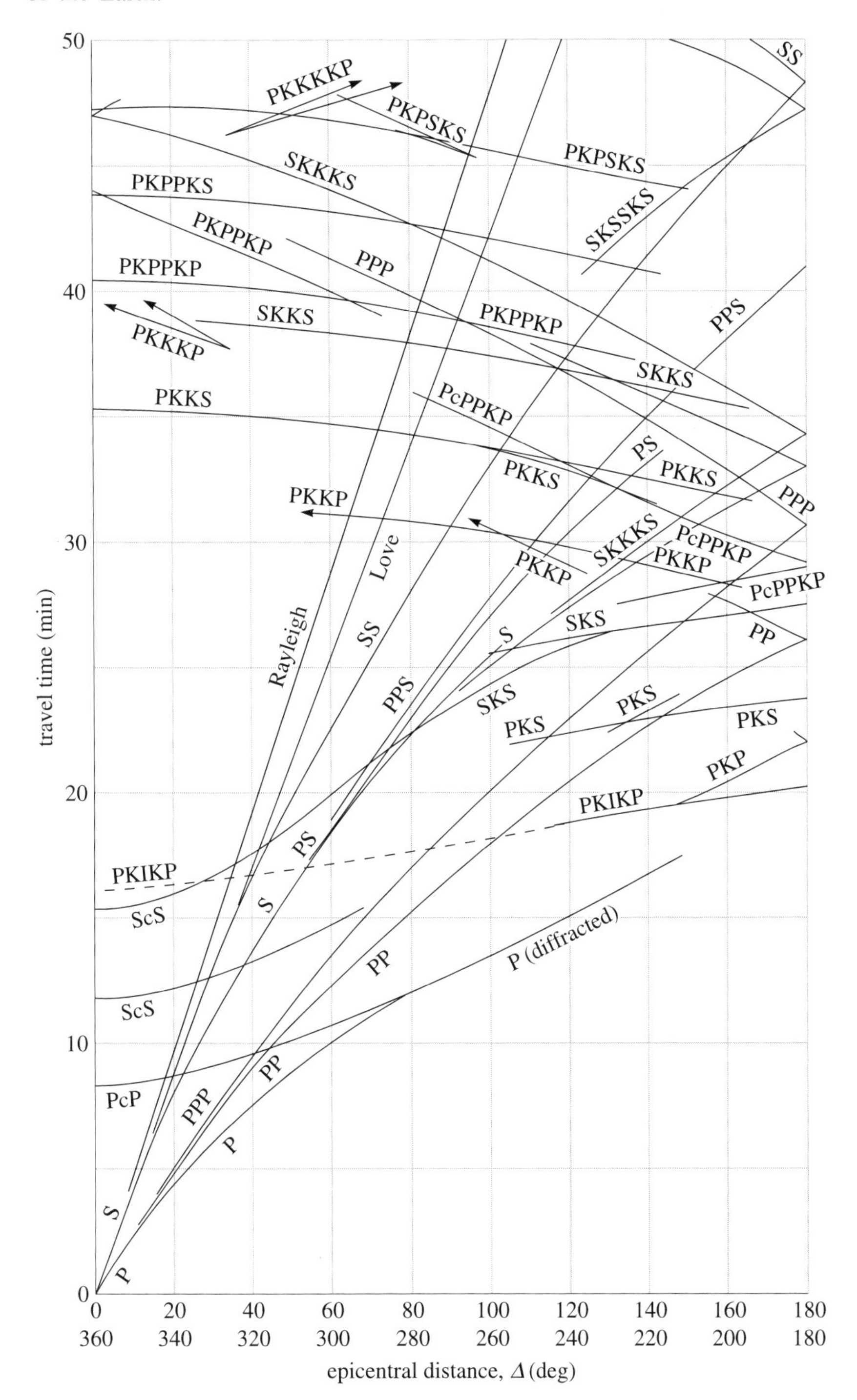

Figure 1.49 The time–distance curves for seismic waves passing through the Earth from an earthquake at the Earth's surface. The distance is expressed as epicentral angle (see Figure 1.29). Note that the times here are in minutes.

1.7.1 The whole Earth

In the study of the seismology of the whole Earth, the seismic sources used are mainly earthquakes, because, with the exception of nuclear explosions, only earthquakes provide enough energy to send waves right through or around the Earth. You will recall that Figure 1.34 showed a selection of those waves. Figure 1.49 shows the time–distance plots for these and other waves through the Earth when the earthquake focus is at the surface. You certainly do not need to remember the details of this diagram, but you should notice two particular phenomena. The first is that, unlike in the plane-boundary case, the time–distance plots are generally curved. The chief reason for this is that within each layer in the Earth, the seismic velocity generally increases with depth. This means that the quantity $1/v$, which represents the gradient of time–distance plots, also varies with depth; and in order for that to be so, the plots must be curved.

The second conspicuous phenomenon on Figure 1.49 is that not all the waves are received by detectors at all epicentral angles up to 180°. In other words, not all the time–distance plots span the complete width of the diagram. The reason for this has to do with the ways certain waves are refracted and reflected inside the Earth; and, indeed, it is those ways that provide information on the Earth's internal structure. How this comes about is illustrated in Figure 1.50, which shows the paths of several waves through all or parts of the Earth. Note that, for discussion purposes, the waves here are not labelled with their standard designations (such as those in Figure 1.34 and 1.49) but with single letters, and that for the moment the inner core is ignored.

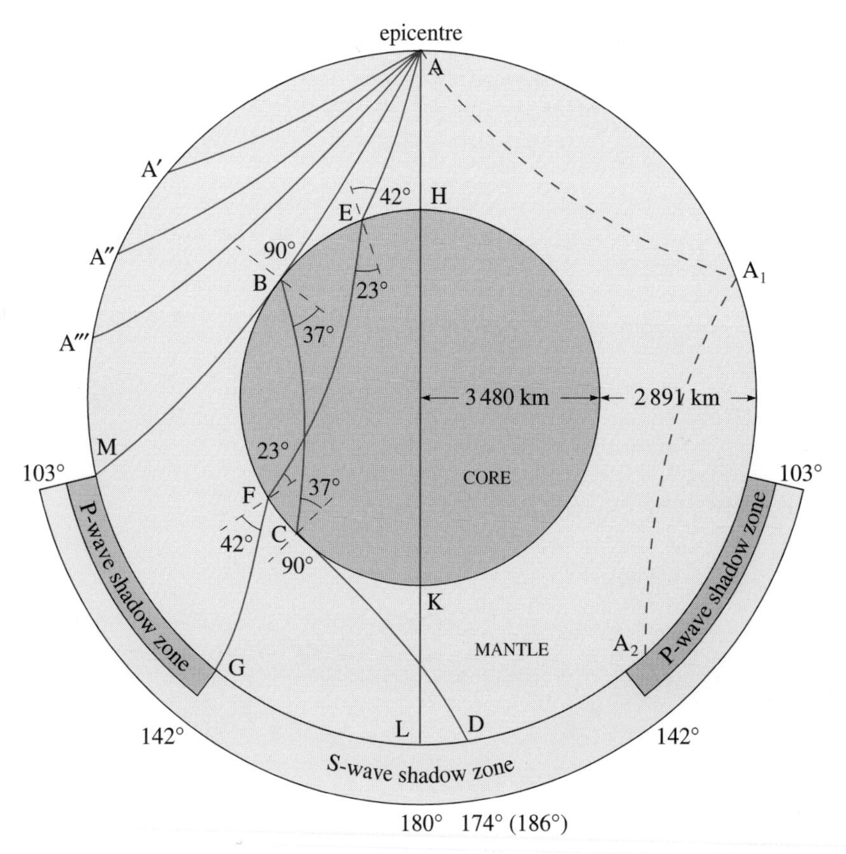

Figure 1.50 The refraction (and reflection) of seismic waves in the Earth. The letters refer to particular waves discussed in the text. Note that on the scale at which this and Figure 1.51 are drawn, the crust is too thin to show up.

Notice, first, that wave paths such as AA′, AA″ and AA‴, which lie entirely in the mantle, are not straight but curved. This, too, is because the velocity of seismic waves within each of the Earth's major layers generally increases with depth. In the plane-boundary simplification of Section 1.6, all the wave paths were straight lines, because within each layer the seismic velocity was constant, and sudden changes in wave-path direction only occurred when there was a sudden change in seismic velocity (i.e. at layer boundaries, or discontinuities). When there is a gradual change in velocity, however, there is an equally gradual change in the wave direction; in other words, there is **continuous refraction**. Waves such as AA′, AA″ and AA‴ are therefore gradually bent back towards the Earth's surface. Waves travelling within the core are likewise curved, and for the same reason. Of course, when waves do reach abrupt velocity discontinuities, they are refracted sharply according to the principles described in connection with horizontal layers. Examples are the waves ABCD and AEFG in Figure 1.50. Note, however, that waves striking discontinuities at incident angles of 0° are not refracted — for example, wave AHKL.

In the context of Figure 1.50, S-waves behave in the same way as P-waves, but with one exception.

❑ Do you recall what that exception is?

■ In Section 1.6.1, we pointed out that S-waves cannot be transmitted through fluids, and Figure 1.3 showed that the Earth's outer core is fluid. S-waves therefore cannot enter the core, which is why there are no S-wave paths comparable to P-wave paths ABCD and AHKL in Figure 1.50.

The fact that S-waves do not enter the core is the chief piece of evidence supporting the view that the outer core is fluid, and is part of the evidence for supposing that the Earth has a core in the first place. To see how this comes about, consider such waves as AA′ and AA″ (regarding them as S-waves), which travel entirely in the mantle. If there is a fluid core through which S-waves cannot pass, it's clear that AA′-type waves cannot lie to the right (as viewed on Figure 1.50) of wave ABM, which re-emerges at the Earth's surface at an epicentral angle of 103° (this applies on each side of the Earth, of course). It follows, therefore, that for direct S-waves from the epicentre (i.e. for waves not refracted at discontinuities) there will be a zone from 103° on one side of the Earth to 103° on the other at which no waves can be received.* The fact that this **S-wave shadow zone** does indeed exist is thus strong evidence in favour of both the core and its fluidity. In practice, some S-waves *are* received in the S-wave shadow zone, but they are only *reflected* waves such as AA_1A_2 in Figure 1.50.

P-waves behave rather differently because they do enter the core, although there is also a **P-wave shadow zone**. Consider again such waves as AA′, AA″ and AA‴, now regarded as P-waves. The limiting wave is again ABM, which glances the core and emerges at the surface of the Earth at an epicentral angle of 103°. Now, however, consider a wave just to the right of ABM (i.e. striking the core's surface at an angle of incidence very slightly less than 90°). This wave (ABCD) will be refracted into the core, travel through it, be refracted back into the mantle, and will emerge at D, which lies at an epicentral angle of 174° on the opposite hemisphere (or 186° if the epicentral angle is measured anticlockwise from the epicentre).

What happens to the other waves passing through the core may appear rather strange at first sight but arises from the geometry of the Earth and the particular seismic velocities in the mantle and core. A P-wave from the epicentre striking the core's surface at an angle of incidence smaller

* But bear in mind that the Earth is spherical, not two-dimensional as shown in Figure 1.50. In reality, therefore, the S-wave shadow zone is a cap-shaped segment of the Earth's surface with a circular circumference.

than that of wave ABCD will emerge at the Earth's surface *between D and M*. Thus whereas a wave with an angle of incidence of 90° at the core's surface emerges at M (ABM) and a wave with an angle of incidence slightly less than 90° emerges at D (ABCD), as the angle of incidence decreases further the point of emergence moves from D towards M — but only as far as G (at an epicentral angle of 142°), represented by wave AEFG. As the angle of incidence gets even smaller, the point of emergence begins to move back towards D until it reaches L (represented by the undeflected wave, AHKL, through the centre of the Earth). The net result of this is that no P-waves at all emerge between 103° and 142° (on both sides of the Earth, of course). This is the P-wave shadow zone.

It should be clear from this that both P-waves and S-waves can be used to define the depth at which the core starts, although only S-waves can be used to demonstrate its fluidity.

❑ Does the behaviour of S-waves as described above demonstrate that all the core is fluid, or just part of it?

■ As far as S-waves are concerned, the two cases cannot actually be distinguished. All that we know is that S-waves cannot enter the core, which need only mean that the core's uppermost layer is fluid. The bulk of the core could be solid, but the S-waves would never know because they could not get in there to find out.

In fact, we know from other evidence that there is a solid inner core and that the fluid outer core is more substantial than merely a fluid uppermost layer. This evidence comes partly from the fact that some P-waves do emerge within the P-wave shadow zone, although they are not waves of the mantle – fluid core – mantle type. Some are simply mantle waves that have been reflected at the Earth's surface (e.g. AA_1A_2 in Figure 1.50) — but there are other, more complicated wave paths, as shown in Figure 1.51. Analysis shows that waves such as ANPQR can only be interpreted as having passed through a solid inner core in which the seismic velocity is higher than in the fluid outer core, although there are also mantle – outer core – inner core – outer core – mantle waves, such as ASTUVW, that do not emerge in the P-wave shadow zone.

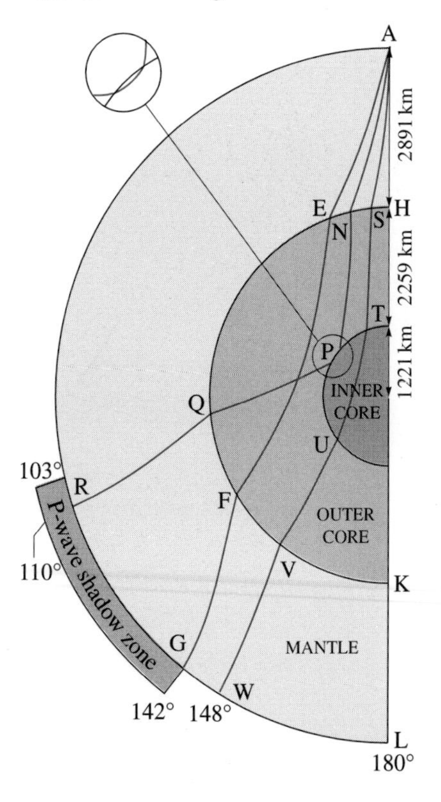

Figure 1.51 The refraction of seismic waves through the Earth's core.

❑ Go back now to look at Figure 1.50. Look specifically at points B and E, where waves are refracted into the core. There is something going on there that you have not come across before. Can you spot what it is, with particular reference to Equation 1.19?

■ According to Equation 1.19,

$$\frac{\sin i}{\sin r} = \frac{v_1}{v_2}.$$

You will recall that in Section 1.6.1, we considered the situation in which waves pass from a lower-velocity medium into a higher-velocity medium. In that case, $v_2 > v_1$, and so (from Equation 1.19) $\sin r > \sin i$ and therefore $r > i$. But you should have spotted that at points B and E in Figure 1.51, $r < i$. This can only mean that $v_2 < v_1$.

❑ What physical interpretation do you put on that conclusion?

■ In passing from the lower mantle into the core, the waves pass from a higher-velocity zone into a lower-velocity zone.

That this is indeed the case is illustrated in Figure 1.52, which combines thousands of observations to provide the best view we have of the variation of P-wave and S-wave velocity throughout the whole Earth. In both P-wave and S-wave curves, there are marked discontinuities, the most marked of which is at a depth of about 2 900 km, interpreted as the mantle–core boundary. From the lowermost mantle to the core, there is a sudden drop in both P-wave and S-wave velocity; indeed, the S-wave velocity drops to zero.

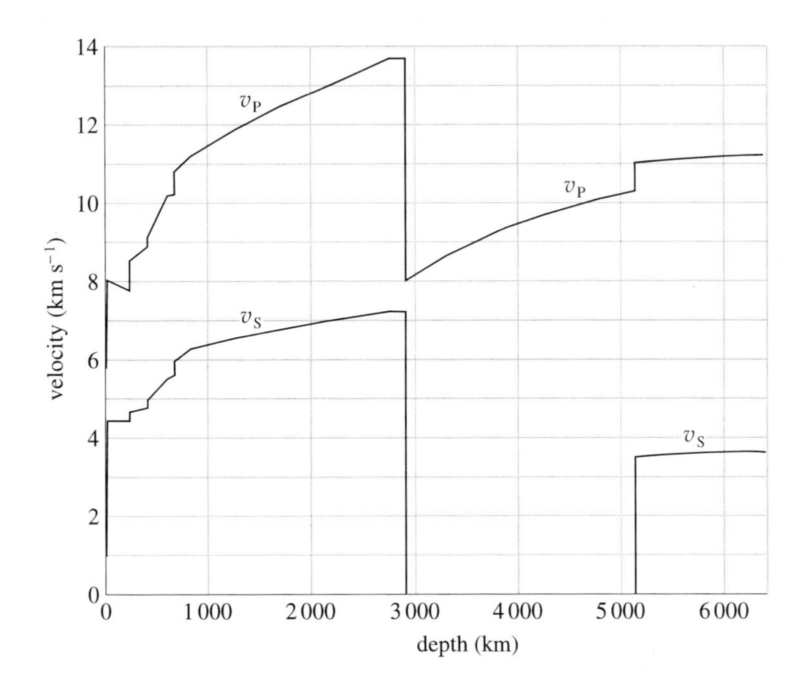

Figure 1.52 The variation of P-wave velocity (v_P) and S-wave velocity (v_S) throughout the Earth.

Below that depth, there is a gradual increase in P-wave velocity throughout the whole core, but a sudden jump at about 5 150 km marks the onset of the solid inner core. There are, of course, no S-waves in the outer core, and so there the S-wave velocity is zero. There is, however, a known S-wave velocity in the solid inner core. You may wonder how this can be, given that S-waves cannot penetrate the outer core to reach the inner core. In fact, you may recall that in Section 1.6 we mentioned, without going into detail about mechanisms, that when P-waves are refracted or reflected they can generate secondary S-waves (and vice versa). P-waves travelling through the fluid outer core and impinging on the surface of the inner core do just that, providing S-waves to travel through the inner core.

The second most conspicuous discontinuities in seismic velocity are those between the crust and the mantle (extreme left of Figure 1.52), although these are not very clear because the crust is so thin that on the scale of Figure 1.52 it barely shows up. Throughout most of the mantle, there is a gradual increase in both P-wave and S-wave velocity. You will notice also that in the uppermost mantle down to a depth of about 700 km there is quite a bit of funny business going on. This is actually extremely important, but we shall defer consideration of it until we look at the mantle in more detail in Section 1.7.5.

In the mean time, what the seismic data have told us is that the primary division of the Earth is into four major layers — **crust**, **mantle**, **outer core** and **inner core** — bounded by major (i.e. conspicuous) discontinuities. The boundary between the crust and the mantle is called the **Mohorovičić discontinuity**, or **Moho** for short, after the seismologist who discovered it. The boundary between the mantle and outer core is sometimes called the **Gutenberg discontinuity**, but more often simply the **core–mantle interface**. The outer core – inner core boundary is occasionally called the **Lehmann discontinuity** after the seismologist who discovered the inner core.

You should now listen to audiovision sequence AV 02, 'Seismic-wave Tracing', which explains how the tracing of seismic waves enables the internal structure of the Earth to be determined. (This lasts about 20 minutes.)

1.7.2 CRUSTAL THICKNESS

The Moho has been observed in many parts of the world, and is now regarded as a worldwide discontinuity. It can be detected using refraction (which is how Mohorovičić discovered it in 1909) or reflection. A particularly nice recording of the Moho using the reflection techniques described in Section 1.6.3 is shown in Figure 1.53. Once beyond the immediate surface layer (0–1 s), there are no obvious reflection signals at all until the horizontal dark line(s) at about 10 s, which geophysicists had no difficulty in recognizing as reflections from the Moho.

ITQ 30

Estimate (a) the depth of the Moho and (b) the thickness of the Moho in Figure 1.53, given that the average P-wave velocity between the Moho and the surface is 6 km s^{-1}.

It may come as a surprise to you to learn that the Moho has a thickness at all, but don't forget that the thickness given by ITQ 30 is negligible on the scale of the whole Earth. The Moho is really a very sharp boundary indeed.

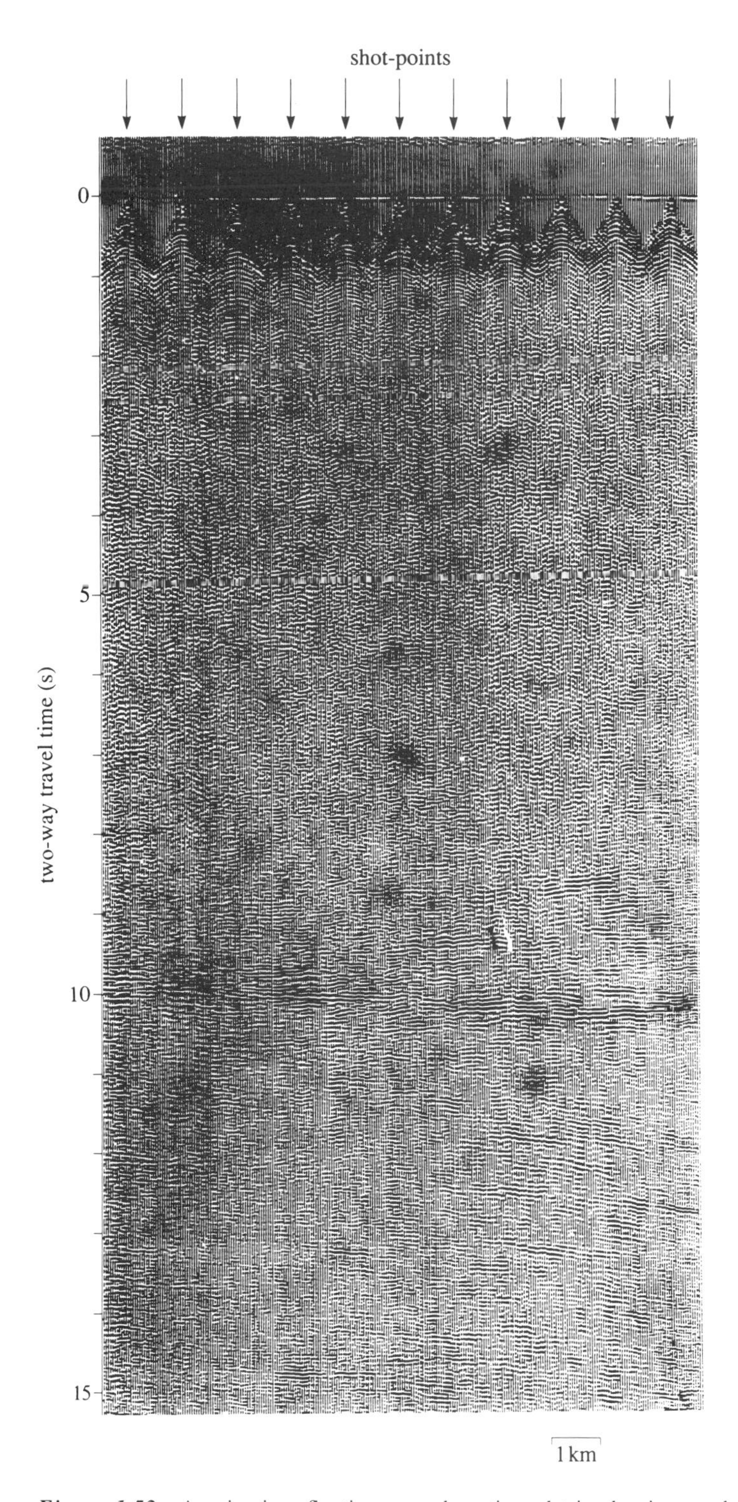

Figure 1.53 A seismic reflection record section obtained using explosions near Mildura, Victoria, Australia, by the Australian Bureau of Mineral Resources.

The average thickness of the oceanic crust is 5–7 km with little variation. The average thickness of the continental crust is 35–40 km but is much more variable. Figure 1.54 shows how the depth to the Moho varies throughout Europe and North America.

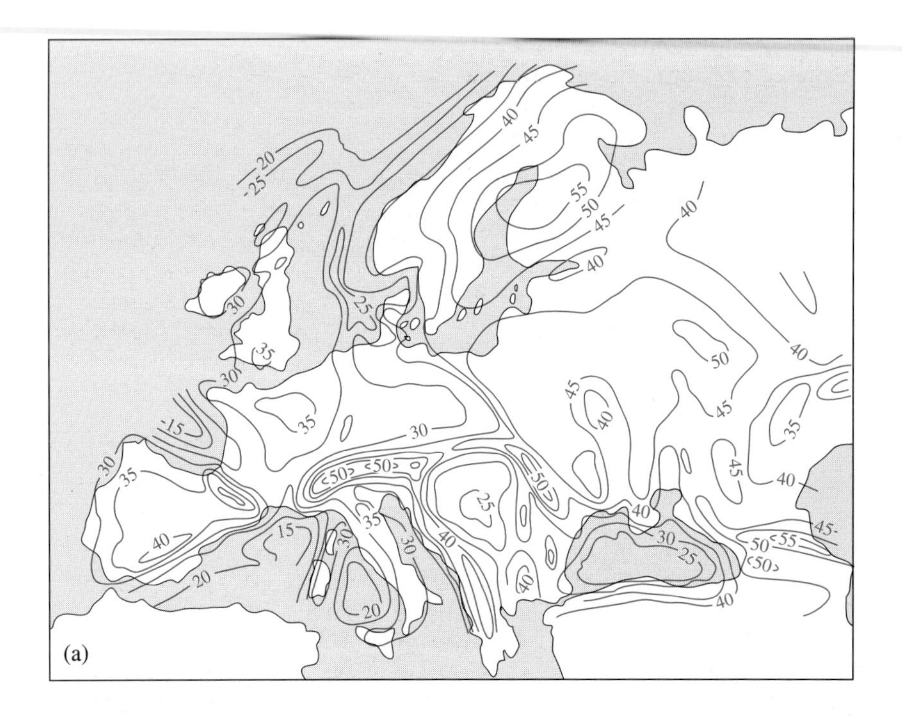

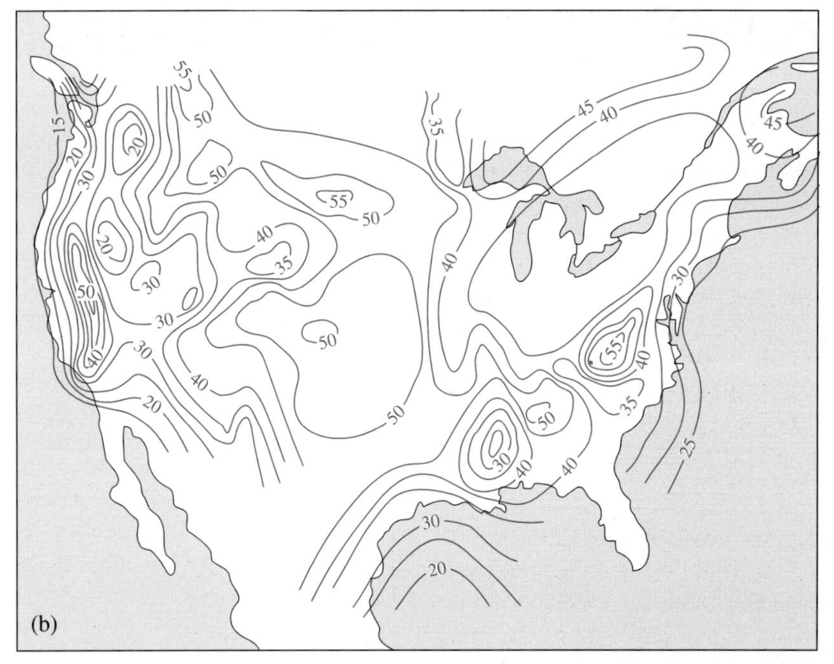

Figure 1.54 Depth of the Moho (measured from sea-level) beneath (a) Europe and (b) the USA. The contours are expressed in kilometres. The two diagrams are not to the same scale. Note that because the depths are measured from sea-level, they do not take account of topography. They therefore represent underestimates of true crustal thickness.

❑ What is the range of crustal thickness demonstrated within continental boundaries on Figure 1.54?

■ The minimum and maximum contours are 20 km and 55 km, respectively. However, don't forget that the crustal thickness (below sea-level) will be rather less than 20 km within the area of the 20 km contours and rather more than 55 km within the 55 km contours.

In fact, these figures rather understate the position on the worldwide scale. While the minimum thickness of continental crust is probably 10–20 km, at the opposite extreme there are certain areas (e.g. beneath the highest parts of the Himalayas) where local crustal thickness exceeds 90 km.

Very often, crustal thickness beneath continents correlates well with surface topography and with geological province (i.e. a region having internally consistent age patterns and/or rock types). One of the regions in which such correlations have been investigated thoroughly is the western USA, illustrated in Figure 1.55. In numerical terms, the average crustal thickness for the various geological provinces are: California coast, 25 km; California Central Valley, 20 km; Sierra Nevada, >40 km; Basin and Range, 25–30 km; Colorado Plateau, 40 km; Rocky Mountains, >50 km; Great Plains, 40–50 km. In other words, each geological province in the western USA appears to have its own characteristic crustal thickness. Generally, the crust is thick beneath young mountain ranges such as the Alps and Carpathians in Europe and the Sierra Nevada and Rocky Mountains in the USA, moderately thick beneath continental interiors, and thin beneath young basins and rifts such as the North Sea and Rhine Graben in Europe and the Basin and Range Province and California Central Valley in the USA.

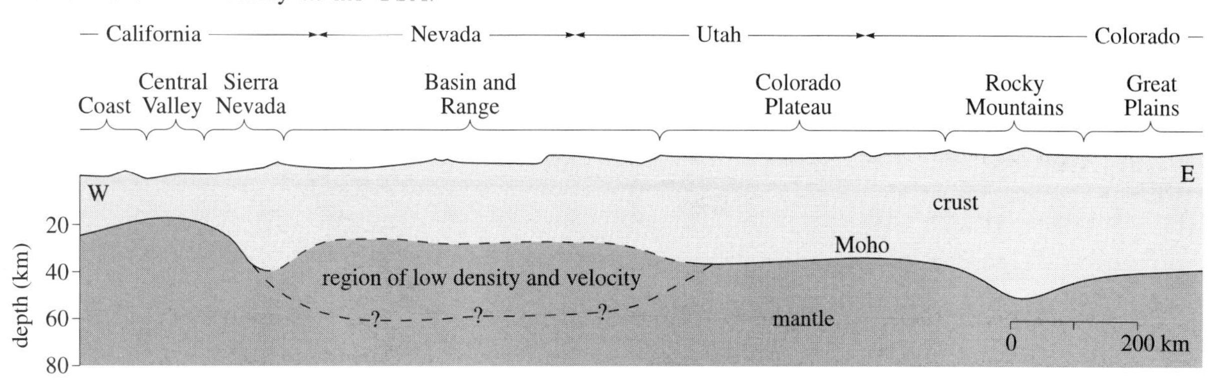

Figure 1.55 Variations in the depth to the Moho along a line from San Francisco, California, to Lamar, Colorado. Crustal thickness is greater under mountainous regions than elsewhere and correlates quite well with the different geological provinces identified here.

1.7.3 The structure of continental crust

Geological observation shows that in many places the Earth's near-surface continental layer is structurally complex and that the complexity can extend down at least to the depth of the deepest borehole (currently about 15 km). But what lies beyond the evidently convoluted uppermost crust? Early seismic experiments gave the impression that the middle and lower crust was structurally bland, and that view was widely held for much of the first half of this century. In fact, there are regions in which deep crustal architecture probably is rather simple. In Figure 1.53, for example, the lack of seismic reflectors above the Moho suggests that that part of the Australian crust is remarkably homogeneous. Elsewhere, however, this is far from the case.

As just one of many hundreds of possible examples, Figure 1.56 depicts a reflection record section from the Wind River Mountains region of Wyoming, USA. What this shows is, first, in the top left-hand corner, reflections from the many near-horizontal layers of sediment in the Green River Basin. The layers are gently folded and extend to a depth of about 12 km (4 s in two-way travel time). Of greater interest, however, is the fact that the fault long known to have cut the surface at B can now be seen to extend far down into the crust and much further than conventional geology could have determined (the diagonal reflector along the line (D) of the sloping arrows). There is also folding and faulting at C and A and, indeed, elsewhere. All this becomes much clearer on the line-drawing interpretation in Figure 1.57, which goes much beyond the region of the record section. Here, the BD fault (now seen to extend to at least 25 km depth) and other major faults are conspicuous, but what the line drawing also brings out is that in the 50–60 km (depth) of crust represented, there are scores of other, lesser reflectors. The existence of such reflectors is not untypical of many regions of deep crust, even in areas such as Kansas where the surface topography and geology are as bland as can be. Precisely what the myriad reflectors mean in terms of structural geology is still a matter of debate; but whatever they are, it can no longer be said that the deep crust is simple.

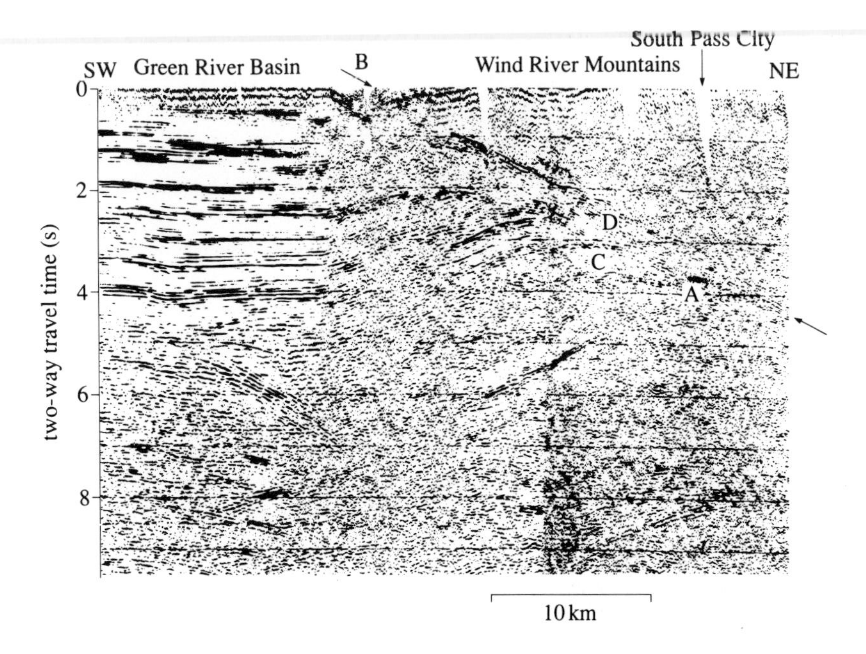

Figure 1.56 A seismic reflection record section from the Wind River Mountains area of Wyoming, USA.

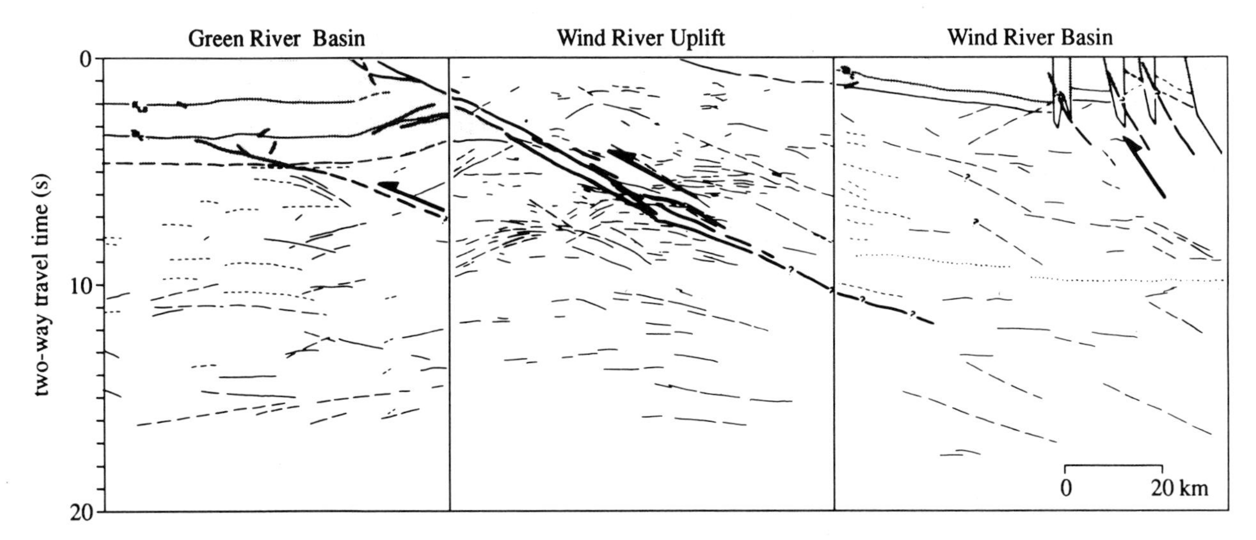

Figure 1.57 A line-drawing interpretation of the record section in Figure 1.56 together with that of the record section continued to the right [NE] (not shown in Figure 1.56). The shaded areas are sediments. The heavy lines are thrusts and faults. The lighter lines are other reflectors.

Refraction data also indicate complexity and, moreover, provide seismic velocities. The direct P-wave (designated P_g) travelling in the crystalline basement of the uppermost crust (i.e. beneath the soil and sedimentary veneer) usually has a velocity of 5.9–6.2 km s^{-1}, depending on the location. The average P-wave velocity in the upper 10–20 km or so of the crust is normally in the range 6.0–6.5 km s^{-1}, whereas that beneath is in excess of 6.5 km s^{-1}. In some regions, there is also a lower crustal layer with a P-wave velocity in excess of 7 km s^{-1}.

What these figures show is that the P-wave velocity generally increases with depth in the continental crust, although probably not often very smoothly. Figure 1.58 shows a selection of six crustal sections as defined by seismic refraction studies. All indicate discontinuities in the crust, although not all the discontinuities are equally clear. Layers with velocities up to about 6.5 km s^{-1} are usually regarded as being part of the 'upper crust' and those with velocities in excess of 6.5 km s^{-1} part of the 'lower crust', although the two terms are often used with rather less objectivity than these definitions would suggest. The P-wave velocity in the uppermost mantle is 7.9 km s^{-1} or greater. Perhaps the most important impression to be gleaned from Figure 1.58 is the remarkable variability of the continental crust, in terms of thickness, of the number of distinct layers and of the velocities within the layers. There is no such thing as 'standard continental crust'.

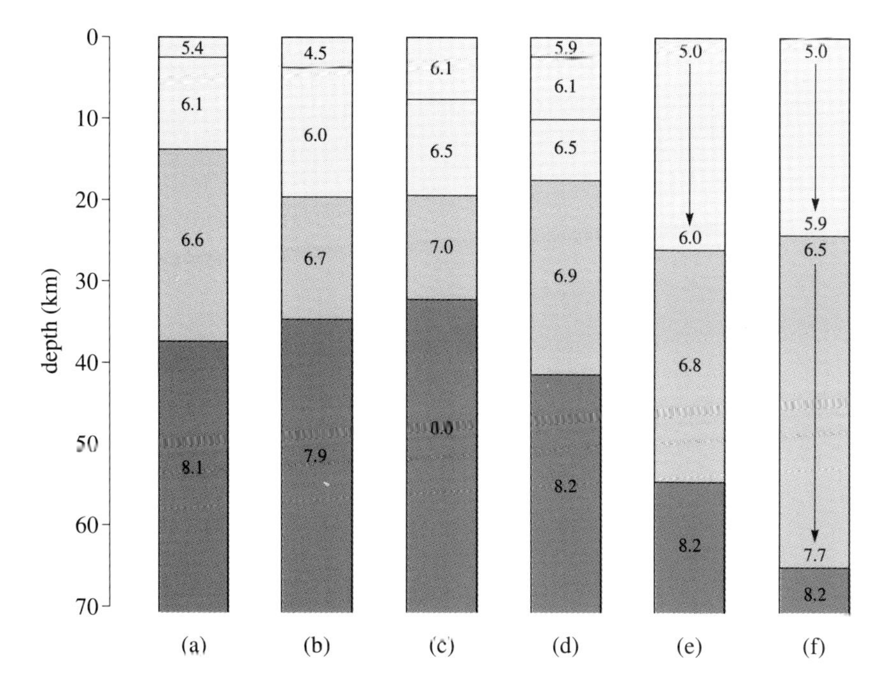

Figure 1.58 Six crustal sections as defined by refraction seismology using P-waves: (a) a stable continental area of Precambrian age in Wisconsin, USA; (b) a modern continental rift environment within Precambrian basement — the Basin and Range Province of the western USA; (c) a 400-Ma-old continent–continent collision zone in northern Scotland; (d) a 100-Ma-old ocean–continent destructive plate boundary in southern California; (e) a modern ocean–continent destructive plate boundary in the central Andean mountains; and (f) a modern continent–continent collision zone in the central Alps. The grey shaded zones are 'upper crust', the lightly coloured zones are 'lower crust' and the more heavily coloured zones are mantle. The arrows in (e) and (f) indicate what appear to be smooth velocity increases between the values given at either end.

1.7.4 THE STRUCTURE OF OCEANIC CRUST

The structure of the oceanic crust is quite different from that of the continents. In part, this is a result of their different thicknesses, but it also reflects the fact that whereas the continental crust is over 3 500 Ma old in places, the oceanic crust now in existence formed entirely during the past 200 Ma. Compared with continental crust, the oceanic crust is surprisingly uniform in thickness and internal structure. Both of these facets are illustrated in a general way by the boxes in Figure 1.59, which compare the average velocity profiles beneath oceans and ancient continental nuclei, or shields.

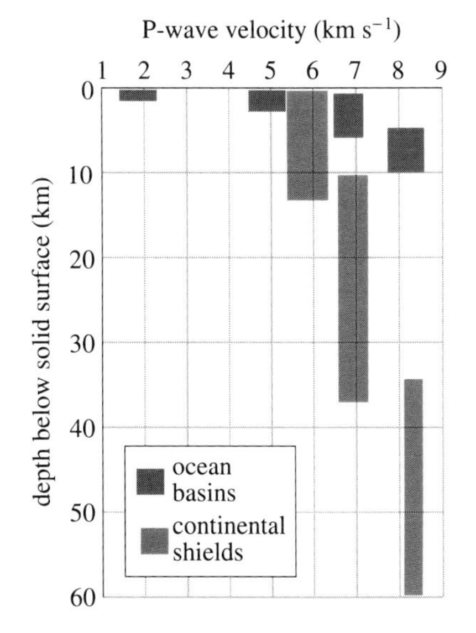

Figure 1.59 Worldwide P-wave velocity averages for two types of crust and upper mantle. Ocean-basin velocities can be divided into four discrete groups, representing four distinct layers in which there is no velocity overlap. Continental velocities fall into three distinct groups, representing the 'upper crust', the 'lower crust' and the upper mantle.

❑ How many distinct layers are there above the oceanic Moho?

■ You should recognize that the fourth layer beneath the ocean basins in Figure 1.59 has P-wave velocities characteristic of the upper mantle ($>7.9\,km\,s^{-1}$), leaving three distinct crustal layers above.

Layer 1, the first ocean-basin crustal layer beneath the water (which, incidentally, has an average depth of 4.5 km and a seismic velocity of $1.5\,km\,s^{-1}$), is easily identified by sampling as the poorly consolidated sediments forming the sea-bed. Its thickness is very variable, ranging from zero at the crests of actively spreading oceanic ridges up to about 3 km near some continental margins. Its average thickness is 0.4 km, and the range of P-wave velocities within it is 1.6–2.5 km s^{-1}. Figure 1.60 shows an example of oceanic structure determined by seismic refraction. Seismic reflection studies, on the other hand, reveal that layer 1 also possesses a fine structure. As Figure 1.61 shows, the layer contains several reflectors, which in parts of the ocean have been directly related to rock types investigated by deep-sea drilling.

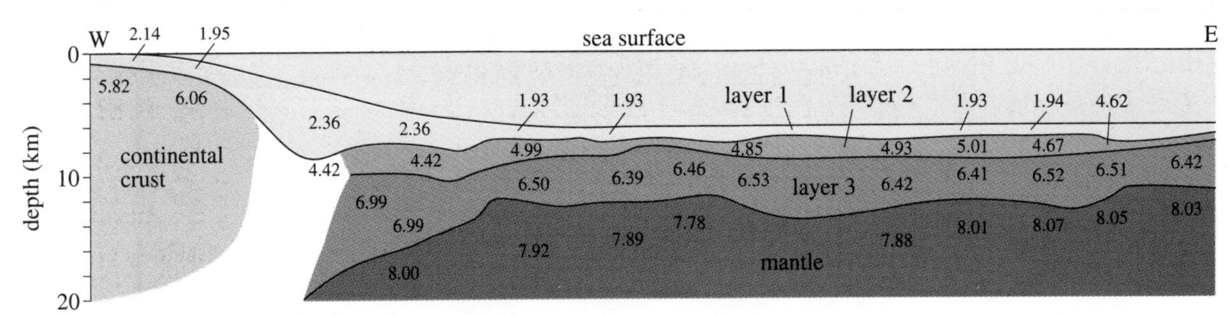

Figure 1.60 An example of oceanic crustal/upper mantle structure based on seismic refraction data, in the Atlantic Ocean east of Argentina. The line runs from 46° S, 60° W on the continental shelf to 43° S, 50° W in the ocean. The small numbers are seismic P-wave velocities in kilometres per second.

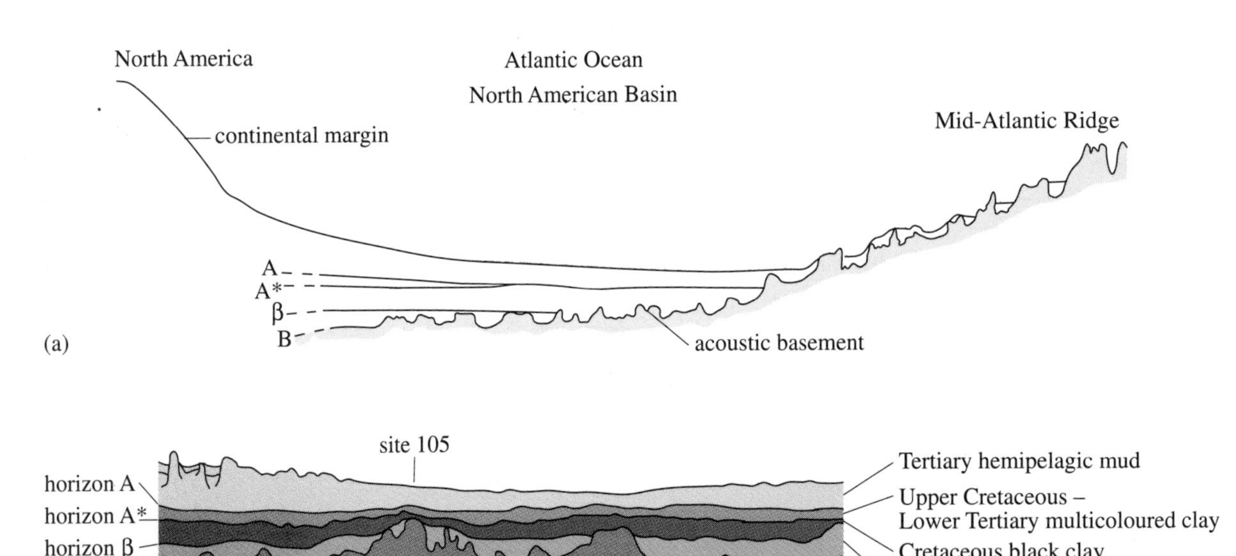

Figure 1.61 (a) Major seismic reflectors in the western Atlantic Ocean, and (b) the corresponding interpretation based on deep-sea drilling data.

Layer 2 of Figure 1.60 is the most variable of the oceanic layers in respect of both thickness (1.0–2.5 km, average 1.4 km) and P-wave velocity (3.4–6.2 km s^{-1}), and is shown by deep-sea drilling to be of igneous origin. There are three sublayers within layer 2. Sublayer 2A (the uppermost sublayer) is found only on actively spreading ridges, where it ranges in thickness from 0 to 1 km. It is mainly highly porous basalt through which seawater circulates and thus generally has a low P-wave velocity of 3.4–3.6 km s^{-1}. Sublayer 2B is also basalt but of much lower porosity and forms the upper part of layer 2 where sublayer 2A is absent (e.g. in ocean basins). Sublayer 2B generally has P-wave velocities in the range 4.8–5.5 km s^{-1}, although values outside this range sometimes occur (e.g. Figure 1.60). Sublayer 2C is also basaltic, is about 1 km thick where it exists and usually has P-velocities in the range 5.8–6.2 km s^{-1}.

Layer 3 is the main oceanic layer, having an average thickness of 5 km. It is certainly igneous and probably mostly gabbro. Its P-wave velocities generally lie within the range 6.4–7.0 km s^{-1}.

1.7.5 THE STRUCTURE OF THE MANTLE

Up to the 1960s, the Moho was regarded as a fairly sharp boundary at which the velocity of P-waves increases abruptly to 8.1 km s^{-1} or higher in an almost homogeneous upper mantle. It is now clear, however, that the upper part of the mantle is less homogeneous than previously supposed. Figure 1.62, for example, shows how the P-wave velocity in the mantle immediately below the crust varies beneath the USA. In parts of the central USA the velocity exceeds 8.3 km s^{-1}, whereas beneath the active mountain regions of the west it falls as low as 7.8 km s^{-1}. It is now more usual to say that the uppermost mantle has a P-wave velocity of 7.9 km s^{-1} or higher, although in certain active regions of the world (e.g. Iceland) the velocity of the top of the mantle can be as low as 7.2–7.4 km s^{-1}.

Figure 1.62 Contours of P-wave velocities (km s^{-1}) in the uppermost mantle beneath the USA.

No less striking than the lateral inhomogeneity of Figure 1.62 are changes in the vertical direction. You will recall that in Figure 1.52, in both the P-wave and S-wave velocity curves, there were squiggles in the upper few hundred kilometres to which we said we would return later. Because of its scale, Figure 1.52 is hard to read in that range, but a more expanded version of the P-wave velocity curve is shown here as Figure 1.63. The first thing to notice is that there are sudden jumps in P-wave velocity at, respectively, about 400 km and about 670 km — in other words, there are discontinuities at these depths. These are generally thought to represent not chemical changes but phase changes — i.e. changes in mineralogical/crystallographic structure of the material without changes in its chemical composition. We shall return to them later in this Block when we come to look at the composition of the Earth's layers.

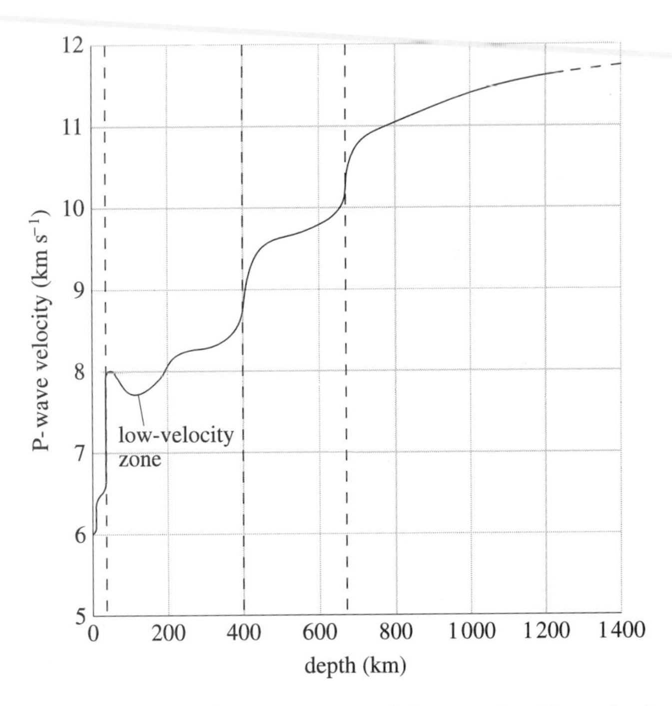

Figure 1.63 The variation of P-wave velocity in the upper 1 400 km of the Earth.

Remaining with the upper part of the mantle, if you look at Figure 1.63 again you will see that P-wave velocity does not everywhere increase with depth but that within the upper two hundred kilometres or so there is a region in which it dips. This is called the **low-velocity zone** (LVZ) and is thought to be due to partial melting. The material involved must be mainly solid, because it transmits S-waves; but theoretical analysis shows that only a small amount of melting (a few per cent) would be sufficient to reduce the P-wave velocity by the amount indicated in Figure 1.63.

The existence of the LVZ provides an alternative way of dividing the Earth. The division of the Earth into crust, mantle and core is essentially chemical; these layers have different chemical compositions. The LVZ within the upper mantle, however, is not a chemical subdivision but one based on physical properties. The material in the LVZ is chemically the same as the material immediately above and below it, but because it is partially molten it has different physical properties — specifically, it is much more able to deform by flow.

At this point, we begin to get into territory which is uncertain and even controversial, and in which the terminology used is confusing and ambiguous because it is not universally agreed. In the last paragraph, we mentioned an alternative method of dividing up the Earth's interior based not on chemical but on physical properties. The general way of expressing this division is to call the layer in the upper part of the mantle that can deform, or flow, the **asthenosphere**. The strong, outermost layer of the Earth above the asthenosphere is then called the **lithosphere**. The lithosphere thus comprises the crust and uppermost mantle, which, despite their chemical differences, are strongly coupled physically. The region of the Earth between the base of the asthenosphere and the surface of the core is called the **mesosphere**. As you will see in Block 2, the lithosphere is divided horizontally into segments, or 'plates', which move and interact in various ways generally referred to as plate tectonics or global tectonics. Suffice it to say here that what gives the lithospheric plates the freedom to move and interact is the existence below of the deformable layer, the asthenosphere.

The question then arises: how, more specifically, is the asthenosphere to be defined? The need for an asthenosphere to account for certain phenomena taking place at the Earth's surface came to be appreciated long before the LVZ had been detected. It was known, for example, that when a segment of the Earth's crust is heavily loaded (e.g. by ice during an ice age) it sinks, and that when the ice is removed the crustal region concerned rises again. There therefore had to be a deformable zone within the Earth just below the rigid outer layer. In short, there had to be an asthenosphere. When the LVZ was discovered, it seemed a godsend, for here was an obvious explanation for the asthenosphere — a region in which there is supposedly partial melting, which would both give the layer an ability to flow and account for the anomalous dip in seismic velocity. It thus became customary for a while to claim that the LVZ and the asthenosphere were coincident — the LVZ *was* the asthenosphere. Later, however, some geophysicists shied away from this equation, which they regarded as an oversimplistic view.

Part of the problem is that whereas the LVZ is now fairly closely defined by seismology (the most recent seismic models of the Earth place its top at 40–60 km on average and its base at about 220 km on average), the zone of partial melting is not. As Figure 1.64 shows, because of increasing pressure with depth, the melting point of mantle material also increases with depth. And so, of course, does the temperature. It turns out, however, that, possibly by chance, the temperature–depth curve intersects, or comes close to, the mantle melting curve somewhere in the Earth's upper few hundred kilometres. This encounter, or close encounter, is apparently sufficient to melt a small part (a few per cent) of the mantle in the appropriate zone, but the limits of that zone are not likely to be very sharp. Moreover, they are likely to differ from place to place. Beneath oceanic ridges, where temperatures are high, the lithosphere will be thin and the top of asthenosphere close to the surface. But in old ocean basins and, especially, beneath continental shields, the opposite is likely to be true.

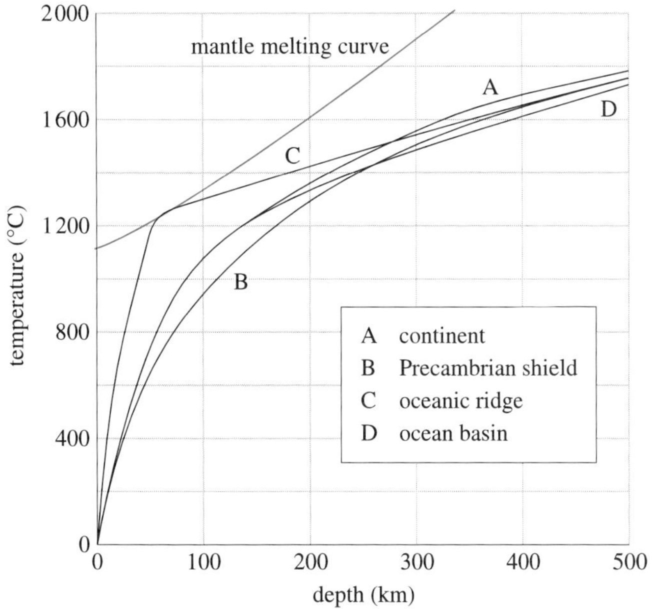

Figure 1.64 The variation of temperature with depth in the upper part of the mantle beneath various continental and oceanic regions (A–D), and the mantle melting curve. Temperature everywhere increases with depth in the Earth; but because pressure also increases with depth, the temperature at which mantle material melts increases faster than does the actual temperature. Nevertheless, within the Earth's upper 200 km or so, the temperature curves approach the mantle melting curve and in one case may even intersect it. This close approach is not sufficient to result in a completely molten uppermost mantle, but could result in partial melting. Note, however, that the temperatures are not known accurately enough to allow precise depth limits to be placed on the supposed zone of partial melting. What is important here is not the detail but simply the concept — that somewhere in the upper part of the mantle the temperature and mantle-melting curves come close enough to produce a zone of partial melting.

Notwithstanding the difficulty of locating the supposed zone of partial melting, a perfectly respectable case can be — and often is — made for saying that the LVZ and the asthenosphere are indeed the same. The asthenosphere thus becomes *by definition* a zone of partial melting in which the material can flow (as opposed to the lithosphere above, which is rigid) and which lies between 40–60 km and about 220 km. (As we shall see later, strictly, these figures apply to the asthenosphere beneath old oceanic lithosphere — i.e. not beneath very young lithosphere or beneath continents). We can then define the **upper mantle** as comprising the subcrustal lithosphere (i.e. that part of the lithosphere beneath the Moho), the asthenosphere and that part of the mantle between the base of the asthenosphere and the 400 km discontinuity. The **lower mantle** is that part of the mantle below the 670 km discontinuity; and the region between 400 km and 670 km is known as the **transition zone**. (This region is called the transition zone, because at depths of 400 km and 670 km, the mineral olivine, the main constituent of the mantle, undergoes phase changes; it remains olivine, but its crystallographic structure changes. The transition zone is thus a region of change between the 'normal' olivine in the upper mantle — as defined above — and the final, high-pressure form in the lower mantle.)

The chief reason why the scheme outlined in the last paragraph is not universally accepted — and the source of much confusion — is that partial melting is not the only, or even the main, cause of deformation, or flow, in the mantle. At high pressures and over long time-scales, solid material can slowly flow by a process known as **solid-state creep**. This is believed to take place throughout some or all of the mantle where, aided by the transfer of heat, it is known more generally as convection (see Section 1.3.3). Whether convection takes place throughout the whole mantle (probably now a majority view) or is limited to that part of the mantle above 670 km (minority view) is still a matter of debate; but both sides do agree that convection takes place above 670 km, which means that it occurs immediately below, and indeed probably within, the partially molten zone defined above as the asthenosphere.

All this raises an interesting issue of perception. If flow can take place as a result of partial melting *and* solid-state creep, why limit the definition of the asthenosphere only to the partially molten zone? What matters, surely, is the behaviour of the material concerned, not the cause of the behaviour. So if flow occurs both in the partially molten zone and in the region below, why not apply the term 'asthenosphere' to the whole region of flow irrespective of the cause of the flow? Of course, you should be able to see that this could lead to an absurd conclusion. If flow takes place throughout the whole mantle, it would place the base of the asthenosphere at the core–mantle boundary. In fact, no one goes as far as that (at least, to the best of our knowledge), but what some geophysicists do now do is place the base of the asthenosphere at the 670 km discontinuity. They then define the whole of the region between the Moho and the 670 km discontinuity as the **upper mantle** [alternative definition] and the region between that discontinuity and the core–mantle interface as the lower mantle. The transition zone for olivine then becomes, somewhat illogically perhaps, part of the upper mantle.

Unless otherwise stated in a particular context, we shall in this Course define the asthenosphere as being coincident with the partially molten LVZ, which places the base of the asthenosphere at a depth of about 220 km. At the same time, we shall regard the upper mantle as the whole of the region between the Moho and the 670 km discontinuity, simply because this has now become the majority usage even among people with diverse views about where the base of the asthenosphere should be placed. The picture that emerges is thus that shown in Figure 1.65.

One issue that we have not yet discussed in any detail, however, is the depth of the *top* of the asthenosphere, which is equivalent to asking where the base of the lithosphere lies, or how thick the lithosphere is. Unfortunately, this is not as simple a matter as it might appear. The obvious point to make is that the lithosphere is unlikely to have the same thickness everywhere. As we mentioned above, beneath oceanic ridges, where temperatures are high, the lithosphere will be comparatively thin, whereas beneath ocean basins and continental shields, where temperatures are lower, the lithosphere can be expected to be thicker. However, even if we consider just one type of region — say, where there is old oceanic lithosphere — the matter is not straightforward.

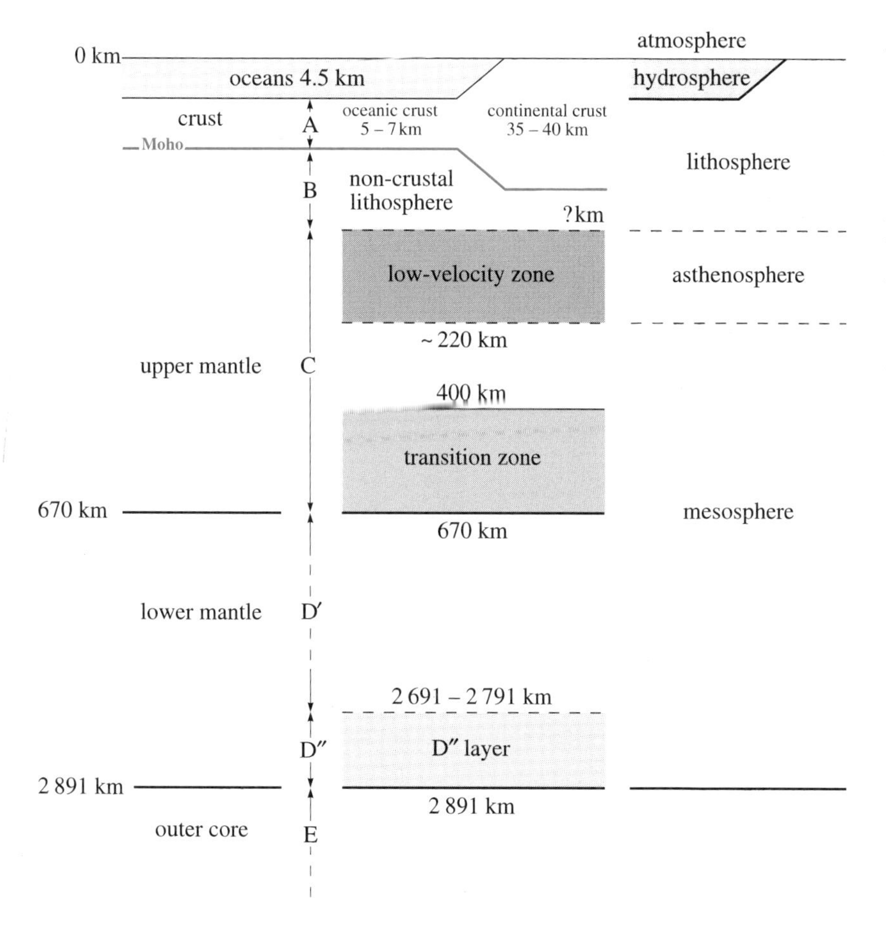

Figure 1.65 The main layers of the Earth's mantle. The designations crust, mantle and core (shown on the left) are based on chemical boundaries. The designations lithosphere, asthenosphere and mesosphere, shown on the right, are based on physical changes. Down the centre is the letter notation sometimes used to refer to the Earth's major layers. Note, however, that not all the detail on this diagram is universally agreed. For example, geophysicists differ over where to place the base of the asthenosphere. Then again, whereas here we regard the transition zone as part of the upper mantle (with the boundary between upper and lower mantle at 670 km), some scientists regard the upper mantle as the layer above the transition zone (i.e. above 400 km) and the lower mantle as the layer below the transition zone (i.e. below 670 km). Others still regard the transition zone as the layer between 400 km and about 1 000 km, a legacy of a time when there was thought to be a major world-wide discontinuity at about 1 000 km. Similar problems occur with the numbers. For example, some geophysicists argue that old oceanic lithosphere reaches a thickness in excess of 100 km. All this confusion merely represents states of ignorance. We burden you with it here for two important reasons. The first is that you may see other notions expressed elsewhere, and we want you to appreciate that others' views are perhaps no more nor less authoritative than ours (although some will be based on outdated ideas). Second, there is genuine confusion/disagreement among the various authorities, and we want you to realize that it exists. It's no use pretending that the details of the Earth's interior are universally agreed 'facts'. This diagram represents the best consensus we have been able to come up with (in 1992) and will be the information we use as required elsewhere in the Course.

On the basis that the asthenosphere is defined as being coincident with the LVZ, the top of the asthenosphere should lie at a depth of 40–60 km beneath old oceanic lithosphere, which is the depth of the top of the LVZ as defined by seismic models. But do these particular seismic data agree with other seismic data and with data from other sources? The top of the LVZ was defined by seismic velocities as plotted in Figure 1.63. An alternative approach as far as the lithosphere–asthenosphere boundary is concerned is to try to determine not the top of the asthenosphere but the base of the lithosphere. The details of the techniques used here are beyond the scope of this Course, but one involves measuring the dispersion of seismic waves and another involves observing how the lithosphere responds elastically to large loads (e.g. an ice sheet, a mountain range, a sediment-filled basin or a volcano) placed on it.

The results of such determinations are shown in Figure 1.66. The seismic-wave dispersion data are summarized by the curve, which suggests that the thickness of the oceanic lithosphere rises to well in excess of 100 km. In other words, there is an alarming difference between this and the seismic-velocity data placing the top of the LVZ at 40–60 km. However, the raw dispersion data are considered by some geophysicists to be misleading. The upper mantle is anisotropic and when the dispersion data are 'corrected' to take the anisotropy into account (open circles and dashed curve in Figure 1.66), they agree much more closely with the data obtained on the basis of elasticity (closed circles) and refraction (triangle) and from seismic-velocity studies of the LVZ. Although the corrected dispersion data give slightly greater values of lithospheric thicknesses than do the elastic-load data, both methods show that the thickness of the oceanic lithosphere increases with the age of the sea floor and both methods suggest that the older oceanic lithosphere has a thickness of 30–60 km, which is in surprisingly good agreement with the top of the LVZ as defined by seismic models. Before we can conclude that the old oceanic lithosphere is indeed 30/40–60 km thick, however, we must examine the thermal data, which we shall do when we have looked at the Earth's heat flow (Section 1.12). In the mean time, you should note that seismic and elastic determinations on land suggest that the continental lithosphere is probably about 150 km thick under shields (the ancient cores of continents), although some studies have given even higher values.

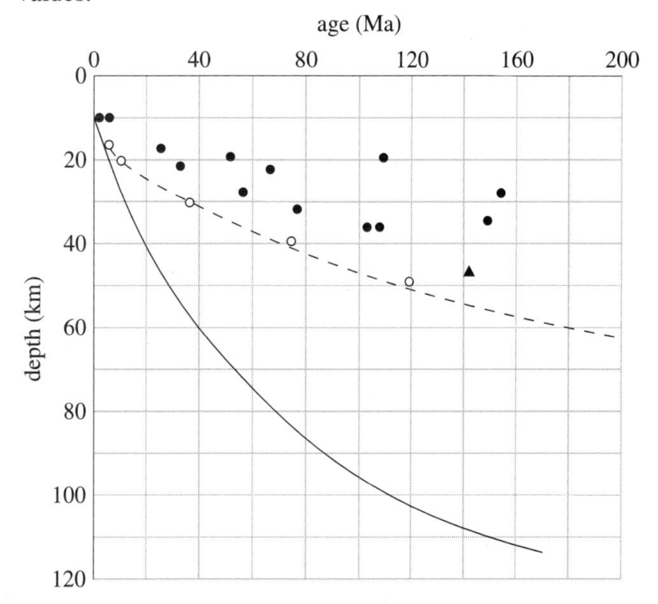

Figure 1.66 Estimates of the thickness of the lithosphere beneath the oceans. The full curve is based on seismic-wave dispersion studies, the open circles (and dashed curve) are the dispersion data corrected for anisotropy, the closed circles are based on loading observations, and the triangle is from a seismic refraction experiment.

The lower mantle appears to be rather bland as far as seismic waves are concerned, with no major discontinuities throughout most of it. There is, however, a layer with a thickness of about 100–200 km at the base of the lower mantle (i.e. above the core–mantle interface) in which the P-wave and S-wave velocities decrease by a few per cent. This is called the **D″ layer** and may be the result of iron diffusing out of the core into the mantle, although it could be a region of anomalously high temperature produced by heat from the core. (Or it could be both.) It's not entirely clear yet whether or not the layer is continuous (i.e. covers the whole of the core's surface); but it seems to be topographically rugged and is shown as such in Figure 1.3 (Section 1.2.2).

1.7.6 Seismic tomography

Hitherto, we have considered seismology as a largely two-dimensional affair (the diagrams of wave paths, for example, have always shown the waves travelling in two dimensions), although in that the Earth is spherical, rather than circular, the third dimension has been implied. During the late 1970s, however, there occurred an advance that made three-dimensional studies of the Earth much more explicit. This was the development of **seismic tomography**, the seismic analogue of CAT-scanning in the medical world (CAT = computer-assisted tomography). In medical tomography, X-rays are passed through the human body in many different directions and the computer-analysed ray paths are converted into three-dimensional images of the density variations in the body's interior. In seismic tomography, the patient is, of course, the Earth, and the probes are large numbers of seismic waves criss-crossing as they pass through the planet.

The application of seismic tomography is complex, but the principles are not. Consider, first, two identical wave paths in the Earth — the waves are of the same type (P-waves, say), the rocks or combinations of rocks through which they pass are the same and at the same temperature and pressure, and the path lengths (source–detector distance) are equal. Under these circumstances, the travel times of the two waves from source(s) to detector(s) will also be equal. Now suppose that a region of the rocks through which one of the waves (but not the other) passes is heated — i.e. a thermal anomaly is placed in the wave's path. The travel times of the two waves will now no longer be equal.

❑ Can you deduce which wave will now have the longer travel time, that passing through the thermal anomaly or that not doing so? (Hint: Think of the implications of Equation 1.13 and what we said about it in Section 1.6.1).

■ If a rock (or any other material) is heated, it will expand and hence its density (= mass/volume) will decrease. The wave passing through the thermal anomaly will thus pass through a lower-density region. Now, Equation 1.13 shows that seismic velocity varies inversely with density so that, on the face of it, seismic velocity will increase as density decreases. But you should recall that we pointed out in Section 1.6.1 that the elastic moduli increase with density and do so, moreover, more rapidly than the density itself increases. So, in fact, seismic velocity increases with increasing density (and vice versa). The seismic wave passing through the thermal anomaly (lower-density region) will therefore slow down slightly as it does so and will thus reach the detector slightly later than the wave not passing through the anomaly (i.e. the wave passing through the anomaly will have the longer travel time).

In the case of the two waves described, it is possible to tell which of them passed through the thermal anomaly simply by observing which arrived the later. What it is not possible to do, however, is to determine *where* along the wave path the anomaly occurred, because the seismic velocity calculated from the travel time is an average along the whole path and not a detailed picture of how the velocity varies along the path. But the position and size of the anomaly can be determined by analysing the travel times of a large number of waves, some of which pass through the anomaly and some of which don't. The principle is illustrated in Figure 1.67. The waves that don't pass through the anomaly have the normal travel times for their respective distances, but those that do are slowed down (if they pass through a lower-density region) or speeded up (if they pass through a higher-density region). The more waves there are and the greater the diversity of travel directions, the more closely will the anomaly (or anomalies) be defined, although the computer power required for the analysis is formidable. Nevertheless, it's possible; and what emerge are maps of 'slow' and 'fast' regions in the Earth's mantle, interpreted, respectively, as lower-density/higher-temperature and higher-density/lower-temperature regions.

Three examples of what can be achieved with seismic tomography are shown in Plates 1.8–1.10.

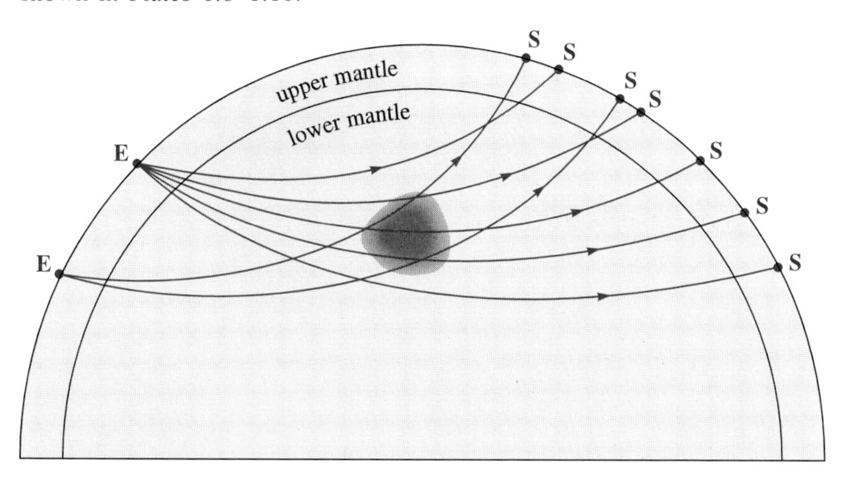

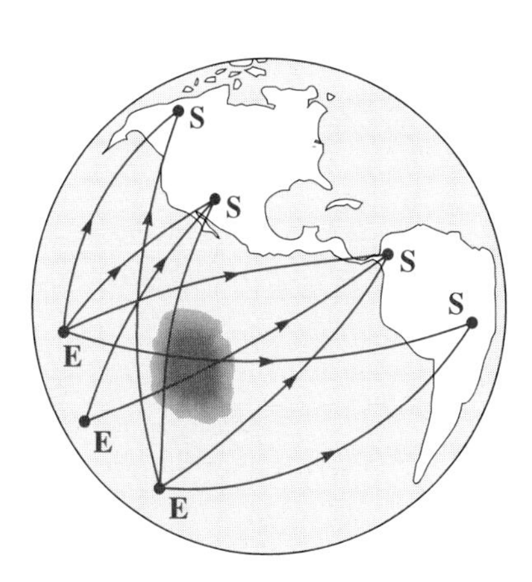

Figure 1.67 The principle of seismic tomography. Information from criss-crossing P-waves (in this case) through the Earth is combined to isolate velocity anomalies such as the one shown here. E are the earthquake sources and S are the seismometers.

1.7.7 Seismic velocities in the core

At the core–mantle boundary, the P-wave velocity drops sharply from about 13.7 km s^{-1} to about 8.1 km s^{-1} (Figure 1.52). The S-wave velocity drops from about 7.3 km s^{-1} to zero. The P-wave velocity then increases again with depth in the outer core to about 10.4 km s^{-1} at its base, jumping to about 11.0 km s^{-1} at the top of the inner core. Within the inner core, there is a slight increase in P-wave velocity to about 11.3 km s^{-1} at the Earth's centre. S-wave velocity rises from about 3.5 km s^{-1} at the top of the inner core to about 3.7 km s^{-1} at the Earth's centre. The velocities in the core thus never rise to those in the lowermost mantle.

On the scale of Figure 1.52, the outer core–inner core boundary appears very sharp, but it is widely believed that the jump in velocity there takes place within a transition zone several hundred kilometres thick. This zone is sometimes called the **F layer**. This region of the Earth is difficult to study, partly because waves have a long way to travel before reaching the inner core (and so their amplitudes are small) and partly because any errors in the velocity profile of the upper parts of the Earth will tend to obscure details at the boundary between the inner and outer cores.

The information summarized in Figure 1.64 was derived entirely from seismology, which, as we pointed out at the beginning of Section 1.5, provides the most important set of techniques for investigating the physical state of the Earth's interior. However, it is not the only way of probing the Earth. In Section 1.8, we turn to another — the study of gravity.

Summary of Section 1.7

The principles governing the refraction and reflection of waves at plane surfaces bounding horizontal layers within each of which the seismic velocity is constant are the same as the principles governing the refraction and reflection of waves at curved surfaces bounding layers within each of which the seismic velocity varies (generally increasing with depth in the case of the Earth). The end results are different, however. In particular, wave paths and time–distance plots for refracted waves in the latter circumstances are generally curved. Moreover, in the case of the Earth, refracted waves are not received at all epicentral angles. There is an S-wave shadow zone beyond epicentral angles of 103° because the Earth has a fluid outer core through which S-waves cannot pass. Indeed, the existence of the shadow zone is the chief evidence for a fluid outer core. There is also a P-wave shadow zone between epicentral angles of 103° and 142° because of the way that waves are refracted at and within the core, in which layer seismic velocities are lower than in the lower mantle. From the analysis of seismic waves, seismologists are able to deduce that the Earth has four major layers — crust, mantle, outer core and inner core — and to estimate the depths at which their boundaries lie. These boundaries are named, in order of increasing depth, the Mohorovičić discontinuity (Moho), the core–mantle boundary or interface (the Gutenberg discontinuity) and the outer core–inner core boundary (the Lehman discontinuity).

The average thickness of the oceanic crust is 5–7 km and that of the continental crust is 35–40 km; but whereas there is little variation in the thickness of the oceanic crust, the continental crust varies considerably in thickness (up to 90 km in places). Crustal thickness very often correlates well with the surface topography and type of geological province, generally being thick beneath young mountain ranges and thin beneath young basins and rifts.

Continental crust is highly variable in nature and structure. Near the surface, this is obvious by inspection, but seismic data also reveal complexity at depth. Reflection data show that in many places the crust contains numerous reflectors; and refraction data indicate that continental crust often has a broadly layered structure with successively deeper layers having successively higher seismic velocities. Crustal P-wave velocities generally lie within the range $5.9\,\mathrm{km\,s^{-1}}$ (top) to $7.0\,\mathrm{km\,s^{-1}}$ (bottom), but there are many exceptions. A velocity of $6.5\,\mathrm{km\,s^{-1}}$ is usually regarded as the boundary between the (loosely defined) 'upper crust' and 'lower crust'.

The much younger oceanic crust has a structure quite different from that of the continental crust, being more distinctively layered and less variable in nature throughout the world. Layer 1 (uppermost), the sediment forming the sea-bed, is very variable in thickness (0–3 km, average 0.4 km) and layer 2, which is basaltic, is also variable in thickness (1.0–2.5 km, average 1.4 km), but layer 3, which is mostly gabbro, is much more uniform (average 5 km). Layer 1 has a fine structure, and layer 2 can be divided into sublayers 2A, 2B and 2C. Successively deeper layers have successively higher seismic velocities or seismic-velocity ranges.

The mantle usually has a P-wave velocity of $7.9\,\mathrm{km\,s^{-1}}$ or higher, although there are a few places in which the uppermost mantle has a somewhat lower velocity. The uppermost mantle is, in fact, laterally inhomogeneous as far as seismic velocity is concerned. There are also velocity discontinuities at depth, chiefly at 40–60 km and 220 km (between which there is a reduction in velocity — the so-called low-velocity zone, or LVZ) and at 400 km and 670 km (both of which discontinuities represent phase changes in the chief mantle mineral, olivine). The LVZ is thought to be partially molten, enabling it to flow and thus giving the segments (plates) of the overlying rigid outer shell of the Earth the freedom to move.

According to one view, the LVZ is the asthenosphere, the crust and mantle segment above being the lithosphere and the rest of the mantle below being the mesosphere. By another view, however, the asthenosphere includes both the LVZ and the mantle below to a depth of 670 km, the latter region being able to flow by solid-state creep. The region of the mantle below 670 km (the lower mantle) may also be able to flow by the creep mechanism. Depending on which view one takes, the upper mantle may be defined as the whole of the mantle between the Moho and the 670 km discontinuity or that part of the mantle between the Moho and the 400 km discontinuity, although there is a tendency these days to adopt the former definition irrespective of the definition of the asthenosphere. Either way, the region between the 400 km and 670 km discontinuities is called the transition zone. In the current state of knowledge, the lower mantle appears to have few discontinuities. However, there is a distinct layer, 100–200 km thick (the D″ layer), immediately above the core, which could be a chemically distinct layer (iron diffusing from the core) or a thermal layer (heat from the core), or both.

Seismic tomography is a technique in which many criss-crossing seismic wave paths through the Earth are analysed to determine the three-dimensional regions in which there are density (and thermal) anomalies.

The P-wave velocity at the top of the fluid outer core is much lower than that at the base of the mantle. It rises with depth in the core (with a jump at the surface of the solid inner core) to the centre of the Earth but never reaches the value at the base of the mantle. There is, of course, no S-wave velocity in the fluid outer core, but in the solid inner core the S-wave velocity is more or less constant at a value lower than anywhere in the mantle.

OBJECTIVES FOR SECTION 1.7

When you have completed this Section, you should be able to:

1.1 Recognize and use definitions and applications of each of the terms printed in the text in bold.

1.28 Describe and interpret the chief characteristics of the refraction and reflection of seismic waves in the Earth and, in particular, how the form of the data differs from that of data from waves refracted and reflected at plane horizontal boundaries separating layers in which the seismic velocity is constant.

1.29 Interpret the plot of seismic velocity with depth in the Earth.

1.30 Quote the approximate values of the average thicknesses of the continental and oceanic crust, and interpret 'contour' maps of crustal thickness.

1.31 Outline the chief seismically determined characteristics of the continental crust, the oceanic crust, the mantle and the core.

1.32 Describe both the chemical and physical internal divisions of the Earth and discuss in general terms the problem of how to define the position of the asthenosphere.

1.33 Summarize the basic principles of seismic tomography.

Apart from Objective 1.1, to which it also relates, the ITQ in this Section tests the Objectives as follows: ITQ 30, Objective 1.28.

You should now do the following SAQs, which test other aspects of the Objectives.

SAQS FOR SECTION 1.7

SAQ 42 (*Objectives 1.1, 1.28, 1.29, 1.30, 1.31, 1.32 and 1.33*)

State, giving reasons where appropriate, whether each of the following statements is true or false.

(a) Both wave paths and time–distance plots for refracted seismic waves passing through the Earth are generally curved, because within each of the Earth's major layers seismic velocity generally increases with depth.

(b) The existence of an S-wave shadow zone at epicentral angles of greater than 103° is evidence of the fluid nature of the outer core.

(c) No S-waves are received within the S-wave shadow zone.

(d) There is a P-wave shadow zone between epicentral angles of 103° and 174°.

(e) At the centre of the Earth, the P-wave velocity is higher than that at the base of the mantle.

(f) Both P-wave and S-wave velocities are almost constant throughout the solid inner core.

(g) The average thickness of the continental crust is at least five times that of the oceanic crust.

(h) Crustal thickness in Britain never exceeds 30 km.

(i) Seismic velocity in the continental crust generally increases smoothly with depth.

(j) Direct sampling shows that layer 1 of the oceanic crust is igneous.

(k) In the mantle immediately below the crust of the USA, P-wave velocity ranges from less than $7.8\,\mathrm{km\,s^{-1}}$ to more than $8.3\,\mathrm{km\,s^{-1}}$.

(l) There is a low-velocity zone (LVZ) in the upper mantle between depths of 40–60 km and about 220 km.

(m) The lithosphere comprises the whole of the crust and that part of the mantle above the asthenosphere.

(n) If the asthenosphere is taken to be coincident with the LVZ, its base must lie at a depth of about 220 km.

(o) Irrespective of where the base of the asthenosphere is placed, flow within the asthenosphere is likely to be due partly to partial melting and partly to solid state creep.

(p) Seismic and elastic-loading data suggest that the top of the asthenosphere lies at a depth of 30–60 km beneath old oceanic crust.

(q) If two waves travel along paths that are identical in all respects except that one wave passes through a thermal anomaly (high-temperature zone) and the other does not, the wave passing through the anomaly will arrive at its respective detector earlier than the other.

(r) There is a sudden jump in P-wave velocity at the outer core–inner core boundary.

SAQ 43 (*Objectives 1.25 and 1.28*)

Use Snell's law (Equation 1.19) to show that a wave originating at the base of the Earth's crust and travelling initially towards the centre of the Earth will actually pass through the centre of the Earth and emerge at an epicentral angle of 180°.

1.8 THE EARTH'S GRAVITY

We have discussed the way in which seismic methods may be used to develop models for the internal structure of the Earth. But how can we be sure that those models are realistic? Some of them can be clarified using other geophysical techniques, such as the measurement of gravity. Indeed, in many circumstances, models for the Earth not supported by a range of different geophysical data are not truly valid. Gravity is a passive technique in that it does not require the release of energy before measurements can be made (seismology, by contrast, often requires use of a controlled explosion). The technique relies on detecting the natural variations in the gravity field caused by the uneven distribution of mass within the Earth.

1.8.1 THE LAW

The law of gravity was first stated by Sir Isaac Newton in 1687. Whether or not there is any truth in the story that an apple fell on his head, he realized that the nature of the force governing the movements of the Moon relative to the Earth was the same as that attracting smaller bodies, like apples and people, to the Earth. Newton's law can be expressed in a simple equation, which relates the force (F) between two particles to their masses (m_1 and m_2) and the distance (d) between their centres of mass:

$$F = \frac{Gm_1m_2}{d^2}. \qquad \text{(Equation 1.32)}$$

The law states that the force (F) experienced by each particle (planet, or apple, or satellite) is directly proportional to the product of the masses of the two objects (m_1 and m_2) and is inversely proportional to the square of the distance (d) between them. The constant G is the **universal gravitational constant**, and the first laboratory measurement of its value was made by Cavendish in an outhouse at his home in London in 1797.

❑ What will the units of the universal gravitational constant G be?

■ By rearranging Equation 1.32, we find that

$$G = \frac{F\ (\text{N})\ d^2\ (\text{m}^2)}{m_1\ (\text{kg})\ m_2\ (\text{kg})}$$

where the terms in brackets are the SI units newtons, metres and kilograms. The metres are squared because distance d in Equation 1.32 is squared. The SI units for G then are

$$\frac{\text{N m}^2}{\text{kg}^2} \quad \text{or} \quad \text{N m}^2\,\text{kg}^{-2}.$$

The value that Cavendish determined was remarkably accurate and in close agreement with the accepted value today of $6.672 \times 10^{-11}\,\text{N m}^2\,\text{kg}^{-2}$.

You may remember from the Foundation course or elsewhere that the force (F) on an object of mass (m) at the Earth's surface is related to the acceleration it experiences; in this case, it is the acceleration due to gravity (g):

$$F = mg. \qquad \text{(Equation 1.33)}$$

Equations 1.32 and 1.33 may be combined to give the force F on an object with mass m caused by its gravitational attraction to the Earth, whose mass we denote as M:

$$F = \frac{GmM}{d^2} = mg. \qquad \text{(Equation 1.34)}$$

In this case, d, the distance between the two centres of mass is simply the radius of the Earth (assuming that the size of our object is negligible compared with the Earth and that the object is on the Earth's surface). Obviously, if the object were above the Earth's surface, as a satellite would be, then d is larger, being the sum of the Earth's radius and the object's height above the surface.

It would seem from our discussion so far that the acceleration due to gravity, g, is a constant all over the Earth, and on a broad scale it is. After all, weight is directly proportional to g and you don't seem to weigh any more or less in Australia than you do in the UK. Sensitive gravity meters — often called **gravimeters** — measure gravity variations of 1 in 10^8 or 1 part in a hundred million and reveal variations in g across the globe. Although small, these tiny differences turn out to reveal much about the interior structure of the Earth. We will now consider the sources of these variations.

1.8.2 SOURCES OF GRAVITY VARIATIONS

If the Earth were a perfectly homogeneous stationary sphere, then the gravitational attraction experienced by objects at its surface would be exactly the same wherever they were. However, the polar radius of the Earth is about 21.5 km less than the equatorial radius, making the Earth a rather squashed sphere, known as an **oblate spheroid**. The distance to the Earth's centre of gravity is less at the pole than at the equator, so that the mass of the Earth is not evenly distributed.

❑ Can you see what difference this will make to the measured value of gravity g?

■ g will be greater at the poles than at the equator because the distance to the Earth's centre is less at the poles.

We can even calculate the difference using Equation 1.34:

$$\frac{GmM}{d^2} = mg.$$

As the mass of the object on the Earth's surface (m) appears on both sides of the equation, it cancels out, leaving

$$\frac{GM}{d^2} = g.$$

From the last equation, you can see that if we measure g and we know d and G, then we can obtain a value for M, the mass of the Earth. The first measurements of the Earth's radius (d) were made about 2 500 years ago by Eratosthenes, who measured the angle subtended by the Sun at two places on the same longitude at the same time and measured the distance between the places. His value of d was within 16% of the real value, which wasn't bad considering he could only measure distance in camel days (1 camel day = 18.5 km). Nowadays, precise measurements of distances and angles are made using satellites.

So the measured value of gravity, g, will depend directly on the mass of the Earth and the gravitational constant, and inversely on the square of the Earth's radius.

The mass of the Earth is about 6×10^{24} kg, the equatorial radius is about 6 378 km and the polar radius is about 6 356.5 km, so we are now ready to calculate the value of g at the equator and poles. Note that since we are using SI units, the distances must first be expressed in metres.

$$g_{\text{pole}} = \frac{6.672 \times 10^{-11}\,\text{N}\,\text{m}^2\,\text{kg}^{-2} \times 6 \times 10^{24}\,\text{kg}}{(6\,356.5 \times 10^3)^2\,\text{m}^2}$$

so $g_{\text{pole}} = 9.908\,\text{N}\,\text{kg}^{-1}$

and since $1\,\text{N} = 1\,\text{kg}\,\text{m}\,\text{s}^{-2}$,

$g_{\text{pole}} = 9.908\,\text{m}\,\text{s}^{-2}$.

ITQ 31

Calculate the value of gravity (g_{equator}) at the equator.

SI units are rather cumbersome when talking about small variations in g, so in honour of Galileo, who was the first person to measure g, the gravity units conventionally used now are gals (Gal), where

$1\,\text{Gal} = 1\,\text{cm}\,\text{s}^{-2} = 10^{-2}\,\text{m}\,\text{s}^{-2}$.

The difference in gravity between the pole and the equator is therefore 6.7 Gal.

This would be the value that a gravity meter would register if there were no other sources of gravity variation on the Earth. However, the Earth is rotating, which produces an outward force at the surface opposing the inward gravitational force.

❑ Can you remember what this force is called?

■ It is the **centrifugal force**.

The Earth is spinning very fast, with a surface speed at the equator of about $464\,\text{m}\,\text{s}^{-1}$, and so compared with the value that g would have on a stationary, perfectly spherical Earth, the observed value is smaller. The values that we just calculated took into account the fact that the Earth is not a perfect sphere, but didn't take account of the centrifugal force produced by the Earth's spin.

The gravitational acceleration on the Earth as a rotating oblate spheroid can be expressed mathematically by the **International Gravity Formula** (IGF), which, as adopted by the International Association of Geodesy in 1967, is

$$g_\lambda = 9.780\,318\,5\{1 + 5.278\,895 \times 10^{-3}\,(\sin\lambda)^2 + 2.346\,2 \times 10^{-5}\,(\sin\lambda)^4\}\,\text{m}\,\text{s}^{-2} \qquad \text{(Equation 1.35)}$$

where λ is latitude in degrees.

You do not need to remember this equation. This formula predicts the latitude-dependent variations in gravity seen across the surface of the Earth at sea-level.

❑ What is the actual variation in gravity according to the IGF going from the pole to the equator?

■ We can calculate this simply by substituting $\lambda = 0°$ for the equatorial value of g, then $\lambda = 90°$ for the polar value into Equation 1.35. So $g_{0°} = 9.780\,318\,5\,\text{m}\,\text{s}^{-2}$ and $g_{90°} = 9.832\,177\,24\,\text{m}\,\text{s}^{-2}$, making the difference $0.052\,\text{m}\,\text{s}^{-2}$ or 5.2 Gal.

This is smaller than the difference we calculated earlier for a non-rotating Earth whereas we would have expected it to be greater because of the effect of the centrifugal force acting against gravity. This is because previously we calculated the value of gravity at the pole assuming the whole Earth had the polar radius (thus *under*estimating g_{pole}), and we *over*estimated g_{equator} because there we assumed the whole Earth to have the equatorial radius. The effect of the centrifugal force is greatest at the equator. In fact, it is more complicated than that as our earlier calculated g_{pole} and g_{equator} are both larger than the IGF values. This is because the Earth's surface is more curved at the equator than at the poles (the assumption that the Earth is spherical with all its mass concentrated at the centre is only true to a first approximation). The purpose of this calculation was to illustrate how, using very crude approximations, the gravity difference between the poles and the equator could be estimated. Equation 1.35 gives the relationship between the acceleration due to gravity at the Earth's surface and latitude, allowing for the rotation and equatorial bulge. This theoretical gravity variation is the field we should observe on a rotating oblate spheroid and it approximates closely to the actual field. But there are other smaller variations caused by irregularities on the surface of and deep within the Earth. These smaller variations are important for geophysicists studying the Earth's gravity field, as departures from the IGF can be analysed to tell us about the fine structure of the Earth.

1.8.3 THE SHAPE OF THE EARTH

In reality, the Earth is neither a perfect oblate spheroid nor a perfect sphere. Obvious examples of this are the deep oceans and tall mountains, which deviate by several kilometres from the smooth surface of the spheroid. Clearly, another shape is needed to describe the real surface of the Earth.

As the oceans cover two-thirds of the planet, and the land is a complex surface with hills and valleys, mean sea-level is a good approximation to the real equilibrium surface of the Earth. On land, this would be the surface of canals open to the sea. At this surface, the value of gravity is the same as on the spheroid (and so increases away from the equator according to the IGF). In other words, at sea-level, gravity obeys the IGF and it is larger in value at the poles than at the equator. The problem is, though, that sea-level is not a smooth surface. This is not simply the effect of waves on the surface caused by the wind and tides, but is due to lateral variations in gravity. The surface of the sea is able to adjust itself vertically to compensate for variations in gravity. If, for reasons that will become apparent later, the value of gravity is larger than it should be according to the IGF at a particular place, sea-level rises until g at the (elevated) sea-surface has its IGF value. Similarly, if gravity is too low, the sea-level falls to accommodate it and g again attains its IGF value. The bumps on the surface are usually expressed as elevation contours above and below the 'theoretical' spheroid surface (Figures 1.68 and 1.69). These elevations are literally the heights of the sea-surface above and below the oblate spheroid. This surface, called the **geoid**, is not theoretical — it is the *real* surface of sea-level at which measured gravity is equal to the theoretical IGF value in the absence of the smaller variations that we come to in the next Section.

A high point on the geoid (relative to the spheroid) represents a surface that is higher than the spheroid surface. Since the value of gravity decreases as the square of the distance from the Earth's centre of mass, then for the geoid surface to be displaced upwards, the value of g on the spheroid surface at that point must be anomalously high. The Earth's surface compensates by moving upwards — reducing g — until the IGF value for that place is reached.

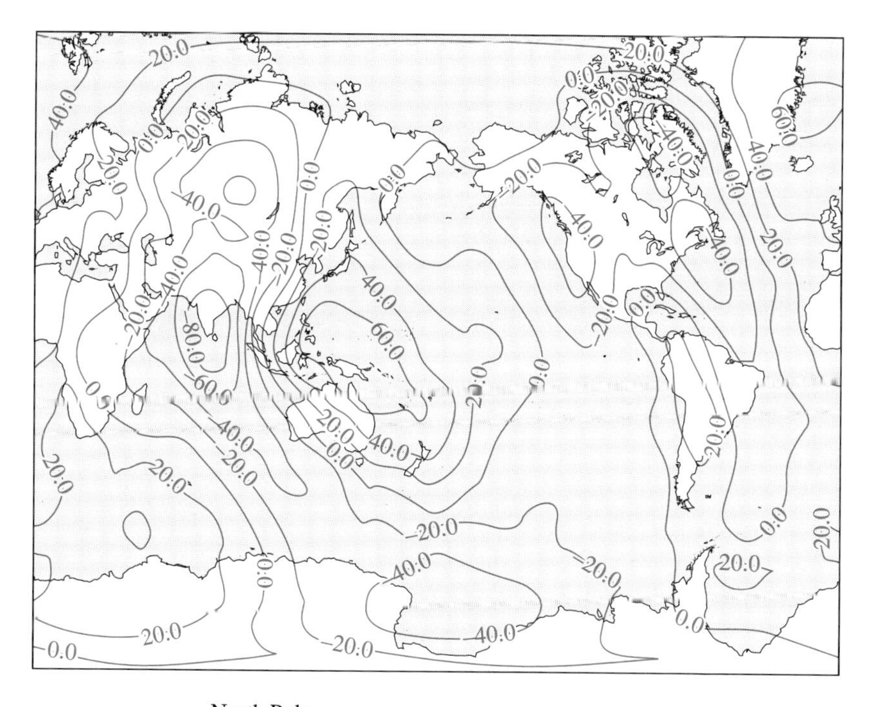

Figure 1.68 The geoid expressed in metres above (positive) and below (negative) the spheroid surface. In general, the value of gravity on the geoid surface is the same as the theoretical value on the spheroid and is given by the IGF. The undulations on the geoid are caused by the uneven distribution of mass within the Earth.

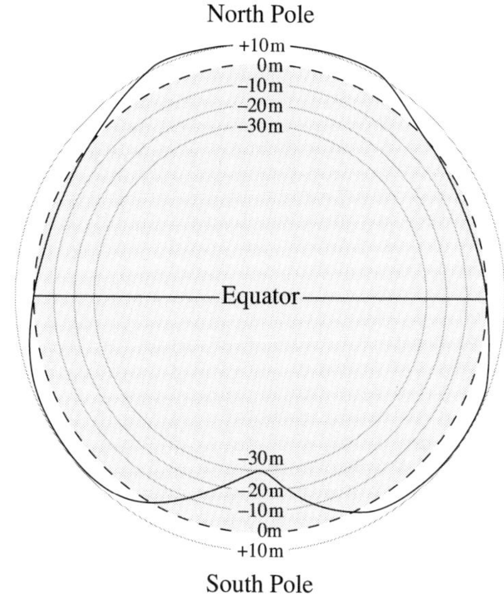

Figure 1.69 The Earth is not symmetrical about the equatorial axis. This exaggerated sketch shows that the Earth's surface (solid line) is higher at the North Pole by 10 m and lower at the South Pole by 30 m than the spheroid surface (dashed line).

❑ Look at Figure 1.68 carefully and see if you can find any correlation between features on the geoid and topographic features such as coastlines, mountain chains and oceans.

■ The oceans do not seem to be related to undulations in the geoid, neither do the mountain ranges on land or the coastlines. The departures of the geoid from the spheroid must therefore reflect deeper structural anomalies within the Earth. These anomalies have been used to provide evidence for the presence of convection cells within the mantle, and we shall discuss these further in Section 1.14.

A massive object buried deep in the Earth would exert a gravitational force at the surface of the Earth, in addition to the usual value of g. Similarly, a large buried body of very low density would exert a net upwards force, because it exerts *less downwards* force than the surrounding 'normal' material. These features, if they were on a large enough scale, would produce hills and valleys, respectively, on the geoid surface.

The most striking feature on the geoid is the apparent hole, or negative anomaly just to the south of India. The surface of the sea is more than 80 m lower here than predicted by the spheroid. A ship sailing from east to west across the hole, would drop by over 80 m and then rise up again on the edge of the hole; but, of course, it would not notice any incline because gravity would not be changing. The only force that the ship experiences (apart from drag by the water) is due to gravity acting perpendicular to the surface of the geoid. This large depression in the geoid is believed to reflect a deep-seated mass deficit, caused by lateral inhomogeneities deep within the mantle. We shall return to this later in the Course.

1.8.4 GRAVITY ANOMALIES

Geophysicists who wish to use a gravimeter as a tool for determining the structure of the shallow part of the Earth are less interested in absolute values of g than in relative values. What they usually want to do is to locate inhomogeneities on a much smaller scale; these cause density contrasts that can be distinguished by making very detailed gravity measurements in the region of interest. The important thing is the *difference* between the gravitational effect of the particular feature and the effect that there would be if no such feature were present. This difference in gravity is the **gravity anomaly** and its value is usually measured in thousandths of gals, or **milligals** (mGal). The gravimeters used for this type of survey work basically consist of a mass on a very sensitive spring. For a constant mass attached to the spring, the higher the value of g, the greater the weight of the mass and the longer the spring becomes. Relative values of g are obtained by comparing the length of the spring at different stations. In principle, therefore, the geophysicist *measures* spring length at many points (S) in the area of investigation and again at a point of reference (P) well away from the area (Figure 1.70). The reference station is, ideally, an international gravity station outside the area of the small-scale anomaly being studied, at which the absolute value of gravity is known accurately. There is a network of such stations throughout the world. The absolute value of gravity is measured using a different sort of gravimeter; some work by measuring the period of a pendulum (which is proportional to the inverse square of g) and others by measuring the acceleration on a falling mass.

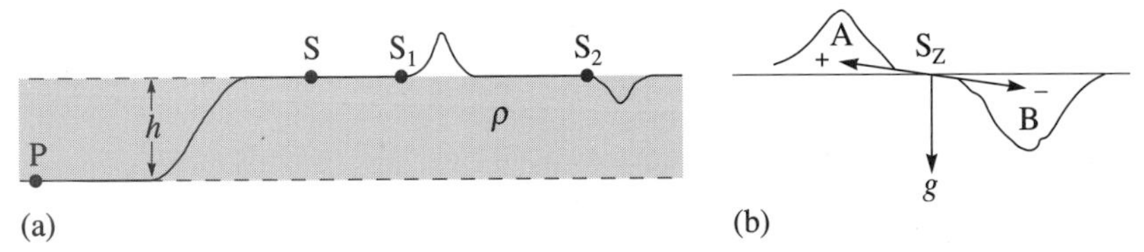

Figure 1.70 (a) Diagram to illustrate the 'corrections' that must be made to the observed gravity anomaly when the stations (S) differ in elevation from the reference point (P). The free-air correction ($\Delta_2 g$) allows for the difference in distance to the centre of the Earth; the Bouguer correction ($\Delta_3 g$) allows for an infinite horizontal slab of material between S and P (shaded); the terrain correction ($\Delta_4 g$) allows for the fact that the material between S and P is not just a flat slab, but consists of hills and valleys. (See text for details.)
(b) The effect of topography on gravity at a point S_Z. The hill (A) exerts an additional attraction in the direction of the + arrow. This has an upward component and so reduces the overall downwards attraction. The value of g at S_Z is thus reduced. Conversely, the valley (B) does *not exert any* attraction at point S_Z There is an overall lack of downwards attraction because of the valley (towards the − arrow). This means that again the value of g at S_Z is reduced.

In fact, the gravity difference between two such points (a site and a reference point) depends on many factors as well as the presence of subsurface density variations. Once the gravity difference Δg between the site (S) and the reference point (P) has been measured, it needs to be 'corrected' in order to eliminate these other effects before it can be used to say anything about the physical state of the Earth beneath the particular area of investigation. In other words, it is necessary to work out what the Δg value *would be* if: (a) the sites were at the same latitude; (b) the sites were at the same topographical height; (c) the attraction of the rocks vertically between the sites were taken into account; (d) the topography of the surrounding region were taken into account. Any difference that remains must be the result of lateral variations in subsurface density. These principal corrections, and the reasons for them, will be covered in turn.

(1) *Latitude*: We have already mentioned that, because of the Earth's shape, gravity increases towards the poles. It follows, therefore, that even if the Earth were perfectly uniform structurally (that is, even if there were no lateral variations in density), there would be a gravity

difference between the site (S) and the reference point (P) if the site and reference point were to differ in latitude. This latitude effect must therefore be removed from the measured value of gravity difference ($\Delta g = g_S - g_P$) before any interpretation may be made in terms of density variations.

The correction is derived directly from the IGF, by firstly calculating g for the reference station (P), and then again for the field station S. The difference between these two calculated values is the latitude correction.

If the site (S) is at a higher latitude than the reference point (P) (that is, S is nearer the pole, where, because of the latitude effect, g is larger than at P) then the latitude correction must be subtracted from Δg. So if we define the **latitude correction** ($\Delta_1 g$) as positive for a site at a higher latitude than the reference point, then *after* the latitude correction, the corrected value for $g_S - g_P$ becomes:

$$(g_S - g_P)_{\text{corrected}} = \Delta g - \Delta_1 g. \qquad \text{(Equation 1.36)}$$

ITQ 32

Calculate the latitude correction if the site (S) is at 10.431° N and the reference (P) is at 10.590° N.

(2) *Elevation*: Equation 1.34 shows that g varies inversely with the square of the distance from the Earth's centre. It follows that g will decrease with increasing height above the Earth's surface, and so if S and P are not at the same elevation this will have an effect on the observed gravity difference.

The elevation correction ($\Delta_2 g$) is simply derived by putting the heights of the reference and field stations into the distance d part of Equation 1.32. The resulting value is reasonably constant over the whole surface of the Earth, and if we define height as being positive above sea-level, then the elevation correction due to the height difference ($h = h_S - h_P$) between S and P becomes:

$$\Delta_2 g = -0.3086h \text{ mGal} \qquad \text{(Equation 1.37)}$$

where h is in metres.

The minus sign indicates that gravity decreases with increasing height, so $\Delta_2 g$ will be negative if S is topographically higher than P. Because gravity is less at S than at P (see Figure 1.70), the *correction* to the gravity difference we measure (Δg) must be positive, so increasing the gravity value at S to the value it would have at the lower elevation of the reference point P. After the elevation correction, the corrected value of $g_S - g_P$ is:

$$(g_S - g_P)_{\text{corrected}} = \Delta g - \Delta_1 g - \Delta_2 g. \qquad \text{(Equation 1.38)}$$

ITQ 33

What is the elevation correction if site S is (a) 10 m above the reference P; (b) 10 m below P; (c) at the same height as P; and (d) 100 m above P?

To make sure that you understand what we have done here, consider the opposite case to Figure 1.70a, where S is *below* P. Ignoring the effects of any anomalies, gravity will be greater at S than at P. This means that to bring the value of gravity at S to the value it would have at the height of P, we must take away an amount appropriate to the height between them. Because the height of S is less than the height of P, we define h to be negative and so $\Delta_2 g$ is positive, from Equation 1.37. This may seem complicated at first sight, but, if you remember that h is *positive where S is at a higher level than P* and then simply use Equations 1.37 and 1.38, you can't go wrong.

If $\Delta_2 g$ is to be known within the geologically useful limit of 0.01 mGal, the difference in elevation (h) between S and P must be known to within about 4 cm. Most gravity surveys must therefore be accompanied by accurate levelling surveys to determine the differences in height between the different measuring sites. The correction for elevation, $\Delta_2 g$, is known as the **free-air correction** because, in calculating it, it was assumed that the only material between the altitudes of S and P was air. The terms 'elevation correction' and 'free-air correction' are interchangeable for the purposes of this Course.

(3) *Material between S and P*: Look again at Figure 1.70a. S lies *above* P, and therefore is further from the Earth's centre. So gravity is less at S than at P. The free-air correction assumes that only air exists between S and P. But, as you can see from Figure 1.70a, there is a slab of rock of density ρ and thickness h lying below S. Because there is rock in this slab and not air, we must make a further correction, known as the **Bouguer correction**, $\Delta_3 g$. This is named after the pioneering geophysicist, Pierre Bouguer (pronounced Boo-gay), who made geodetic measurements in the Andes in about 1740 using the gravitational attraction of a plumb bob to an adjacent mountain to measure G.

ITQ 34

(a) What effect will the slab of rock in Figure 1.70a have on the gravity measured at S?

(b) Do you think the Bouguer correction ($\Delta_3 g$) will increase or decrease the measured gravity difference (Δg) between S and P in Figure 1.70a?

The Bouguer correction is the gravitational effect of an infinite horizontal slab of density (ρ) and thickness (h). So:

$$\Delta_3 g = 2\pi G\rho h \text{ m s}^{-2} \quad \text{or} \quad \Delta_3 g = 4.1921 \times 10^{-5}\rho h \text{ mGal} \qquad \text{(Equation 1.39)}$$

where h is in metres and ρ is in kg m^{-3}.

If S is above P ($h_S - h_P$ is positive), then $\Delta_3 g$ is the additional gravity at S due to the rocks between S and P. In cases where P is above S ($h_S - h_P$ is negative), then gravity at S is less than at P due to the rocks between them, and $\Delta_3 g$ is negative. Thus the free-air correction and the Bouguer correction for a given station *always* have opposite signs.

The corrected value of $g_S - g_P$ after the Bouguer correction becomes:

$$(g_S - g_P)_{\text{corrected}} = \Delta g - \Delta_1 g - \Delta_2 g - \Delta_3 g. \qquad \text{(Equation 1.40)}$$

Going back to the case shown in Figure 1.70a, where S is above P, the additional gravity at S due to the rocks between S and P makes the measured gravity difference (Δg) too large and so, as argued in ITQ 34, $\Delta_3 g$ must be subtracted from Δg.

(4) *Terrain*: Gravity at a site such as S_1 in Figure 1.70a will be lower than at S because the hill adjacent to S_1 will exert an attraction whose net effect as seen from S_1 will be upwards. Similarly, gravity at S_2 will also be lower than at S because the valley has removed an attracting mass from below the level of S_2. The principles involved here are illustrated further in Figure 1.70b and you should read the caption carefully.

❑ To eliminate the effects of topography around a measuring site (S), will the correction to the measured gravity difference (Δg) between S and P be positive or negative?

■ Because g_S is reduced by the effects of both hills and valleys, the correction must be *added* to Δg to make $g_S - g_P$ what it would have been if the topography had been flat. So the terrain correction is always positive.

In fact, the **terrain correction** ($\Delta_4 g$) is worked out by summing the effect of all hills and valleys in the neighbourhood of each gravity site, using standard tables which take into account height, density and the decreasing effect on g_S with distance.

The free-air correction allows for height differences between stations and the reference point; the Bouguer correction allows for rock between the stations, but it does so by assuming that an infinite slab of density ρ and height h exists between the stations. The terrain correction 'removes' the effect of this slab where it is not in fact there — i.e. in valleys and hills (Figure 1.70)

Once the latitude, free-air, Bouguer and terrain corrections have been determined, the final corrected gravity difference between S and P may be calculated, simply using Equations 1.35, 1.37 and 1.39 and tables for $\Delta_4 g$:

$$(g_S - g_P)_{\text{corrected}} = \Delta g - \Delta_1 g - \Delta_2 g - \Delta_3 g + \Delta_4 g. \qquad \text{(Equation 1.41)}$$

This value should, in principle, now be free of all effects except those due to density variations below the surface of the region under investigation — the small-scale anomalies to which we referred earlier. The only correction that needs to be applied at P is the terrain correction, since station P is at the reference latitude and height.

The choice of anomaly

Once all the corrections have been made, the final corrected value of gravity for each field station, ($g_S - g_P$), is plotted on a map and contoured to produce a gravity map of the area. The gravity values on the map are not absolute values, but are values expressed in mGal relative to the IGF. Ideally, the reference station is in a region of zero anomaly, but often this is not the case, so gravity anomalies are denoted as positive or negative relative to a particular place. A gravity map of this type is called a **Bouguer anomaly map**, and the anomalies on it are **Bouguer anomalies**, which should not be confused with the Bouguer correction, which is just one of the many corrections that are applied to raw gravity data before interpretation is possible.

Another type of gravity map which is used mostly for marine and satellite data is the **free-air anomaly map**, which as its name implies is the anomaly obtained after making only the latitude and free-air corrections. The Bouguer and terrain corrections are not applied to data in this case. This is because measurements are all made at the same height, and there is no terrain to worry about.

There is an important and subtle difference between these two types of anomaly. The Bouguer anomaly is used for most land surveys, and the corrected data have the values that would have been measured if there were no hills and valleys causing anomalous effects, and if the field stations were all at the same height. The height to which the data are corrected, the reference station height, is often the average height of the region, but may even be sea-level. The job of the geophysicist is to interpret the Bouguer anomaly in terms of density variations beneath the ground surface. A negative Bouguer anomaly, then, would be interpreted in terms of a body or bodies beneath the ground surface, of density rather lower than the density (ρ) that had been adopted as the average for the area in Equation 1.39. The final model produced to account for the gravity anomaly is calculated for densities relative to this value of ρ (see Figure 1.71).

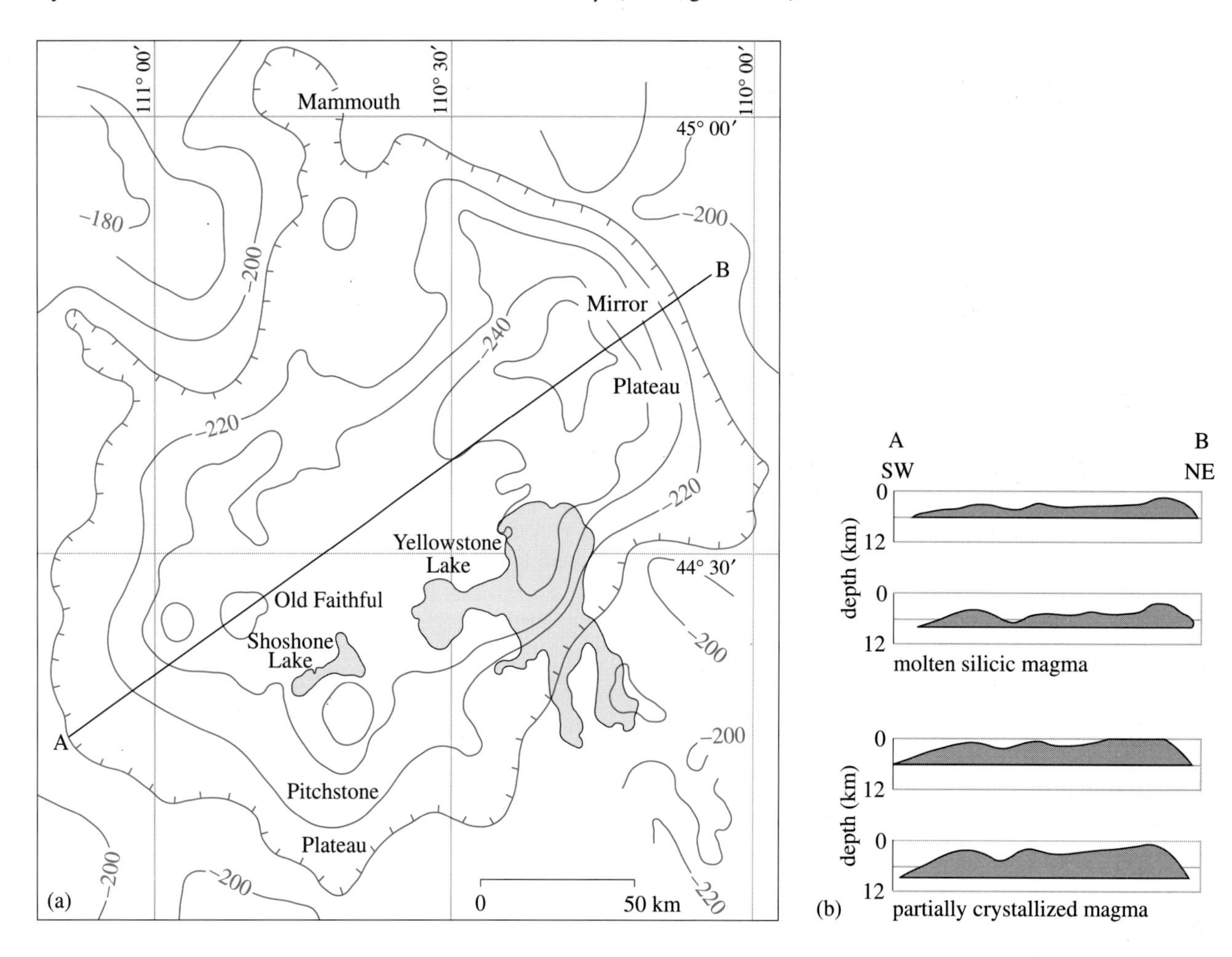

Figure 1.71 (a) Bouguer anomaly gravity map of the Yellowstone Plateau in Wyoming. This is an active volcanic region, characterized by a huge caldera — a depression formed in association with explosive eruptions. The contour interval is 10 mGal (10^{-4} m s^{-2}) and the gravity anomaly is clearly associated with the caldera wall (line with tick marks).
(b) Several models can be used to account for the observed anomaly. Each model and several geologically less reasonable ones would produce the negative anomaly inside the caldera. Two types of model are shown here, assuming a plane base for simplicity at 6 km (top diagram of each pair) and 8 km (bottom diagram) depth. The anomaly could be caused by a reservoir or chamber of molten magma 450 kg m^{-3} less dense than the surrounding country rock or by a chamber of partially crystallized magma only 220 kg m^{-3} less dense than the surroundings. Note that the smaller the density contrast between an anomalous body and the surroundings, the closer it needs to be to the surface in order to affect the gravity field. On the scale of this diagram, the topography appears as a flat plane.

The Bouguer gravity anomaly, then, gives information on the density distribution beneath the ground surface. Anomalies occur where regions beneath this surface have densities different from the surrounding material and/or different from the density used to calculate the Bouguer anomaly.

In contrast, the free-air anomaly gives information on the actual gravity field in the area of interest corrected only for height variations, and includes the effects of local terrain and the material (if any) vertically between the field stations and the reference station. In other words, the free-air anomaly gives information on the total *mass* distribution beneath the surface. This will be illustrated when we discuss isostasy in the next Section.

ITQ 35

A scientist measures gravity at a site (S) at latitude 30° N and finds the value to be 9.70 mGal higher than at a reference point (P) that lies 2′ to the south and 10 m below the site. Assume the density of the rocks forming the crust between the site and the reference point is 2 670 kg m^{-3}. (*Note* An angle of 2′ (2 minutes) is $(\frac{2}{60})°$, and you can ignore the effects of the topographic correction $\Delta_4 g$ in this ITQ.)

(a) What is the corrected gravity difference between the site and reference point?

(b) Which is the most important correction (in size)?

(c) Bearing in mind that the Earth's gravity is approximately 9.81 m s^{-2}, what percentage of this value is the (corrected) gravity difference between the site and the reference point?

ITQ 36

What would be the corrected value of the anomaly in ITQ 35 if the sites had been at 30° S, with the reference point still 2′ to the south, but this time 10 m below the reference point? Take it that Δg was again +9.7 mGal.

1.8.5 ISOSTASY

At the beginning of Section 1.8.2, we found that

$$g = \frac{GM}{d^2}.$$

This equation can be rearranged and the mass (M) of the Earth substituted as follows:

$$M = \text{density of the Earth } (\rho) \times \text{volume of the Earth } (V) = \rho \times V.$$

Now for a sphere which, to a first approximation, the Earth is:

$$V = \frac{4\pi d^3}{3}$$

so we can combine these equations:

$$g = \frac{GM}{d^2} = \frac{G\rho V}{d^2} = \frac{G\rho 4\pi d^3}{3d^2}$$

which simplifies to:

$$g = \frac{4\pi G\rho d}{3}. \qquad \text{(Equation 1.42)}$$

So if g is measured at a point on the Earth's surface below which ρ is anomalously low, g will be anomalously low — and by 'below' we mean anywhere between the surface and the centre of the Earth. Likewise, if g is measured at a point below which ρ is anomalously high, g will be anomalously high. This correlation may also be described in terms of mass rather than density. Volume for volume, a region of low density will possess less mass; and conversely, a region of higher density will possess more mass. Thus an anomalously low value of g at a point implies that below that point there is a **mass deficiency**, and an anomalously high value of g implies a **mass excess**.

It is useful to trace the ancestry of this idea, which really stems from gravity measurements. Between 1735 and 1745, Pierre Bouguer led an expedition to the Andes in which he attempted to compare the vertical gravitational pull of the Earth with the horizontal gravitational attraction of a mountain. The method involved observing the deviation of a long plumbline from the vertical (obtained using astronomical observations) due to the sideways pull of the mountains. He thought that he could calculate the mass of the mountain from the density of surface rocks and the volume of material above sea-level. Later analysis showed that the deviation of his plumbline was much less than expected on the basis of the Andes' topography and density. Much later, when similar effects were found near other mountains (particularly during Himalayan levelling surveys by Sir George Everest and others), it was realized that the mass deficiency beneath the mountain chains, which accounts for the reduced deflection of the plumblines, is approximately equal to the mass of the mountains themselves. This observation is reminiscent of Archimedes' principle: 'a floating body displaces its own weight of water'. The displaced material beneath a mountain has the same weight as the mountain. So a mountain chain can be compared with a floating iceberg or cork. This requires the rigid surface layers of the Earth (lithosphere) to float on a deformable denser layer (asthenosphere). The asthenosphere therefore moves to compensate for the weight of a mountain.

In 1889, C. E. Dutton coined the term 'isostasy' to describe this compensation phenomenon. The principle is that there is a certain **level of compensation** above which all columns of material having the same cross-sectional area must have the same weight. Thus if there is apparently an excess mass at the Earth's surface (a mountain range, for example) this must be compensated for by a mass deficiency below the surface feature but above the level of compensation. In ideal circumstances, there would be a state of **isostatic equilibrium** (or balance) in which all mountainous areas have low-density material beneath them just sufficient to balance the mountains.

Having described the principle of isostasy, the next step is to examine *models* for the mass distribution at depths that are consistent with observations. There are two theories here, illustrated in Figure 1.72, both of which seem to contain important elements of truth. Both of them date from 1855.

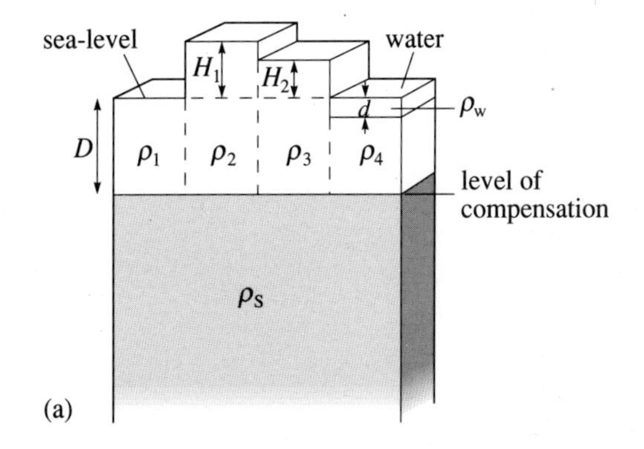

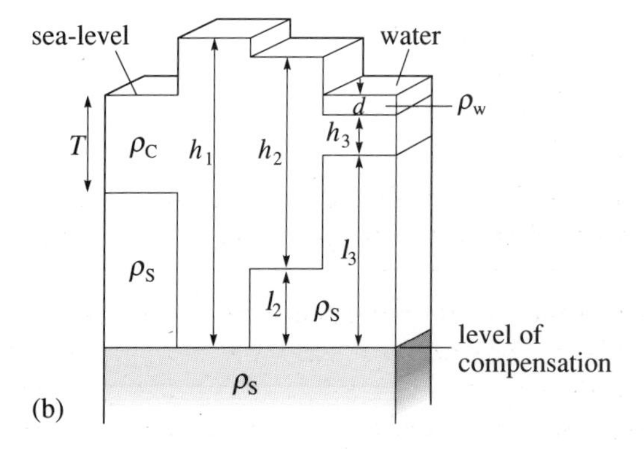

Figure 1.72 Isostatic compensation according to the theories of (a) Pratt and (b) Airy. In (a), each column has the same cross-sectional area and depth to base. The height of the column up from the level of compensation depends on the height of the ground or depth of water, and the density of each column can be different. In (b), all the columns have the same cross-sectional area and density but although the column tops again represent ground or sea-level, the bottom of each column can be at a different depth. The level of compensation is deeper than the deepest column.

(1) **Pratt's hypothesis:** In 1855, Archdeacon J. H. Pratt suggested that the level of compensation lies at a constant depth (D) below sea-level around the Earth (see Figure 1.72a) and that the material above the level of compensation varies in density laterally, according to the elevation of the overlying topography. Thus mountains are made of and are underlain by material of anomalously low density which extends down to the level of compensation, and ocean basins, by contrast, are underlain by material of relatively high density. For isostatic equilibrium, the total weight of each column (Figure 1.72a) must be the same. The volumes of the columns are different because, although they have the same cross-sectional area, they have different heights. So for all columns in isostatic equilibrium:

density × volume = mass = constant

and if we compare columns of unit cross-sectional area:

density × height = constant

or, in general

$$\rho_c h + \rho_w d = \text{constant} \qquad \text{(Equation 1.43)}$$

where ρ_c is the density of columns, which varies laterally (see caption to Figure 1.72a), h the thickness of columns ($H + D$ for columns on land, $D - d$ for columns under the sea), ρ_w the density of seawater (if any), and d the depth of seawater (if any).

ITQ 37

Are the columns shown in Figure 1.73a all in isostatic equilibrium? What would happen if the density of the tallest column was increased to $2\,700\,\text{kg}\,\text{m}^{-3}$? (*Hint* Use Equation 1.43 and take the density of seawater to be $1\,000\,\text{kg}\,\text{m}^{-3}$).

(2) **Airy's hypothesis**: Sir George Airy suggested, also in 1855, that although the level of compensation lies at a constant depth below sea-level around the Earth (see Figure 1.72b), the material above it forms a low-density crust of thickness T overlying a higher-density substratum. The crust and substratum each have uniform but different density, and the boundary between them is an exaggerated mirror image of the surface topography. In other words, isostatic equilibrium is reached by variations in the depth of the crust–substratum boundary rather than, as in Pratt's hypothesis, by lateral variations in density.

According to Airy's hypothesis, isostatic equilibrium requires that for columns of a given cross-sectional area above the level of compensation:

$$\rho_c h + \rho_s l + \rho_w d = \text{constant} \qquad \text{(Equation 1.44)}$$

where ρ_c is the density of column (which is constant), h the thickness of column (which varies at both the top and the bottom), ρ_s the density of substratum (which is constant), l the thickness of substratum (which varies only at the top), ρ_w the density of seawater (if any), and d the depth of seawater (if any).

ITQ 38

The columns shown in Figure 1.73b for Airy's model are in isostatic equilibrium. What would happen if the water in column F dried up?

Aspects of both models are consistent with observations of the structure of the lithosphere. For example, high mountains are indeed underlain by crust that is thicker than normal, forming a 'root', and oceans are underlain by crust that is thinner than normal, as in Airy's model. On the other hand, ocean crust is more dense than continental crust, as in Pratt's model. Moreover, in reality there is a substratum, equivalent to the mantle part of the lithosphere. So, although the two models are often treated in isolation, properties of both are required to explain fully the principle of isostatic compensation recovery where the lithosphere–asthenosphere boundary is the level of compensation.

ITQ 39

If all the columns in the Airy model in Figure 1.73b are in isostatic equilibrium, (a) what is the value of the constant in Equation 1.44, and (b) what would be the effect of doubling the width of column C?

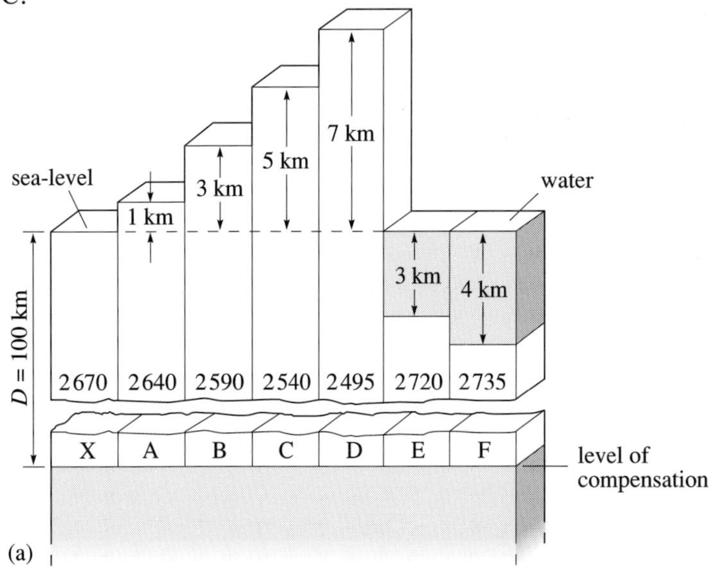

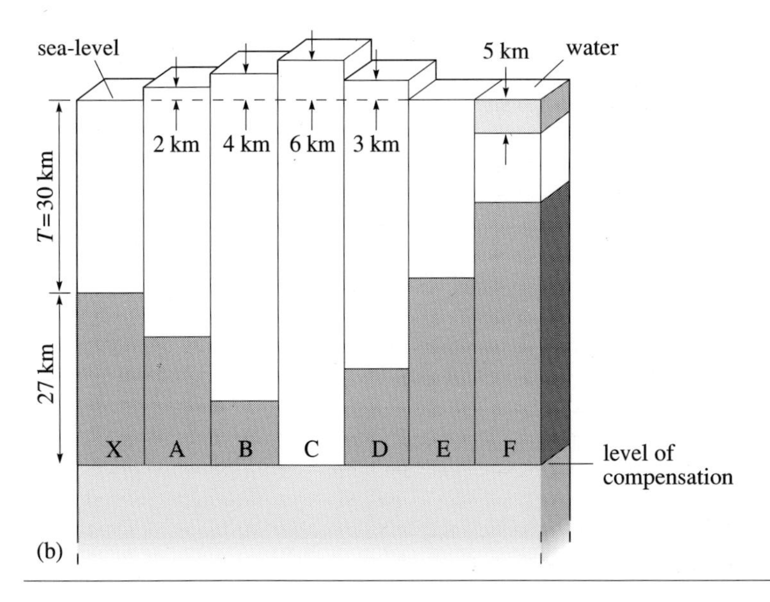

Figure 1.73 Crustal columns for use with ITQs 37 to 39.
In (a), the four-figure numbers are densities in kg m^{-3}; the other numbers represent distances to sea-level.
In (b), the substratum density is 3 270 kg m^{-3} (shaded), and the crustal density is 2 670 kg m^{-3} (unshaded). Take the density of seawater to be 1 000 kg m^{-3}.

It is clear from Equations 1.43 and 1.44 that both the Airy and Pratt hypotheses are consistent with the principle of isostasy. Moreover, it is clear from ITQs 37–39 that these equations can be applied to models for the densities of columns to determine whether isostatic equilibrium operates. In practice, however, it is often more difficult to 'test for isostatic equilibrium' than you might think from ITQ 39. For one thing, there is the problem of choosing the right densities for blocks of crust, and then, of course, there is the problem of choosing the most appropriate model and the level of compensation.

Now, we can try to test for isostasy using gravity measurements because, on a regional scale (hundreds of kilometres), there should be *no* gravity anomaly over a region in isostatic equilibrium. In other words, there should be no mass excess or deficiency above the level of compensation. But as you will realize from the end of Section 1.8.3, it really depends on the way you choose to look at the gravity data, because the presence of a large-scale negative Bouguer anomaly in a mountainous region (as in

Figure 1.71) does not necessarily mean that it is out of isostatic equilibrium. This is because the Bouguer anomaly is telling us that there is a mass deficiency caused by a region of anomalously low density beneath the surface, after the gravitational effect of the mountains has been corrected for. That mass deficiency may exactly compensate for the extra mass of the mountains above, giving us a cross-section as envisaged by Airy (Figure 1.72b) and no *net* mass excess or deficiency (in other words, isostatic equilibrium). Say we use Airy's hypothesis to make a model with estimated values of T and density in a root zone, or Pratt's hypothesis to model values of D and laterally changing density (T and D are defined in Figure 1.72), then we can calculate the value of the Bouguer anomaly that should result from each model. This can be compared with the Bouguer anomaly that has been observed and, if the two are roughly equal for geologically reasonable models, then the area is likely to be near isostatic equilibrium. To make this easier to deal with, we define an **isostatic anomaly** which, in cases of isostatic equilibrium, should, of course, be zero, as follows:

$$\text{isostatic anomaly} = \text{Bouguer anomaly corrected to a reference point at sea-level} - \text{predicted anomaly of modelled root zone below sea-level needed to satisfy Airy's or Pratt's hypotheses}$$

(Equation 1.45)

Unlike the Bouguer and free-air anomalies, isostatic anomalies contain an element of prediction; for example, that of the size of the root zone. But the predicted anomaly of the model and the corresponding isostatic anomaly are different depending on which hypothesis is used to make the calculation. So it should be possible to use isostatic anomalies to decide between Airy's and Pratt's hypotheses. In practice, however, this is not always easy because models based on either hypothesis may give isostatic anomalies that do not differ greatly. The problem is illustrated for various degrees of compensation in Figure 1.74.

The fact that the Bouguer anomaly in Figure 1.74a and 1.74b is large and negative shows that there is a mass deficiency beneath the mountains and that at least some isostatic compensation (sinking) must have occurred. The gravitational effects of subsurface structural models based on Airy's hypothesis (with T = 20 and 30 km) and Pratt's hypothesis (with D = 80 km) have been calculated. The densities used in the Pratt model columns were $3\,330\,\mathrm{kg\,m^{-3}}$ to either side of the mountains and $2\,850\,\mathrm{kg\,m^{-3}}$ beneath the mountains. The same two densities were used for the substratum and crust in the calculations based on Airy's model. Isostatic anomalies based on these three models are recorded in Figure 1.74a.

❑ Look again at Figure 1.74: what do the three isostatic anomaly profiles suggest about crustal structure beneath mountains?

■ The first thing to note is that all three profiles are parallel and horizontal, running along the zero line, suggesting that the area is near isostatic equilibrium. In fact, by taking $T = 30\,\mathrm{km}$ for Airy's hypothesis, the isostatic anomaly is precisely zero, indicating perfect equilibrium. Remember that T is the thickness of the crust in the Airy model, whose surface lies at sea-level: this 'preferred' model is the one shown in cross-section in Figure 1.74a.

It should not have escaped your attention that in making this interpretation we have pulled what, on the face of it, is a fast one. We may be agreed that a zero isostatic anomaly indicates perfect isostatic equilibrium; but have we not fiddled the books by choosing the value of T that will give just the zero isostatic anomaly required? In short, quite apart from the fact that we have assumed the two values of density shown in Figure 1.74, we have chosen an arbitrary value of T that gives us perfect isostatic equilibrium. So in what sense have we tested for isostatic equilibrium?

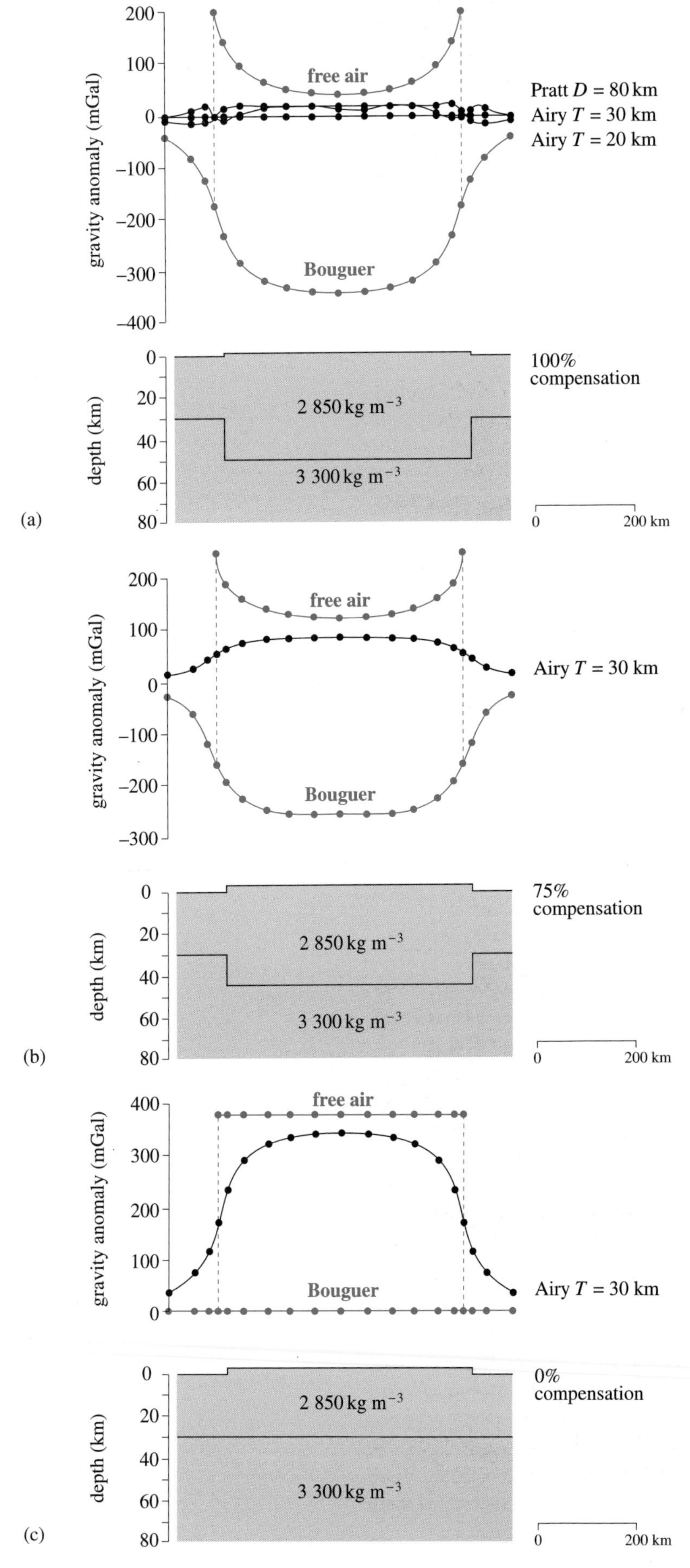

Figure 1.74 Bouguer, free-air and isostatic anomalies over a schematic mountain range. The mountain range and the material beneath it have densities of $2\,850\,\mathrm{kg\,m^{-3}}$ and $3\,300\,\mathrm{kg\,m^{-3}}$, respectively.
(a) The large negative Bouguer anomaly indicates the presence of a large region of low-density material beneath the mountain (indicating isostatic adjustment); the small free-air anomaly (except at the edges of the structure where the steep topography causes anomalies) indicates no anomalous mass is associated with the mountain; the isostatic anomalies are practically zero, indicating isostatic equilibrium. Thus the mountain is said to be 100% compensated.
(b) This is the same mountain chain, but this time it is only 75% compensated — isostatic equilibrium has not yet been reached. The isostatic and free-air anomalies are therefore positive — reflecting the excess mass of the mountain. The Bouguer anomaly is not quite so negative as in (a) because of the smaller amount of low-density material accumulated in the mountain root.
(c) The same mountain range, but this time it is not compensated at all. There is no Bouguer anomaly because the anomalous mountain is completely corrected for in the Bouguer and terrain corrections and there is no buried density anomaly. The anomalous mountain mass produces large positive free-air and isostatic anomalies.

This brings us back to the question of what is a 'reasonable' model for the subsurface structure. The values of T and D chosen to compute the isostatic anomalies were 'reasonable' in the sense that they could be predicted quite closely and, of course, are roughly what might be expected from previous experience. This experience, in turn, is based on information other than that from gravity surveys — seismic data, for example. For instance, in the case of Figure 1.74a, had it been necessary to adopt a T value of, say, 300 km to produce a zero isostatic anomaly, it would have been quite clear that the system could not be in isostatic equilibrium because, in Equation 1.45, the predicted anomaly would have been unreasonable. This is because we know from seismic data that the thickness of the Earth's crust is far less than 300 km; and so T = 300 km is not a valid Airy model.

This highlights a very important fact about gravity surveys: *structure cannot be determined uniquely from gravity alone*. For example, in spite of the strong presumption in favour of the 'root' structure shown in Figure 1.74a, there is no definite proof from gravity alone that this is correct. All we can say is that the structure we have deduced is consistent with the observed Bouguer anomaly. But for any observed anomaly, there are, in fact, an infinite number of different structures that are consistent with it. In terms of isostatic models, some mountains may have a low-density root (Airy), others may have low-density columns down to the level of compensation (Pratt), and the real situation may, in most cases, be some combination of the two.

There are many cases where isostasy does not give complete equilibrium because, once a geological system has been disturbed, it may take a considerable time (typically tens of thousands of years) for it to respond fully and to restore isostatic equilibrium. For example, during the last glaciation Scandinavia was loaded with ice, and the 'excess' mass of ice caused the region to sink. When the ice retreated (about 10 000 years ago), Scandinavia began to rise again and is still rising today — that is, it has not yet returned to isostatic equilibrium. This is a useful example to consider further. When the mass of ice formed on Scandinavia, it upset the isostatic equilibrium because the total mass above the level of compensation increased with respect to non-glaciated areas. In order to restore the equilibrium, Scandinavia sank, the implication here being that material must somehow have flowed out laterally beneath the region in order to ensure that the mass above the level of compensation did not stay in excess. By the same token, of course, now that the ice has melted Scandinavia is now rising, and material must be flowing in beneath it. This lateral movement of material implies flow at, or just below, the level of compensation. It is this flow, within the asthenosphere, that allows the isostatic movements of oceanic and continental lithosphere to occur. The isostatic rebound of Scandinavia is still going on. Over the short term (geologically speaking), regions may well be just approaching isostatic equilibrium, the process of **compensation** not yet being complete. Figure 1.74b and 1.74c show the gravitational effects of the same mountain chain as Figure 1.74a before compensation (0% compensated — Figure 1.74c) and after 75% compensation (Figure 1.74b).

Lateral flow within the asthenosphere allows the surface of the Earth to rise or fall in order to reach isostatic equilibrium. This is not a one-way process, because ice cover does not last forever, and although the surface sinks under the weight of ice, it rises again after the ice has retreated. Similarly, tall mountains are eroded more rapidly than plains and as mass is removed from a mountain, so the surface will rise to restore isostatic equilibrium. The level of compensation is identified as the top of the asthenosphere in which plastic flow of material occurs slowly at depth in order to maintain isostatic equilibrium.

Gravity observations suggest that most of the Earth's major surface features are in rough isostatic equilibrium but they cannot be used to deduce unambiguously what form the subsurface structure takes. Seismic refraction studies do enable structure and thickness (and thus the geometry) of anomalous bodies to be determined. Together, seismic and gravity studies give us a very powerful tool for investigating the Earth and we will introduce you to this combined approach in Section 1.9.

Finally, a word about scale: the geoid height variations we examined in Figures 1.68 and 1.69 are broad (up to thousands of kilometres across) and probably reflect small differences in density at depth on a similarly large scale. The Bouguer anomalies that we have just been discussing are of medium scale (up to a few hundred kilometres across), reflecting the scale of mass excesses and deficiencies sufficient to cause isostatic readjustment. On a smaller scale still (up to a few tens of kilometres across), the gravity anomalies used to investigate shallow crustal structures represent mass excesses and deficiencies of insufficient size to cause isostatic movements. These are the sort of anomalies of interest for example to oil companies, which use gravity to locate likely oil-bearing rocks. These are detectable by their relatively low density relative to surrounding crystalline basement rocks. Volcanologists also use gravity measurements to map out magma chambers and conduits beneath volcanoes (such as Yellowstone) and to monitor volcanic activity using the density contrasts between solid lava, molten magma and volcanic gases.

SUMMARY OF SECTION 1.8

Gravity varies over the Earth's surface, increasing generally from equator to poles. This is firstly due to the equatorial bulge; an object at the Earth's surface is further from the Earth's centre of mass if it is at the equator than if it is at the pole so the gravitational attraction at the equator is less than at the pole. Secondly, centrifugal forces act to oppose the internal gravitational force due to the Earth's mass; these also are at a maximum at the equator. The International Gravity Formula (IGF) (Equation 1.35) describes the theoretical variation of gravity over the Earth's surface taking these two effects into account. The spheroid is a theoretical surface on which the value of gravity is given by the IGF.

Departures from the IGF on a *larger* scale can be mapped from satellites, leading to a definition of the geoid (Figure 1.68), which is the real surface where the value of gravity is the same as the value predicted by the IGF. The geoid is contoured in heights above and below the spheroid surface. Where there are geoid highs, gravity at the spheroid radius from the Earth's centre is stronger than the spheroid gravity value and, where there are geoid lows, gravity is weaker than the spheroid value. The 'bumps' and 'dents' and asymmetry of the geoid do not relate simply to surface topographic features of the Earth but to inhomogeneities deep within the mantle.

Gravity anomalies are deviations from the IGF on a smaller scale, due to lateral variations in the crust or shallow lithosphere. The data recorded at various sites in a gravity survey are compared with gravity at a reference point. But first they must be corrected for the effects of latitude difference, height difference and rock density (the latitude, free-air and Bouguer corrections) between the sites and reference point and, finally, for topographic effects. A fully corrected gravity anomaly is usually called a Bouguer anomaly and this reflects inhomogeneities below the surface.

Large negative Bouguer gravity anomalies over mountain ranges are due to a mass deficiency (the effects of topography having been removed in the calculation of the anomaly). Such mountain ranges must therefore have anomalously low-density material beneath them which 'compensates' for the excess mass which has been corrected away. This leads to the concept of isostasy, which states that an equilibrium exists whereby different columns of material of different heights all have the same mass above a level of compensation. Thus when mountains are eroded from above, they will rise until they reach more typical crustal thicknesses and when continental areas are glaciated, they will sink. Following melting of the ice that covered it during the last Ice Age, Scandinavia is, at present, a region with a negative Bouguer anomaly that cannot be explained by models with mass deficiency below sea-level. It is therefore a region with a negative isostatic anomaly and is rising to achieve isostatic equilibrium.

Pratt's hypothesis for isostasy proposes that the columns all have different but uniform densities down to the level of compensation so that column height is simply a function of density (Figure 1.72a). Airy's hypothesis proposes that each column comprises an upper and a lower zone; the same two densities apply to all columns so the difference in height of the columns depends on the relative amounts of low-density and high-density material (Figure 1.72b). Below the level of compensation plastic flow occurs to maintain equilibrium. Models based on these two hypotheses may be evaluated by matching their predicted gravitational effect to that of the Bouguer anomaly for a region thought to be in isostatic equilibrium (that is, with a zero isostatic anomaly — Equation 1.45). Unique solutions using gravity data alone are difficult to obtain, but, with the added information provided by seismic data, satisfactory models of shallow-Earth structure can be devised.

OBJECTIVES FOR SECTION 1.8

When you have completed this Section, you should be able to:

1.1 Recognize and use definitions and applications of the terms printed in the text in bold.

1.34 Summarize the major variations in gravity over the surface of an oblate spheroid, including the basis of the International Gravity Formula (IGF), and explain the difference between the spheroid and the geoid.

1.35 Understand that gravity anomalies due to near-surface inhomogeneities related to the differing densities of rocks are small but significant departures from the IGF 'theoretical' gravity values.

1.36 Perform simple calculations to correct 'raw' gravity survey data, enabling subsurface inhomogeneities to be quantified.

1.37 Describe the differences between Bouguer, free-air and isostatic gravity anomalies.

1.38 Explain the principle of isostasy indicating the use of a 'level of compensation' in Pratt's and Airy's models of isostasy, and perform simple calculations based on these models.

Apart from Objective 1.1, to which they all relate, the ITQs in this Section test the Objectives as follows: ITQ 31, Objective 1.34; ITQs 32–36, Objectives 1.35, 1.36 and 1.37; ITQs 37–39, Objective 1.38

You should now do the following SAQs, which test other aspects of the Objectives.

SAQs for Section 1.8

SAQ 44 (*Objectives 1.1, 1.34, 1.37 and 1.38*)

State, giving the reasons where appropriate, whether each of the following statements is true or false:

(a) A contour map that shows large-scale variations from the spheroid as heights above and below the theoretical IGF surface is a map of the geoid.

(b) Free-air and Bouguer anomalies are equivalent, but the former are measured at sea and the latter on land.

(c) Pratt and Airy models are used to describe isostatic rebound after ice caps have retreated.

SAQ 45 (*Objective 1.36*)

(a) In a gravity experiment carried out at about 45° N, the site investigated lies 200 m to the west and 20 m higher than the reference point. If there is no material (apart from air) in the vertical space between the site and the reference point and if there are no topographic effects, calculate the correction that must be applied to the observed gravity anomaly.

(b) Calculate the final correction that must be made if crustal material of density 2 670 kg m^{-3} lies between the site and the reference point.

SAQ 46 (*Objectives 1.34, 1.37 and 1.38*)

Match each of the items (i)–(iv) with one of the items A–D.

(i) An isostatic anomaly

(ii) The spheroid

(iii) The latitude correction

(iv) The IGF

A is determined by estimating the anomaly of a low-density root zone beneath a mountain range and subtracting it from the Bouguer anomaly.

B is used to predict the value of gravity at sea-level at any point on the Earth assuming that the Earth is a homogeneous rotating oblate spheroid.

C is the theoretical, or mean sea-level, surface of an ideal homogeneous Earth, whose gravity is calculated from the IGF.

D is the first correction to be applied to make the observed gravity difference between field stations and the reference station comparable prior to the evaluation of a Bouguer anomaly.

SAQ 47 (*Objective 1.38*)

Two columns having the same cross-sectional area, shown in Figure 1.75, have the following characteristics:

Column A: l = 20.0 km, h = 38.6 km

Column B: l = 27.0 km, h = 30.0 km

(a) If the density of the crust (ρ_c) is 2 670 kg m^{-3} and that of the substratum (ρ_s) is 3 270 kg m^{-3}, are the two columns in isostatic equilibrium according to Airy's hypothesis?

(b) What must be the height of column A above sea-level?

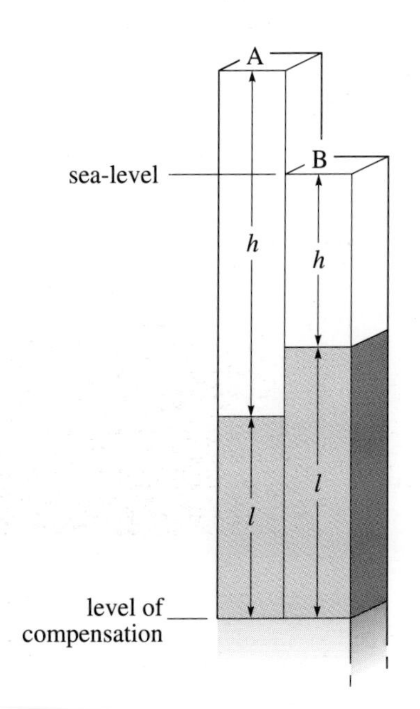

Figure 1.75 Two columns comprising crust (unshaded) and substratum (shaded) of height h and l, respectively; for use in the calculation in SAQ 47.

1.9 GRAVITY–SEISMIC INTERPRETATIONS

We stated earlier the view that a model could not be considered to be valid unless it was supported by more than one type of geophysical data. In Section 1.7, we looked at crustal structures based largely on the results of seismic work. As you will now realize after reading Section 1.8, there are many advantages in combining gravity and seismic interpretations. We shall now consider some examples.

1.9.1 OCEANIC RIDGES

Oceanic ridges will be discussed in detail in Block 2, but for now you just need to know that they are particularly significant features from the point of view of sea-floor spreading and plate tectonics. They are the tectonically active regions along which new oceanic lithosphere is being produced. Figure 1.76 is a purely schematic view of the Atlantic ocean crust and upper mantle, based on P-wave velocities. Here the mantle immediately beneath the ridge zone is both shallower and has lower P-wave velocities than normal mantle. The free-air gravity anomaly across the same oceanic ridge zone is shown above the profile. At sea, the free-air anomaly is measured directly at the sea-level surface and represents the total variation of mass (including seawater) below that surface. The free-air anomaly in Figure 1.76 is relatively flat and near zero across the ridge. In oceanic areas, where there is no material above water to worry about, the free-air anomaly gives us a direct test of isostasy. The lack of a free-air anomaly over oceanic ridges indicates that there can be no mass excess or deficiency beneath the sea-level surface and down to the level of compensation.

❑ What does this tell us about isostasy at oceanic ridges?

■ Zero free-air anomaly is an indication of isostatic equilibrium.

Since the ridge is a topographic high, it *must* be underlain by relatively low-density material in order to keep the overall mass constant across the ridge. As we shall see in Blocks 2 and 3, oceanic ridges are the sites of hot, rising material and the low density is caused by thermal expansion.

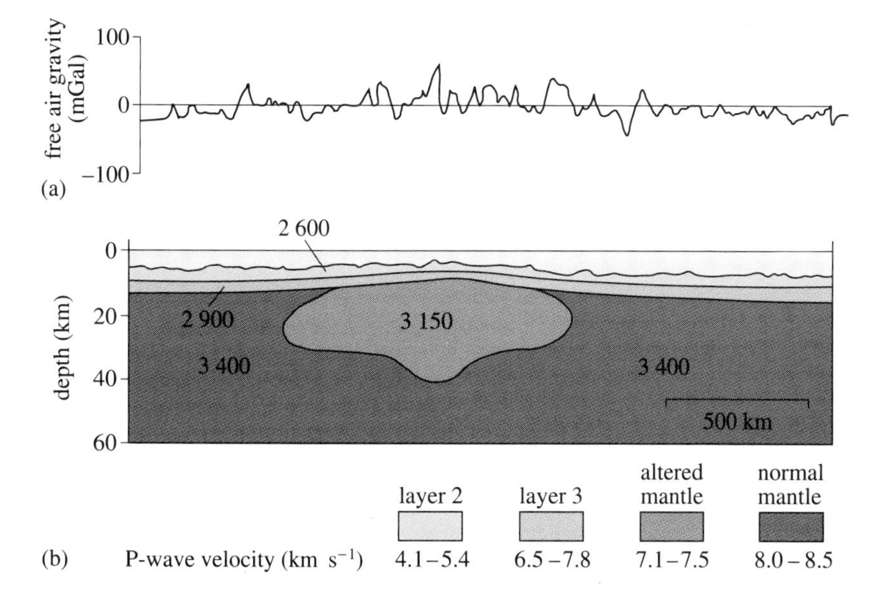

Figure 1.76 (a) Free-air anomaly across the Mid-Atlantic Ridge. (b) A structural/density model that satisfies the gravity and seismic data. The large low-density region beneath the ridge is needed in the model to counteract the positive effect of the ridge topography. The numbers are densities in kg m^{-3}.

1.9.2 Continental margins

There are two general types of continental margin: the **inactive aseismic (passive) margin** typified by those around the Atlantic that do not form plate boundaries, and the **seismically active margin** (for example, around the Pacific) where oceanic crust is being subducted. These concepts will be explained and developed fully in Block 2. For now, though, you just need to realize that aseismic or passive margins are usually close to isostatic equilibrium, but the active margins are the sites of large isostatic gravity anomalies.

The typical active margin has a trench along its oceanic side and either a mountain range or an island arc along its landward side. The strong, negative free-air anomalies, like those over the Sumatran Trench (Figure 1.77) and the Puerto Rican Trench (Figure 1.78), occur in narrow strips over the traces of most trench zones.

- ❑ What do you think that negative free-air anomalies over oceanic trenches must indicate about their state of isostatic equilibrium?
- ■ From what we have said above about free-air anomalies over oceans, it should be clear that trenches are not in isostatic equilibrium and that there must be net mass deficiencies in the oceanic lithosphere below trenches.

Figure 1.78b shows a combined gravity–seismic model. The model produces a negative anomaly over the trench, so the trench is *not in isostatic equilibrium*. To make sure that you appreciate this, notice that the trench is a zone of net negative relief (Figure 1.78b), so in contrast with an oceanic ridge zone, it would have to be underlain by a mass excess to be in isostatic equilibrium. In fact, to satisfy the observed gravity and seismic data, it must be underlain by a mass deficiency giving a small low-density 'root' immediately below the trench (Figure 1.78b). *So trenches are regions of negative isostatic anomaly*, as labelled in Figure 1.77.

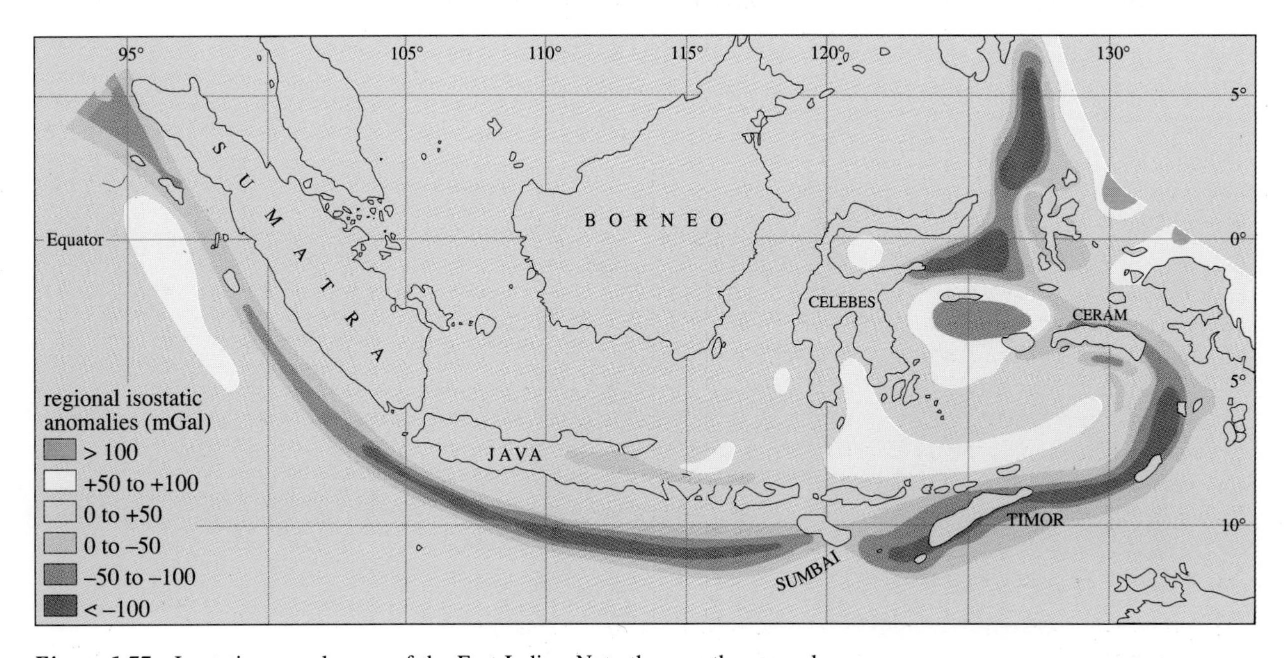

Figure 1.77 Isostatic anomaly map of the East Indies. Note the way the anomaly follows the line of the trench.

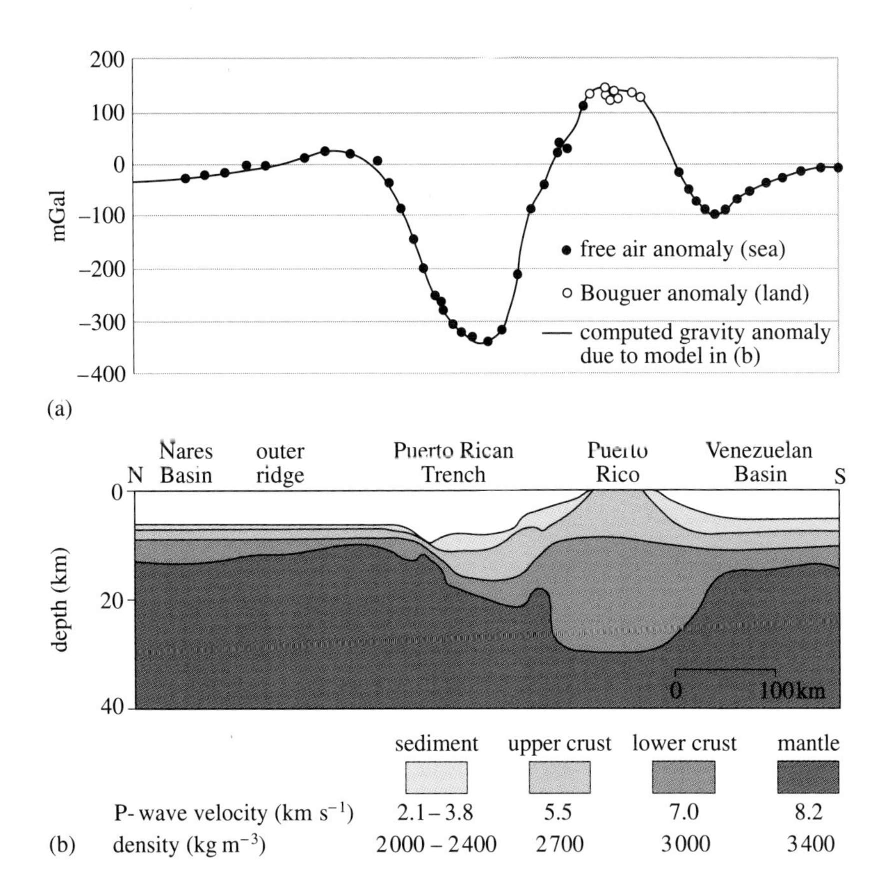

Figure 1.78 (a) Free-air and Bouguer gravity profile across the Puerto Rico arc–trench system. (b) A structural density model which satisfies the observed gravitational and seismic data.

Finally, you may have noticed that the island arcs behind the trench zones in Figures 1.77 and 1.78 have positive gravity anomalies. The positive Bouguer anomaly shown in Figure 1.78a can be accounted for by the combined gravity–seismic model in Figure 1.78b. The positive anomaly (presence of excess mass) is caused by the presence of more relatively dense upper crustal material (such as lavas) than of lower density sedimentary material close to the surface.

But is Puerto Rico in isostatic equilibrium?

You could have argued that Equation 1.45 is satisfied because the model shown gives predicted gravity equal to the Bouguer anomaly, but, as Equation 1.45 makes clear, that would be true only if the column below Puerto Rico were of equal mass to adjacent oceanic crustal columns that are in isostatic equilibrium, so satisfying Airy's or Pratt's hypothesis. Generally, arcs are regions of positive isostatic anomaly (for example, many of the island-arc regions in Figure 1.77). So, many island arcs are underlain by mass excesses despite their low-density 'roots' (Figure 1.78b). One possible reason may be that they are regions of magmatic activity: material is being added to the crust in the form of igneous rocks faster than the crust subsides isostatically. Another possible reason is that the subducting lithosphere below island arcs is cold and has a rather higher density than the surrounding mantle, again giving excess mass.

SUMMARY OF SECTION 1.9

The combined constraints of gravity and seismic data for oceanic ridge zones, trenches and island arcs lead to contrasting conclusions. Oceanic ridge zones have zero free-air anomaly, are in isostatic equilibrium and stand higher than the surrounding ocean floor simply because they are underlain by material of anomalously low density. Trench zones, on the other hand, have negative free-air and isostatic anomalies and are regions of relative mass deficiency; island arcs often have (smaller) positive isostatic anomalies indicating excess mass. As far as trenches are concerned, this is undoubtedly linked to the sinking of oceanic lithosphere into the mantle, at such active margins, although the mass deficiency cannot explain the sinking. This is because the material over trench zones would have to *rise*, not sink, to reach isostatic equilibrium. This suggests that some much more important force operates at subduction zones which drives oceanic lithosphere down into the mantle, keeping it out of isostatic equilibrium. We will develop these ideas further in Block 2.

You should now look at video VB 02, 'Under the Volcano', which describes work on volcanoes in Costa Rica and Italy (Sicily), and then listen to audiovision sequence AV 03, 'Volcano Geophysics', which examines volcano monitoring. (The video lasts about 74 minutes, and the audiovision sequence lasts about 24 minutes.)

Objectives for Section 1.9

When you have completed this Section, you should be able to:

1.1 Recognize and use definitions and applications of the terms printed in the text in bold.

1.39 Show how, together with seismic data, Bouguer, free-air and isostatic anomalies can be used to test for isostatic equilibrium, especially over ocean ridges and trenches.

You should now do the following SAQ.

SAQ for Section 1.9

SAQ 48 (*Objectives 1.1 and 1.39*)

Complete the following paragraphs, using some of the words given below. You may need to use some words more than once.

Low P-wave _________ is indicative of partially molten material beneath the surface. Oceanic ridges are characterized by _________ P-wave velocity, negative _________ anomaly and zero _________ anomaly. These features are therefore in _________ equilibrium and have _________ material underlying raised topography.

An active continental margin typically has a _________ on the seaward side and a mountain range or _________ on the other side. There is usually a _________ isostatic anomaly over the trench and a smaller positive _________ anomaly over the accompanying island arc. The isostatic anomaly reflects a mass _________ over the trench and the Bouguer anomaly a mass _________ over the arc or mountain chain.

Words: Positive, negative, high, low, dense, light, Bouguer, isostasy, isostatic, partially molten, frozen, crystallized, liquid, solid, fast, slow, trench, ridge, island arc, mountain chain, deficiency, excess, volume, mass, density, velocity, bulk modulus.

1.10 THE EARTH'S DENSITY

If we are to understand how the Earth works, we need to know what it is made of. If we could measure or estimate the density of the Earth, we could then compare this information with the densities and behaviour of plausible materials to estimate the chemical composition and, therefore, the nature of the material present at depth.

It would be relatively easy to consider the material deep within the Earth if it were the same as the material exposed at the surface. Since we know the Earth's mass and volume, we can determine its *average density*, which is 5 520 kg m^{-3}. Compared with the densities of surface rocks, which lie in the range 2 000–3 100 kg m^{-3}, this density is rather large. As the crustal rocks are of lower density than that of the Earth as a whole, we conclude that part of the Earth's interior must have a density substantially greater than 5 520 kg m^{-3}.

At first sight, it appears to be an easy matter to determine the variation of density within the Earth. The variations of P-wave and S-wave velocities within the Earth are now known quite accurately (Figures 1.52 and 1.63); and seismic-wave velocities depend on density. Therefore, it should be possible to determine density directly from velocity. Unfortunately, however, this is not so, for other unknown quantities are involved. We know the relationship between density and velocity for a given substance:

$$v_P = \sqrt{\frac{K + 4\mu/3}{\rho}} \qquad \text{(Equation 1.16)}$$

and

$$v_S = \sqrt{\frac{\mu}{\rho}} \qquad \text{(Equation 1.17)}$$

where v_P and v_S are the P- and S-wave velocities, ρ is the density, μ is the rigidity modulus and K is the bulk modulus (defined in Section 1.6.1).

❑ Can you see from these equations why there is no unique solution for density?

■ The problems is that we have two equations and three unknown quantities. We can square Equations 1.16 and 1.17 and eliminate μ to give:

$$v_P^2 = \frac{K + 4\mu/3}{\rho} \quad \text{and} \quad v_S^2 = \frac{\mu}{\rho}$$

$$\text{so} \quad \mu = (\rho v_P^2 - K)\frac{3}{4} \quad \text{and} \quad \mu = v_S^2 \rho$$

$$\text{so that} \quad (\rho v_P^2 - K)\frac{3}{4} = v_S^2 \rho$$

$$\text{and} \quad \frac{K}{\rho} = v_P^2 - \frac{4v_S^2}{3}. \qquad \text{(Equation 1.46)}$$

But this gets us little further since we have no information on how bulk modulus varies with depth. At the same time, however, whatever estimates for the density–depth profile of the Earth's interior do emerge from such a procedure, they must conform to certain other well-known physical quantities. So before we take Equation 1.46 any further, it is useful to introduce those quantities.

The Earth's mass: The mass of the Earth has been determined; we gave it earlier as about 6×10^{24} kg, but more precisely it is 5.977×10^{24} kg. Whatever conclusion we come to about density variations in the Earth, therefore, the sum of the density × volume (= mass) products for all zones of the Earth must add up to 5.977×10^{24} kg.

Rearranging Equation 1.34, the mass of the Earth (M) is:

$$M = \frac{gd^2}{G}$$

where G is the gravitational constant, d is the Earth's radius and g is the acceleration due to gravity at the Earth's surface. As G and d are known independently, the mass of the Earth is determined by measuring g. However, although g depends on M, it gives no information at all about how the mass is distributed with depth. In fact, g would be the same at distance d whether the Earth's mass were all concentrated in a pin-head volume at the Earth's centre or whether it were concentrated in a thin outer shell of an Earth with a hollow core.

The Earth's moment of inertia: A fundamental property of rotating bodies which also depends on the internal distribution of mass is the **moment of inertia**. To appreciate this, first consider a body of mass M, moving in a straight line with velocity v; it is said to have a **momentum** of $M \times v$. However, if the body stayed in one place and merely rotated, it would have no momentum in the linear sense, though each particle within the body would be moving with respect to the axis of rotation. This is because each particle moves in a direction perpendicular to the imaginary line, or radius, joining it to the axis of rotation.

To clarify these ideas, consider a body such as that in Figure 1.79, which is rotating about O at a constant rate in the direction of the arrow. The *body as a whole* has no linear velocity (in the normal sense of straight line movement). But it does have an **angular velocity** (ω), defined as the angle through which the body rotates in unit time. Suppose the body rotates through an angle of θ radians ($2\pi/360$ radians = 1 degree) in time t, then

$$\text{angular velocity} = \omega = \frac{\theta}{t}. \qquad \text{(Equation 1.47)}$$

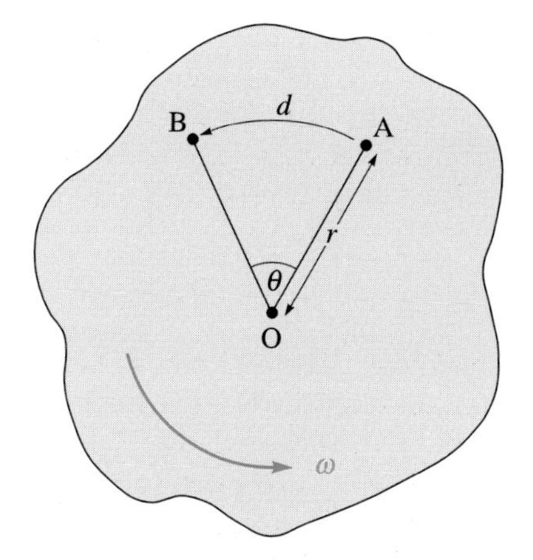

Figure 1.79 Circular motion of a body centred on an axis of rotation through O. A particle at a radial distance r from O starts at A and moves through a distance d to B by rotation through an angle θ, with angular velocity ω. The angular momentum about O is $mr^2\omega$, where m is the mass of the particle. The angular momentum of the whole body is the sum of the angular momenta of all the particles in the body.

Now consider a particle of mass m at A which is distance r from O. In time t, A moves to B through distance d (Figure 1.79), and it can be shown that the linear velocity of the particle is $r\omega$ and that its **angular momentum** is $mr^2\omega$. (Note that you need not try to remember these formulae.) Now the whole body that is rotating is made up of many small particles, each of mass m, and so the total angular momentum of the body is the *sum* of all the $mr^2\omega$ terms. The moment of inertia, I, is defined as the *sum* of mass × radius^2 for all the component parts of a large rotating body. So:

$$\text{angular momentum} = I\omega. \qquad \text{(Equation 1.48)}$$

From perturbations in the orbits of satellites, we can calculate the moment of inertia; the Earth, a rotating body, has a moment of inertia of 8.07×10^{37} kg m^2. This is only 83% of the value it would have if it were of uniform density throughout, and this implies that there is a concentration of mass near the Earth's centre. Therefore, in the same way that we deduced for the Earth's mass, any density model must also be consistent with the moment of inertia, or the model cannot be valid. But moment of inertia is more useful than mass because it says something about the *internal distribution of mass*, through the inclusion of the r^2 term.

So any density variation models devised must be consistent with both M and I. But how are such models devised in the first place? There are, in fact, several ways, beginning with random suggestions. The simplest possibility, for example, is that the Earth's density is uniform throughout; but that can be ruled out partly by the comparison of near-surface density with average density and partly because such a model does not give a correct I. A model in which density increases uniformly with depth also fails the I test and is unlikely to be correct in any case in the light of our knowledge that the Earth has discrete layers. But there are so many models to guess at that the game could go on for ever without success.

A better approach was devised by L. H. Adams and E. D. Williamson in 1923. They derived a third equation so that the third variable (K), the bulk modulus could be eliminated from Equation 1.46. They assumed that the only reason that density increases with depth inside the Earth is that the overlying layers press down on the material below. This is called **self-compression** (described earlier in Section 1.3), and is easily illustrated if you imagine the Earth to be made up of an infinite number of extremely thin spherical shells. The increase in pressure (ΔP) caused by moving in from a distance $r + \Delta r$ to r is derived simply from the equation for pressure (P) which depends on density (ρ), gravity (g) and height (h):

$$P = \rho g h.$$

So in the case of a shell of thickness Δr, the change in pressure across it is

$$\Delta P = -\rho_r g_r \Delta r \qquad \text{(Equation 1.49)}$$

where ρ_r is the density at radius r, and g_r is the gravitational acceleration at radius r. The minus sign indicates that pressure increases with decreasing radius r. By considering all values of r with the variation of the bulk modulus with depth, and adding these effects together, Adams and Williamson were able to derive Equation 1.50 which describes the change of density ($\Delta\rho$) with change of depth (Δr) assuming that density is related only to depth (amount of compression):

$$\frac{\Delta\rho}{\Delta r} = \frac{\rho G m}{r^2(v_P^2 - v_S^2/3)}. \qquad \text{(Equation 1.50)}$$

Equation 1.50 is known as the **Adams–Williamson equation,** after its originators. We do not expect you to use this equation — only to realize that it is important because it describes the variation of density with depth (left-hand side) as a function of the density, mass and seismic velocities (ρ, m, v_P, v_S) at a particular radius (r) from the Earth's centre. This means that it can be applied by taking known or estimated values on the right-hand side (for example, those at the Earth's surface). Using the value of $\Delta\rho/\Delta r$ so deduced, it allows values of ρ and thus m to be calculated, at a different depth (r), where v_P and v_S are known. In turn, this process can be continued to greater and greater depth until ρ is known at all depths.

❑ But what about the assumption that went into the Adams–Williamson equation? Do you think that density increases in the Earth are simply due to self-compression by the weight of material above?

■ You may have realized that the presence of discontinuities in the Earth, where density is thought to change sharply because of changes in the composition of the material, is going to confuse the issue. The equation can only be applied *within* layers inside the Earth where the only variations are due to self-compression of material with uniform chemical composition.

But there is another limitation, to do with the rate at which temperature increases with depth.

❑ What happens to the temperature of a material when the pressure on it is increased? (Think about what happens when you squeeze air through a bicycle pump.)

■ All things being equal, the temperature increases as the atoms are squashed more closely together, and, conversely, if you allow the material to expand or escape, it will cool again. If no heat is added or removed, the amount by which the temperature increases, as the pressure on a material increases, is known as the **adiabatic temperature increase**.

❑ Now what will happen to the temperature at a given depth if, inside the Earth, the pressure increases with depth only due to the weight of overlying material — that is, due to self-compression?

■ Temperature will increase with depth and, again, assuming that no heat is added or removed at any depth, the temperature increase will be adiabatic. We describe the rate of temperature increase with depth in terms of a temperature gradient (temperature/depth) and the particular case of adiabatic self-compression in the Earth as an **adiabatic temperature gradient**. The Adams–Williamson equation assumes that the temperature increase with depth in the Earth is simply adiabatic (this would be about a $0.25\,\text{K}\,\text{km}^{-1}$ rise in temperature for most rocks).

ITQ 40

As an exercise to clarify the meaning of adiabatic temperature gradient, what would be the effect on the *density* at a particular depth in the Earth if heat was added there, giving a rise in temperature, so that, above that depth, there was an above-adiabatic (or **super-adiabatic**) temperature gradient?

Notice, therefore, that higher temperatures than predicted from adiabatic self-compression lead to lower density at a particular depth, and, in general, higher temperature gradients lead to lower density gradients. Such super-adiabatic conditions occur in the Earth because more thermal energy is produced by decay of long-lived radioisotopes than is required to maintain adiabatic conditions. Inside the Earth, the situation is potentially unstable and the material at depth, being of lower density (and, therefore, lighter) than it would be under adiabatic conditions, will tend to rise, to replace denser, cooler material above.

This is the start of convection; if the material is able to respond plastically (that is, it is able to move), it will tend to convect in order to remove the excess temperature gradient, leading towards the establishment of stable, adiabatic conditions with the denser material below the lighter material.

The Adams–Williamson equation (Equation 1.50) can be used to describe the variation of density with depth in a layer (a) which is of uniform composition, and (b) in which changes in physical properties, particularly temperature, are due only to increasing compression with depth. A different approach to the density problem was developed in the late 1960s and early 1970s by Frank Press of the Massachusetts Institute of Technology. He adopted a statistical procedure, known as the **Monte Carlo inversion method** (so-called because it starts by 'guessing' the result), which does not depend on the Adams–Williamson assumptions. Several million random density models were generated by computer. They were then tested against known geophysical quantities such as the Earth's mass, moment of inertia and seismic-wave velocities.

Figure 1.80 gives the variation of density in the Earth. The 2 500–3 000 kg m^{-3} values of the crust give way to about 3 300 kg m^{-3} in the mantle above the asthenosphere. At a depth of 1 000 km the density is about 4 600 kg m^{-3}, increasing to 5 400 kg m^{-3} at the base of the mantle. There is then a jump to 9 900 kg m^{-3} at the top of the outer core, increasing to 12 200 kg m^{-3} at the base of the outer core. The inner core has a density of about 12 800 kg m^{-3}.

Once the density and seismic velocity distributions are known, the elastic moduli distributions can be calculated. But, more usefully, gravity and then *pressure* inside the Earth can be determined (Figure 1.80). Although, not surprisingly, pressure increases steadily towards the Earth's centre, gravity remains almost constant through the mantle. This is because a large proportion of the Earth's mass (about 31%) is concentrated in the core, which occupies only 16% of the Earth's volume.

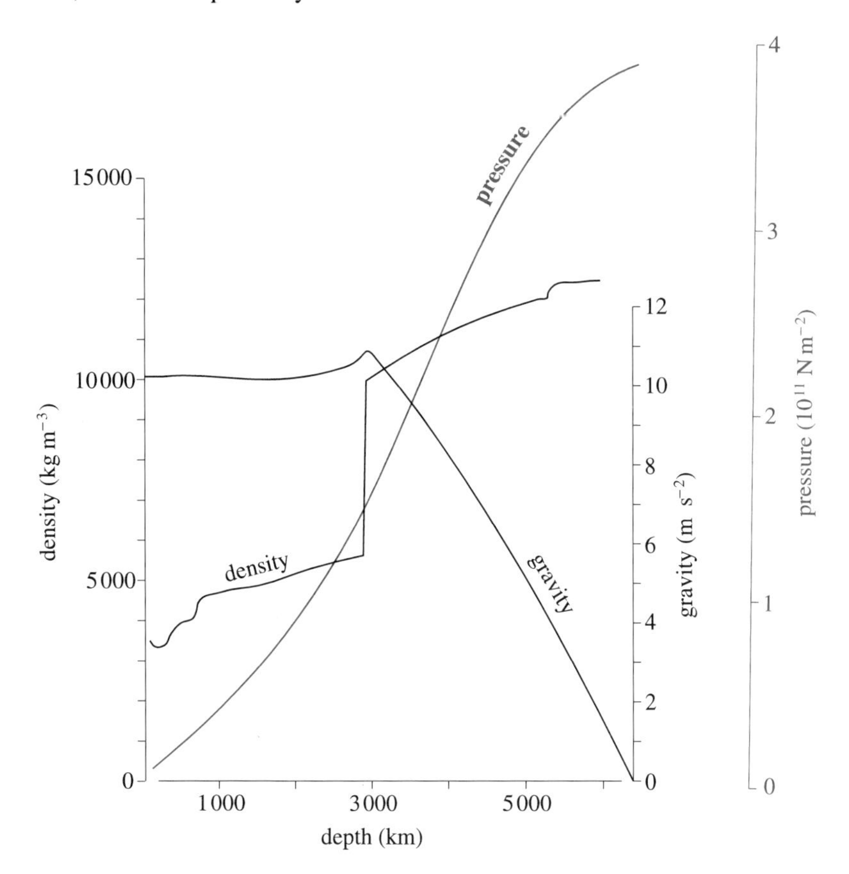

Figure 1.80 Density, pressure and gravity in the Earth.

Summary of Section 1.10

The densities of surface rocks are much lower than the average density for the whole Earth. There are two contrasting approaches to the problem of the relationship between density and depth within the Earth. The Adams–Williamson method assumes that density increases only due to self-compression. This works well within layers, but the model breaks down at seismic (and density) discontinuities. The Monte Carlo inversion method does not rely on any assumptions and the model is adjusted until it fits the observed geophysical data.

Objectives for Section 1.10

When you have completed this Section, you should be able to:

1.1 Recognize and use definitions and applications of the terms printed in the text in bold.

1.40 Describe the methods used to model the Earth's density–depth distribution, including the importance of non-adiabatic temperature gradients and describe the main features of the density–depth curve.

Apart from Objective 1.1, to which it also relates, the ITQ in this Section tests the Objectives as follows: ITQ 40, Objective 1.40.

You should now do the following SAQ, which tests other aspects of the Objectives.

SAQ FOR SECTION 1.10

SAQ 49 (*Objectives 1.1 and 1.40*)

State, giving reasons where appropriate, whether each of the following statements is true or false:

(a) The Adams–Williamson equation allows the density within a particular layer of the Earth (which can be a thick region such as the upper mantle) to be computed from known density elsewhere at other depths within that layer, provided the composition of the layer is uniform and that there is an adiabatic temperature gradient across the layer.

(b) The Monte Carlo inversion method is used to guess the Earth's mass and moment of inertia.

(c) The most significant increase in density at depth in the Earth occurs at the core–mantle boundary.

1.11 THE COMPOSITION OF THE EARTH'S LAYERS: A BRIEF INTRODUCTION

In Section 1.4.2, we came to the conclusion that the Earth probably has a chondritic overall composition, at least as far as refractory elements are concerned. If it does, we can tell immediately what the most abundant refractory elements in the Earth are in terms of numbers of atoms.

❑ Assuming the Earth to be chondritic overall in respect of refractory elements, what are the four most abundant refractory elements in the planet? (See Section 1.4.2.)

■ The data to answer this question are best displayed in Figure 1.16, where the most abundant elements are clearly those in the top right-hand corner. Of the seven most abundant elements (O, Mg, Fe, Si, C, N and S), three (C, N and S) are volatiles (see Figure 1.5). Oxygen (O) is also a volatile under certain circumstances, but it also readily combines with silicon and other elements to form (refractory) silicates. The four most abundant refractory elements in the Solar System, in meteorites and, by implication, in the Earth, are therefore O, Fe, Si and Mg.

Remember, however, that Figure 1.16 shows relative abundances of the elements in terms of the number of atoms of each element for every 10^6 atoms of Si. This is not the same as the relative abundances in terms of mass, for different atoms have different atomic masses. (The **atomic mass** of an element is the mean mass of an atom of that element expressed in atomic mass units. The atomic mass unit, in turn, is based on the carbon isotope ${}^{12}_{6}C$, which is assigned a mass of 12 atomic mass units. The atomic mass scale is thus an arbitrary scale of relative masses, although, in absolute terms, an atomic mass unit is about 1.66×10^{-27} kg. On this scale, the atomic masses of the four elements mentioned above are O = 16.00, Fe = 55.85, Si = 28.09 and Mg = 24.31.)

ITQ 41

For every 10^6 Si atoms in the Earth (assuming a chondritic Earth), there are about 1.2×10^6 Fe atoms. So the relative abundance of Fe compared to Si in terms of number of atoms is $1.2 \times 10^6/10^6 = 1.2$. Calculate the relative abundance of Fe compared to Si in terms of mass.

Clearly, the relative abundances of the elements expressed in terms of number of atoms are numerically different from those expressed in terms of mass. Nevertheless, when calculations such as that in ITQ 41 are carried out for other elements, it becomes clear that even in terms of mass the four most abundant refractory elements in the Earth — assuming a chondritic composition — remain O, Fe, Si and Mg.

But what of the real Earth? Do we have sufficient data to determine what the most abundant elements in the Earth are *without* assuming the planet to be chondritic? As it is not possible to sample the interior of the Earth directly, indirect methods of estimating overall composition must be used. One approach here is to make use of seismic shock-wave data combined with the mixing of meteorite compositions — a technique involving two steps.

Step 1: Experiments on sending shock waves* through minerals in the laboratory show that there is a linear relationship between the seismic-wave velocity in a material and the material's mean atomic mass. Comparison with average seismic-wave velocities in the Earth then suggests that the Earth material has a mean atomic mass of about 27, made up of mantle material with a mean atomic mass of 22.4 and core material with a much higher mean atomic mass of 47.0.

Step 2: Assuming the Earth to be constructed from material similar to that of meteorites, it would be a neat solution to the problem of determining the Earth's major constituent elements if there were a single class of meteorite material with an average atomic mass of exactly 27. But there is not. Carbonaceous chondrites, for example, have mean atomic masses of 23.4–24.0, ordinary chondrites 24.4, enstatite chondrites 25.6 and iron meteorites 55.0. To assemble an 'Earth' with a mean atomic mass of 27, it is therefore necessary to mix meteorite classes in such proportions as to give the 'correct' result and also to satisfy other constraints. For example, the Earth thus assembled must have a core : mantle mass ratio equal to that of the real Earth (8 : 17). Unfortunately, these constraints are not tight enough to enable a unique solution to be obtained. There are thus various different combinations of meteorite materials that conform with the constraints, three of which are shown in Figure 1.81.

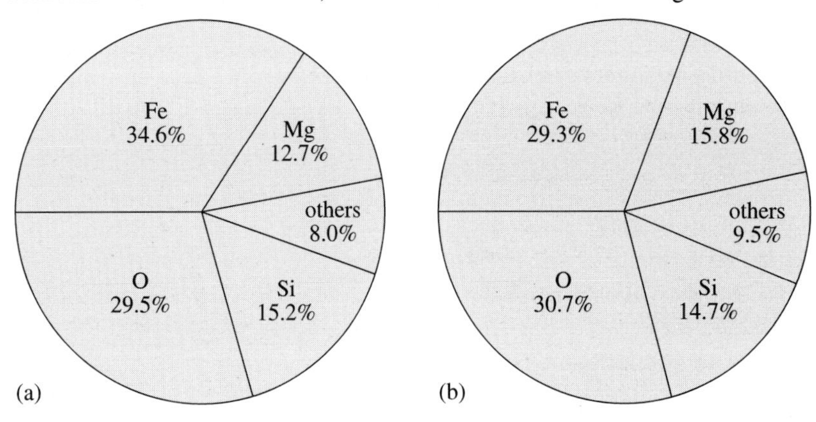

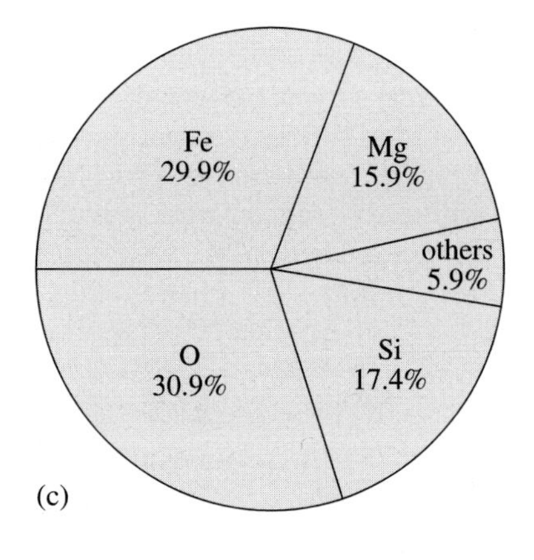

Figure 1.81 Three models of the bulk composition of the Earth. The figures represent mass per cent; thus, for example, in model (a), 34.6% of the Earth by mass is iron, 29.5% is oxygen, etc. These models are constructed by mixing various types of meteoritic material in such proportions as to produce an Earth with a mean atomic mass of 27 and a core : mantle mass ratio of 8 : 17. Model (b), for example, is obtained by mixing 40% C1 carbonaceous chondrite (see Figure 1.15), 50% ordinary chondrite and 10% iron meteorite. Of course, models other than these would be possible; the three shown are just plausible examples.

The one definite conclusion to be drawn from Figure 1.81 is that, although we do not know the bulk composition of the Earth in terms of mass per cent of elements with any great accuracy, all the meteorite-mixing models do agree that O, Fe, Si and Mg are the four most abundant elements in the Earth, accounting for more than 90% of the mass of the planet. To that extent, the models are in encouraging conformity with the chondritic Earth model (CEM). However, there is a rather obvious snag. The models in Figure 1.81 are not themselves actually chondritic, for they consist not solely of carbonaceous chondrites but of *mixtures* of meteorite types. We therefore have an apparent paradox here. If any one of the three models in Figure 1.81 is a true representation of the overall composition of the Earth, the Earth cannot be chondritic. On the other hand, all three models predict that the four most abundant refractory elements in the Earth are the same as the four most abundant elements according to the CEM. The only possible solution to this conundrum is that predicting the four most abundant refractory elements in the Earth is not a very sensitive test of the CEM; many models, whether chondritic or not, give the same answer. We therefore turn next to the compositions of the Earth's major layers in the hope that *they* will provide data for a more valid test of the CEM.

---

* In shock-wave experiments, high-pressure shock waves of a few microseconds duration are passed through a material, enabling certain physical properties — such as seismic velocity and density — to be measured almost instantaneously at that pressure.

1.11.1 Continental crust

Even if the bulk composition of the Earth were known very precisely, such knowledge would provide only very broad constraints on the compositions of the planet's individual layers. How, then, is it possible, if at all, to deduce the compositions of the layers (crust, mantle, outer core, inner core), given that they cannot be sampled directly?

A promising approach on the face of it would be to use geophysical data. Different materials have different seismic velocities; so in principle it should be possible to measure (in the laboratory) the velocities in various materials at the temperatures and pressures characteristic of different depths in the Earth, compare them with the actual velocities at those depths, and thus identify the materials by comparison. A similar approach involves density. The variation of seismic P-wave velocity with density for a wide variety of rocks has been determined experimentally, with the result shown in Figure 1.82. This diagram is known as the **Nafe–Drake curve** after the two scientists who produced it. So in principle it should be possible to measure seismic velocity (e.g. in a near-surface rock formation), estimate the density of the rock from the Nafe–Drake curve, compare the density thus obtained with the densities of known materials, and hence identify the rock.

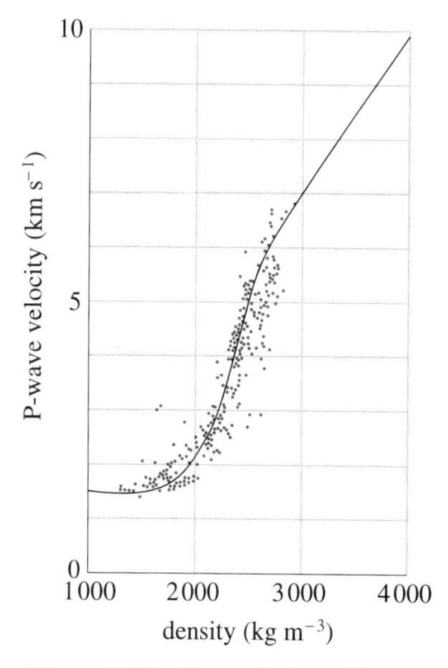

Figure 1.82 The variation of P-wave velocity with density (the Nafe–Drake curve). The experimental points are based on sediments and sedimentary rocks. The upper extension for igneous rocks is continuous but differs in gradient. There is considerable scatter in the plotted data, which means that an attempt to use the curve in any individual case (e.g. to determine density from seismic velocity) could be subject to errors of up to about 100 kg m^{-3}. Note that the Nafe–Drake curve applies only to rocks of the uppermost crust.

ITQ 42

The seismic P-wave velocity in a near-surface rock formation is found to be 6.0 km s^{-1}. Estimate the density of the rock using the Nafe–Drake curve.

ITQ 43

Table 1.5 lists the densities (ranges and averages) of some common sedimentary, igneous and metamorphic rocks. Use the table to determine the type of rock making up the formation in ITQ 42.

Table 1.5 Rock densities (in kg m^{-3}) under near-surface pressures.

Rock	Density range	Average density
SEDIMENTARY		
alluvium	1 500–2 000	1 980
sand	1 400–2 300	2 000
clay	1 300–2 600	2 210
sandstone	1 600–2 760	2 350
shale	1 560–3 200	2 400
limestone	1 740–2 900	2 550
IGNEOUS		
rhyolite	2 350–2 700	2 520
andesite	2 400–2 800	2 610
granite	2 500–2 810	2 640
diorite	2 720–2 990	2 850
basalt	2 700–3 300	2 990
gabbro	2 700–3 500	3 030
peridotite	2 780–3 370	3 150
METAMORPHIC		
quartzite	2 500–2 700	2 600
schist	2 390–2 900	2 640
granulite	2 520–2 730	2 650
marble	2 600–2 900	2 750
slate	2 700–2 900	2 790
gneiss	2 590–3 000	2 800

However, while neither seismic-wave velocities nor densities can be used by themselves to determine composition, *composition must be consistent with them*. In ITQs 42 and 43, for example, it is clear that the rock concerned could be neither alluvium (density too low) nor peridotite (density too high). In other words, while geophysical data cannot be used to determine composition uniquely, they do act as important constraints, enabling some compositions to be excluded, as we shall see in a moment in respect of the lower continental crust.

The continental crust has been formed from mantle material throughout the Earth's history by a series of processes, including melting, crystallization, metamorphism, erosion, deposition and subduction. These processes are the subject of Blocks 3 and 4; so here we shall look at crustal composition in just a very general way.

We mentioned in Section 1.4.2 (see Table 1.3 in particular) the problem of determining the average chemical composition of the crust, even though the uppermost crust at least is open to direct sampling. Not to put too fine a point on it, every authority one is ever likely to consult will provide a different answer, even when they claim to have combined data to produce an 'average' crustal composition. Three such 'averages' for the whole continental crust are shown in Table 1.6.

Table 1.6 Estimates of the chemical composition of the whole crust. Note that because the crust and mantle are highly oxidized, it is customary when presenting their chemical compositions to express major elements in terms of oxides (in mass per cent) and minor elements as elements (in parts per million, ppm).

	Estimate 1	**Estimate 2**	**Estimate 3**
SiO_2	60.2	63.3	57.3
TiO_2	0.7	0.6	0.9
Al_2O_3	15.2	16.0	15.9
Fe_2O_3	2.5	1.5	—
FeO	3.8	3.5	9.1
MgO	3.1	2.2	5.3
CaO	5.5	4.1	7.4
Na_2O	3.0	3.7	3.1
K_2O	2.9	2.9	1.1
H_2O	1.4	0.9	?
Rb	95	85	?
Sr	340	350	?
Ba	670	1150	?
U	3	1.8	?
Th	100	6	?
Ni	50	20	?

Estimates such as those in Table 1.6 are obtained partly by analysing near-surface rocks, some of which are thought to have been formed in the lower crust and subsequently uplifted to the surface, and partly by 'mixing' rock types (e.g. basalt and granite) in proportions such as to give a result that appears to mimic the real crust. Although they differ in detail, they could be said to be broadly similar. Similar differences/similarities are also observed when data for the upper and lower continental crusts are tabulated separately.

Despite the chemical differences, continental upper crust appears to have an average composition similar to that of the rock granodiorite, a coarse-grained igneous rock consisting mainly of the minerals quartz [SiO_2], plagioclase [$(Na,Ca)Al(Si,Al)Si_2O_8$], potassium feldspar [$KAl(Al,Si)_3O_8$], biotite [$K(Mg,Fe^{2+})_3(Al,Fe^{3+})Si_3O_{10}(OH)_2$] and hornblende [$(Ca,Na)_{2-3}(Mg,Fe^{2+},Fe^{3+},Al)_5(Al,Si)_8O_{22}(OH)_2$].*

The seismic velocities in the upper crust are consistent with a granodiorite composition, which means to say that if the upper continental crust were indeed granodiorite, it would have the seismic velocities actually observed. You should not imagine, however, that the continental upper crust is a homogeneous layer of granodiorite. Indeed, we already know from observation that the uppermost continental crust is very heterogeneous, that sedimentary rock buried during thrusting can now be found deep in the crust and that parts of the continental crust comprise oceanic crust that has been thrust up onto the continents or accreted to their edges. All that can be said is that the continental upper crust appears to have the composition of granodiorite on average.

If that seems rather tentative, the composition of the lower continental crust is even more uncertain. The first thing to be said is that the lower crust cannot have the average composition of granodiorite. If it were simply granodiorite that had been self-compressed by burial, it would have a lower seismic-wave velocity than it actually does. In other words, the seismic velocities in the lower continental crust are higher than those to be expected on the basis of granodiorite under lower-crustal conditions of pressure and temperature. There must therefore be a real chemical difference between the lower and upper continental crust. In fact, the lower continental crust is now thought to be granulite, a metamorphic rock with a composition broadly similar to that of metamorphosed basalt, having a granular texture and consisting mainly of feldspar [$MAl(Al,Si)_3O_8$, where M = K, Na, Ca, Ba, Rb, Sr or Fe], pyroxene [$ABSi_2O_6$, where A = Ca, Na, Mg or Fe^{2+} and B = Mg, Fe^{2+}, Fe^{3+}, Fe, Cr, Mn or Al] and garnet [$A_3B_2(SiO_4)_3$, where A = Ca, Mg, Fe^{2+} or Mn^{2+} and B = Al, Fe^{3+}, Mn^{3+}, V^{3+} or Cr]. However, just as the upper continental crust cannot simply be homogeneous granodiorite, nor can the lower continental crust simply be homogeneous granulite. There is evidence from **xenoliths** ('foreign bodies' — in this case, pieces of the lower crust broken off and brought to the surface by magmas rising from the mantle), for example, that the lower crust is compositionally complex, with significant regional variations.

1.11.2 THE MANTLE

Clues to the composition of the mantle have been obtained from seismic velocity/density studies, from laboratory high-pressure/temperature experiments on materials thought likely to be candidate mantle constituents, from geochemical modelling based on meteorite compositions, and from studies of **ophiolites**.

Ophiolites, or, more accurately, ophiolite sequences (because they comprise a sequence of layers) are sections through the oceanic crust and upper mantle that have been uplifted and exposed at the Earth's surface by

* In formulas such as these, the items in brackets and separated by commas are alternatives. Thus, for example, $NaAlSi_3O_8$ is an example of plagioclase, as is $CaAl_2Si_2O_8$. In other words, crystals of the mineral plagioclase do not all have the same composition but belong to a series of minerals with various compositions, albeit ones conforming with the general formula. By the same token, a rock such as granodiorite is not defined on the basis of a unique composition but is a series of rocks with variable mineral content. You do not need to remember these complicated chemical formulas. Why, then, you might ask do we give them? Simply to give you some idea of what the rocks and minerals *are*. 'Plagioclase', for example, is just a name that could mean anything. Quoting the formula does at least give you some idea of what the beast 'plagioclase' is — i.e. a complex silicate consisting of certain particular elements — even if you do not need to remember the formula in detail.

being thrust onto land (e.g. in Cyprus, the Oman and many other places). In other words, they consist of series of layers which correspond to the seismic layers of the oceanic lithosphere (see Section 1.7.4). Thus the upper layer is sediment, the second layer is lava, and so on. What is important in the present context, however, is not the layers thought to represent oceanic crust but the lowermost layer, thought to represent the uppermost mantle. This invariably comprises mainly peridotite, a type of rock consisting predominantly of the mineral olivine [$(Mg,Fe)_2SiO_4$].

Additional evidence for supposing the upper mantle to comprise mainly peridotite comes from **olivine nodules**, pieces of olivine-rich (>80% olivine) material found as xenoliths in the basaltic lavas of oceanic islands. These are thought to have been derived directly from the upper mantle by being torn away from the walls of magma conduits as magma rose to the surface. **Kimberlite pipes** are also thought to have been derived directly from the upper mantle. These are steep-walled vertical pipes of (largely) peridotite that occur in Kimberley, South Africa, and elsewhere, where they are mined for diamonds. Kimberlite pipes are believed to have been formed by explosive volcanism involving gas–solid mixtures with little or no liquid; and their origin in the mantle is supported by their diamond content, for the diamonds could only have formed at the high pressures and temperatures characteristic of the upper mantle. Finally — in case you are wondering — peridotite has a P-wave velocity of 8.1 km s^{-1}, which matches the P-wave velocity in the uppermost mantle (see, for example, Figure 1.60).

If the upper mantle is indeed mainly peridotite, it becomes possible to characterize it in chemical terms. The major-element geochemistry of the mantle based on the type of peridotite (garnet peridotite) thought to form the bulk of the mantle is shown in column 1 of Table 1.7. However, it took a long time for Earth scientists to reach a consensus on peridotite; and in the mean time, attempts had been made to deduce mantle composition by other methods — for example, by mixing various meteorite classes, by analysing magnesium-rich silicate rocks now found at the Earth's surface, and by designing theoretical mixtures. Examples of mantle composition thus derived are shown in columns 2–7 of Table 1.7.

Table 1.7 Estimates of the composition of the mantle based on garnet peridotite (column 1) and other methods (columns 2–7). Figures are mass per cent.

	1	2	3	4	5	6	7
SiO_2	45.1	44.5	44.2	45.2	46.0	48.1	43.2
Al_2O_3	3.3	2.6	2.7	3.5	3.6	3.0	3.9
FeO	8.0	8.7	8.3	8.4	8.6	12.7	9.3
MgO	38.1	41.7	41.3	37.5	38.1	37.1	38.1
CaO	3.1	2.2	2.4	3.1	3.1	2.3	3.7
Na_2O	0.4	0.3	0.3	0.6	0.6	1.1	1.8

Despite minor differences, the seven compositions in Table 1.7 are remarkably similar, which, bearing in mind the different ways in which they have been derived, suggests that they provide a fairly accurate view of mantle geochemistry. It is thus clear that more than 90% of the mantle by mass can be accounted for by the system SiO_2–MgO–FeO, that no other oxide exceeds 4%, and that more than 98% of the composition of the mantle can be expressed in terms of just six oxides.

Strictly speaking, all that we have said so far about mantle composition relates only to the upper mantle, for that is where the evidence comes from. Much less is known about the lower mantle. It is often assumed that the lower and upper mantles are chemically similar but that, as the depth increases, the minerals adopt higher-pressure forms, with particularly conspicuous phase transformations taking place at the 400 km and 670 km discontinuities. However, it is by no means clear that this assumption is fully justified; indeed, there is some evidence that it is not. For example, if we assume that the figures for FeO in Table 1.7 apply to the whole mantle and then add together that iron, the iron in the core and that in the crust, we end up with a total abundance of iron that is less than there should be if the Earth is chondritic in iron. So if the Earth *is* chondritic in iron, where is the 'missing' iron? The only possible place for it to be is in the lower mantle, which would mean that the concentration of iron in the lower mantle is higher than in the upper by a factor of about two. As we discover more about the lower mantle we are likely to find that not only does it differ from the upper mantle in terms of composition but that it is also much less homogeneous than it appears in our current state of ignorance.

As for the phase changes in the mantle, laboratory experiments show that, at pressures and temperatures corresponding to a depth of about 400 km in the Earth, olivine transforms into a form in which the atoms are packed more closely, known as **spinel**, with a consequent increase in density of about 10%. This largely accounts for the discontinuity at 400 km. There is then a further phase change at about 670 km where the spinel structure transforms into an even higher-pressure form known as **perovskite** (and residual magnesium oxide) with another density increase of about 10%. The whole of the lower mantle, which means most of the mantle, thus largely comprises minerals with a perovskite structure. Incidentally, phase changes (e.g. at 400 km and 670 km) take place over depth ranges of a few tens of kilometres. Thus they give rise to seismic discontinuities which are not very sharp, and certainly less sharp than those formed by chemical changes (e.g. the Moho).

1.11.3 THE CORE

At several points throughout this Block, we have asserted, without adducing much evidence, that the Earth's core consists largely of iron. In fact, there is no direct evidence at all — only circumstantial. The latter includes the following:

(1) If the Earth is chondritic in iron (see Figure 1.16), the core must be predominantly iron, for there is otherwise far too little iron in the combined crust and mantle. Using similar arguments, the core must also contain a few per cent nickel.

(2) There are only two basic types of meteorite (see Section 1.4.1) — stony and iron (plus, of course, the stony–iron hybrids). If meteorites are to be taken seriously as indicators of Earth composition and if the silicate stony meteorites are analogous to the Earth's mantle, it seems likely that the iron meteorites — which, you will recall, are thought to be derived from the cores of differentiated parent bodies — correspond to the Earth's core (which implies that the core should also contain a small proportion of nickel).

(3) The only known mechanism by which the Earth's rapidly varying magnetic field can be generated is by motions within the fluid outer core, and that requires that the outer core be highly conducting (i.e. metallic). The only metal likely to be available in sufficient quantity is iron.

(4) Shock-wave experiments on materials at core pressures show that the only elements, apart from iron, having densities and seismic velocities capable of matching those of the core are such metals as titanium (Ti), vanadium (V), chromium (Cr), cobalt (Co) and nickel (Ni), none of which (as suggested by solar abundances) are available in sufficient quantities to be the core's main constituent(s).

Seismic velocities and densities obtained from shock-wave experiments on iron are rather higher than those in the outer core, which suggests that the outer core cannot be pure iron but must be an alloy of iron containing 5–15% of a less-dense element. And, of course, if the core contains some nickel, which has an atomic mass greater than that of iron, the proportion of the less-dense element would have to be slightly higher. There has been, and still is, considerable debate about what the lighter element (or elements) might be. Most can be ruled out on geochemical grounds or because they are unlikely to be available in sufficient abundance, but that still leaves oxygen (O), silicon (Si), sulphur (S), carbon (C), hydrogen (H) and potassium (K) as serious contenders.

Oxygen

At low temperatures, FeO (one of the oxides of iron) is non-metallic and does not mix with iron. However, high-pressure/temperature experiments on FeO show that at core pressures and temperatures it becomes metallic and does readily mix with iron. Moreover, other laboratory experiments show that liquid iron and iron alloys react vigorously with solid oxides and silicates, suggesting that the Earth's core might react with the lowermost mantle and thus derive oxygen from it. Some geochemists thus believe that the presence of oxygen (7–13% by mass) in the core is highly likely.

Silicon

During the 1960s and 1970s, silicon was seen as a very strong contender as the light element in the core, partly because it is such an abundant element in the Earth, partly because it readily alloys with iron, partly because traces of a natural Fe–Ni–Si alloy called perryite are found in some types of iron meteorite, and partly because small quantities of silicon are dissolved in the iron found in some chondrites. However, it was later discovered that the incorporation of silicon into the core would require an unduly complicated hypothesis of Earth formation involving the outgassing and escape of huge amounts of carbon monoxide (CO), for which there is no evidence. The idea that the core contains a significant quantity of silicon is thus now out of fashion, although a very small amount of silicon in the core is not impossible.

Sulphur

Part of the evidence in favour of sulphur as the core's chief light element is that all chondritic meteorites, irrespective of their state of oxidation, contain substantial amounts of sulphur in the form of sulphides, chiefly FeS. A second piece of evidence is shown in Figure 1.83, which illustrates how certain elements are depleted in the mantle as compared with carbonaceous chondrites.

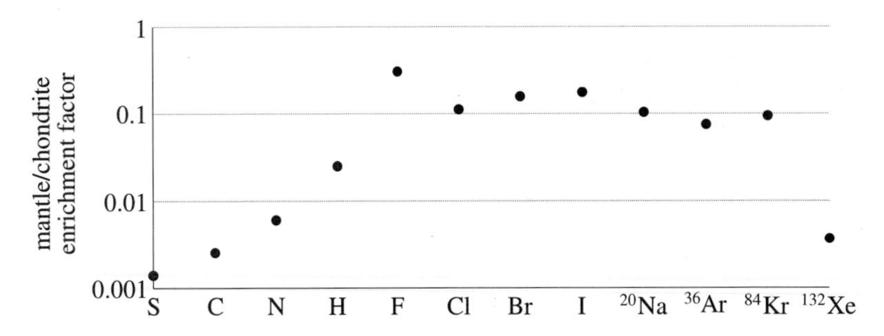

Figure 1.83 The depletion of some volatile elements in the Earth's mantle compared with carbonaceous chondrites. (The horizontal axis of the diagram is only to enable the elements to be separated from one another.)

❑ Examine Figures 1.83 and 1.5 (Section 1.3) with particular reference to the elements sulphur (S), chlorine (Cl) and bromine (Br). Can you spot any inconsistency between the behaviour of S on the one hand and Cl and Br on the other?

■ Figure 1.5 shows that Cl and Br are more volatile than sulphur — i.e. they vaporize at lower temperatures (darker squares for Cl and Br; lighter squares for S). However, Figure 1.83 shows that S is relatively more depleted in the mantle than are Cl and Br. So it would appear, on the basis of the mantle alone, that the Earth has ended up with relatively more of the more volatile elements Cl and Br than of the less volatile element S!

One possible line of argument is that such a situation is unlikely — that, given that sulphur is the less volatile element, it should be less depleted, or at least no more depleted, in the Earth than are chlorine and bromine. If that premise be accepted, it follows that, if the missing sulphur is not in the mantle, it must be in the core. However, sulphur *is* a volatile; and there really is no absolute rule that says that, when the Earth accreted, sulphur must have behaved in a similar way to other volatiles, especially given the unpredictable effects of giant impacts. There could, for example, have been a special reason — which we don't understand yet — for sulphur's behaving differently. Moreover, there are some geochemists who argue that the problem doesn't exist anyway, because the abundance of sulphur in the mantle could have been underestimated. Sulphur remains a fairly strong contender as the core's light element, although it's difficult to find firm evidence in its favour.

Carbon

Carbon has been very little studied in the context of the core. However, as you will see by looking at Figure 1.83, the volatility arguments applying to sulphur apply no less equally to carbon. Moreover, carbon readily alloys with iron. Some geochemists therefore argue that carbon is at least as likely as sulphur to be the core's light element.

Hydrogen

At pressures of above $3.5 \times 10^9\,\mathrm{N\,m^{-2}}$, hydrogen and iron react to form iron hydride (FeH), a compound which is unstable, and indeed unsynthesizable, at lower pressures. FeH has been found to be stable up to pressures of at least $62 \times 10^9\,\mathrm{N\,m^{-2}}$ and thus, by implication, would probably be stable at core pressures of about $3.3 \times 10^{11}\,\mathrm{N\,m^{-2}}$. As hydrogen was very common in the early Solar System, it must therefore be considered a prime contender for the role of the core's light element.

Potassium

The alkali metals potassium (K), rubidium (Rb) and caesium (Cs) are known to be depleted in the crust and upper mantle relative to chondritic abundances. One possible explanation of this is that the depletion is just part of the general depletion of volatile elements in the Earth — in other words, that the Earth simply contains less than chondritic abundances of K, Rb and Cs. An alternative possibility, however, is that these elements really are present in the Earth in chondritic abundances but are mostly 'hidden' in the core, in which case up to 75% of the Earth's potassium could be in the core. Speculation along these lines gains some support from the fact that potassium, while not a notably chalcophile element, can nevertheless combine with sulphur to form the sulphide K_2S and that there is theoretical evidence for supposing that potassium has an even greater affinity for sulphur at high temperatures and pressures. *If* potassium does have an affinity for sulphur under core conditions and *if* sulphur is present in the core in significant quantities, then, so the argument goes, the sulphur going into the core could have taken most of the Earth's potassium with it. A sulphide of iron and potassium, know as djerfisherite, is also found in small quantities in some meteorites, where it is often closely associated with troilite (FeS).

The possible presence of potassium in the core also has implications for the generation of the Earth's magnetic field. To generate the magnetic field in the outer core it is necessary to have some way, which could be thermal, of producing motion in the iron conductor.

❑ Why should the presence of potassium in the core have thermal implications?

■ Because one of the isotopes of potassium (^{40}K) is one of the four major heat-generating long-lived radioactive isotopes in the Earth (see Section 1.3.3).

^{40}K could therefore be the heat source required for generation of the magnetic field. However, it is not absolutely necessary to have a radioactive heat source in the core, for the core motions could be produced in other ways. They could, for example, simply be a consequence of the Earth's rotation. Alternatively, there is a widely supported hypothesis to the effect that the Earth's solid inner core is growing by the gradual solidification of the liquid outer core, a process that would release heat. The requirement for potassium in the core is thus not compelling from the geomagnetic point of view.

It's probably fair to say that at the time of preparation of this Course (1992), oxygen is the most favoured contender as the core's light element, closely followed by sulphur (and thus carbon?) and then potassium. However, as the core cannot be sampled, all speculation about the light elements must be treated with great caution. Don't forget, too, that the core could contain not one, but a mixture of, light elements.

And finally, what of the inner core? Here is a selection of quotations:

(1) 'The inner core, however, has a seismic velocity and density consistent with a composition of pure iron'. *Global Tectonics*, Kearey and Vine, Blackwell Scientific Publications, 1990.

(2) '... the IC [inner core] is not pure Fe, i.e., like the OC [outer core] it must contain some light component as well'. *Deep Interior of the Earth*, Jacobs, Chapman & Hall, 1992.

(3) 'In contrast to the outer core, the seismic velocity and density inferred for the inner core are in good agreement with the shock wave data for pure iron'. *The Interior of the Earth: Its Structure, Constitution and Evolution*, Bott, Edward Arnold, 1982.

(4) 'From density data, shock compression studies and meteorite evidence, the Earth's core comprises ... an inner solid region ... composed of iron–nickel alloy (probably roughly 20% Ni, 80% Fe)'. [Note: Nickel is denser than iron.] *The Inaccessible Earth*, Brown and Mussett, Allen and Unwin, 1981.

(5) 'Within the uncertainties the inner core may be simply a frozen version of the outer core, Fe_2O or FeNiO, pure iron or an iron–nickel alloy'. *Theory of the Earth*, Anderson, Blackwell Scientific Publications, 1989.

❑ What conclusion do you draw from these quotations?

■ That we haven't the faintest idea whether the inner core is denser than, less dense than or equal in density to pure iron, and thus whether it contains just iron, iron plus a less dense element or iron plus a more dense element.

1.11.4 THE CHONDRITIC EARTH MODEL REVISITED

You will recall that when we left the chondritic Earth model (CEM) in Section 1.4.2, we said we could return to it when we had more information about the compositions of the Earth's individual layers. Well, here we are. Do the compositions of the crust (Section 1.11.1), mantle (Section 1.11.2) and core (Section 1.11.3) add up to a chondritic Earth or not? In fact, it turns out to be more difficult to put the CEM to the direct test than you might suppose, especially in respect of minor elements, but it is possible to go some of the way, with encouraging results.

Major elements

As far as the major rock-forming elements are concerned, the crust can be ignored, because it accounts for less than 1% of the mass of the Earth. In other words, by ignoring the crust, which is rather complex, we shall not be introducing any large error in putting the CEM to the test. Let's also assume for the moment that the core is pure iron and, instead, concentrate on the mantle. We cannot, of course, ignore the core, because it accounts for a major part of the Earth's mass; we shall return to it later. In the mean time, how does the mantle compare to the CEM as represented by carbonaceous chondrites?

Table 1.8 shows, in column A, the major-element composition of the mantle as represented by garnet peridotite (this is the same as column 1 of Table 1.7) and, in column B, the average major-element composition of carbonaceous chondrites.

Table 1.8 The compositions of (A) the mantle as represented by garnet peridotite and (B) average carbonaceous chondrite. The figures are mass per cent.

	A	B
SiO_2	45.1	33.25
TiO_2	0.2	0.09
Al_2O_3	3.3	2.27
Cr_2O_3	0.4	0.47
MgO	38.1	22.04
FeO	8.0	14.39
MnO	0.14	0.30
CaO	3.1	2.16
Na_2O	0.4	1.02
K_2O	0.03	0.09
FeS	—	23.20

❑ Is the implied comparison in Table 1.8 realistic in terms of the bulk composition of the Earth?

■ No, because it ignores the fact that a large part of the Earth's mass is represented by the core. Thus if the abundance of oxide X in the mantle is x in mass per cent, and assuming that there is no X in the core, the abundance of X in the bulk Earth — i.e. as a proportion of the whole mass of the Earth — will be less than x.

To convert mantle abundances to bulk-Earth abundances is a matter of simple arithmetic. As the mass of the Earth is about 5.9×10^{24} kg and the mass of the mantle is about 3.9×10^{24} kg, the ratio of the mass of the mantle to that of the whole Earth is $3.9 \times 10^{24}/5.9 \times 10^{24} = 0.66$. Thus for oxide X,

$$\text{mass per cent of X in bulk Earth} = \text{mass per cent in mantle} \times 0.66.$$

For SiO_2, for example, mass per cent in mantle = 45.1 (Table 1.8); so mass per cent in bulk Earth = $45.1 \times 0.66 = 29.8$.

ITQ 44

(a) Complete Table 1.9 by calculating the bulk-Earth compositions of each of the oxides shown.

Table 1.9 Abundance of elements in the bulk Earth (for use with ITQ 44). Figures in mass per cent.

		B
SiO_2	29.8	33.25
TiO_2	0.13	
Al_2O_3	2.18	
Cr_2O_3	0.26	
MgO	25.15	
FeO	5.28	
MnO	0.09	
CaO	2.05	
Na_2O	0.[illegible]6	
K_2O	0.02	

(b) Now plot the bulk-Earth abundances you have estimated in Table 1.9 against the corresponding chondritic abundances (Table 1.8, column B) using the log–log paper in Figure 1.84. You have seen similar plots as Figures 1.16 and 1.17, although in those cases they referred to elements, and the abundances were in terms of atoms per 10^6 atoms of silicon. Draw in the line representing equal abundances in the bulk Earth and carbonaceous chondrites.

(c) Bearing in mind that oxides depleted in the bulk Earth with respect to carbonaceous chondrites will plot below the line and that oxides enriched in the bulk Earth with respect to carbonaceous chondrites will plot above the line, what does your graph indicate about the depletion/enrichment of the refractory major elements Si, Mg, Al and Ca?

The one major element we have not dealt with satisfactorily here is iron, which Figure 1.129 shows to be severely depleted in the bulk Earth. Remember, however, that the bulk-Earth abundances were calculated on the assumption that there is none of the relevant elements in the core, which is not true in the case of iron. The calculations involving iron are complicated, and we do not propose to go into them here. However, when the iron in the core is taken into account it becomes clear that the Earth as a whole is not depleted in iron and thus the Earth is as chondritic in iron as it is in Si, Mg, Al and Ca.

So does the exercise represented by ITQ 44 demonstrate the Earth to be chondritic or not? This depends how we look at the issue. On the one hand, the analysis we have just carried out covers only six (including oxygen) of the 90 naturally occurring elements. Don't forget, however, that these six elements represent more than 95% of the Earth's mass. For the most important elements, therefore, the Earth does indeed appear to be chondritic.

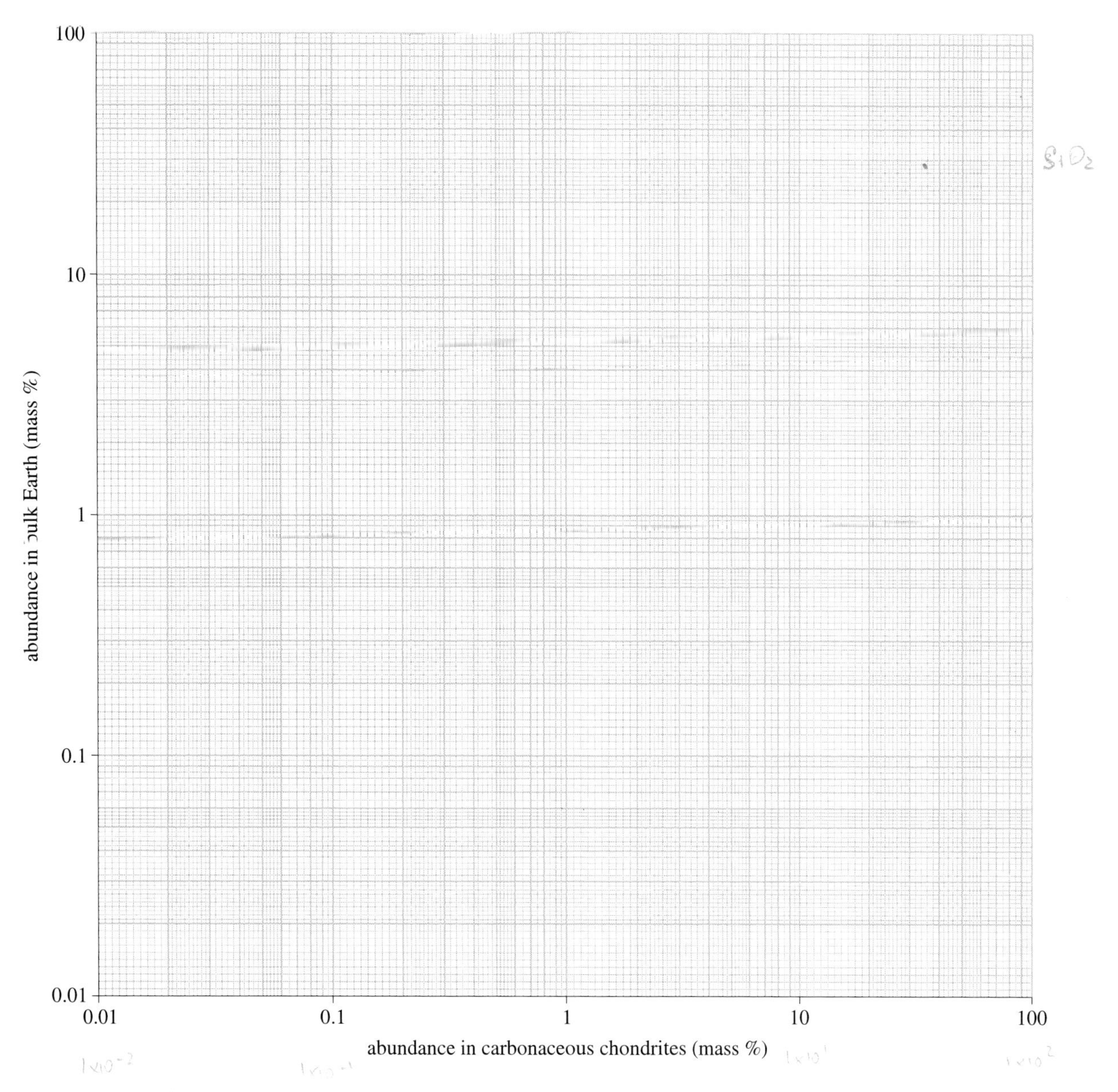

Figure 1.84 Relative abundances of oxides in the bulk Earth and carbonaceous chondrites (for use with ITQ 44).

Minor elements

Testing the CEM in respect of the minor and trace elements is extremely difficult, and for most, if not all, such elements, impossible. For one thing, we can no longer ignore the crust, because we know that certain minor/trace elements are now concentrated there. For example, elements such as K, Rb, Sr, Ba, Y and Zr have large ions that do not fit easily into the lattices of mantle minerals, and so when mantle material melts such elements are preferentially extracted and carried up into the crust in magmatic fluids. Secondly, many trace elements do not form minerals in their own right but are tucked away in the crystal lattices of other minerals. Thirdly, many minor elements are volatile; and we have already seen how difficult it is to predict how individual volatiles are likely to behave. Fourthly, we cannot be sure which minor elements have ended up in the core. And finally, minor/trace elements are, by definition, not very abundant; so at the very least it is very difficult to determine the true concentrations of many of them, especially if they are unevenly distributed.

Potassium offers a good illustration of some of these problems. Figure 1.129 shows quite clearly that if only the K in the mantle is taken into account, the Earth appears severely depleted in K with respect to carbonaceous chondrites. We also know, because we understand some of its geochemical characteristics, that much of the K originally in the mantle is now likely to have migrated into the crust. But it is extremely difficult to estimate just how much K there is in the crust, which can be done only by making a variety of assumptions that may or may not be valid. In fact, the best estimates of the maximum concentration of K in the crust, when added to the supposed concentration (e.g. from garnet peridotite) in the mantle, still show the Earth to be depleted in K. But is the Earth really depleted in K — quite possible, because don't forget that K is a volatile (see Figure 1.5) — or could the 'missing' K be hiding in the core (see Section 1.11.3)? We simply do not know.

More generally, one way of illustrating the position with regard to minor and trace elements is to plot their supposed abundances in the mantle in terms of their enrichment factors. The **enrichment factor** (**EF**) of an element is simply the abundance of the element in the mantle divided by its abundance in chondrites. Thus if the mantle contains 0.35 ppm of element X and chondrites contain 0.43 ppm of the element, the EF is 0.81 or, in other words, element X is enriched in the mantle relative to chondrites by a factor of 0.81. (An enrichment factor of less than 1 indicates depletion.)

Figure 1.85 illustrates the EFs for a number of minor and trace elements, where the EFs are expressed with respect to Cl carbonaceous chondrites. (You should notice that this gives results slightly different from those in Figure 1.83, where the EFs are expressed with respect to carbonaceous chondrites in general.) Examine Figure 1.85 and then attempt to answer the following questions.

Figure 1.85 Some minor/trace elements' enrichment factors (EFs) in the mantle relative to Cl carbonaceous chondrites. (The elements' EFs are separated from one another horizontally for clarity.)

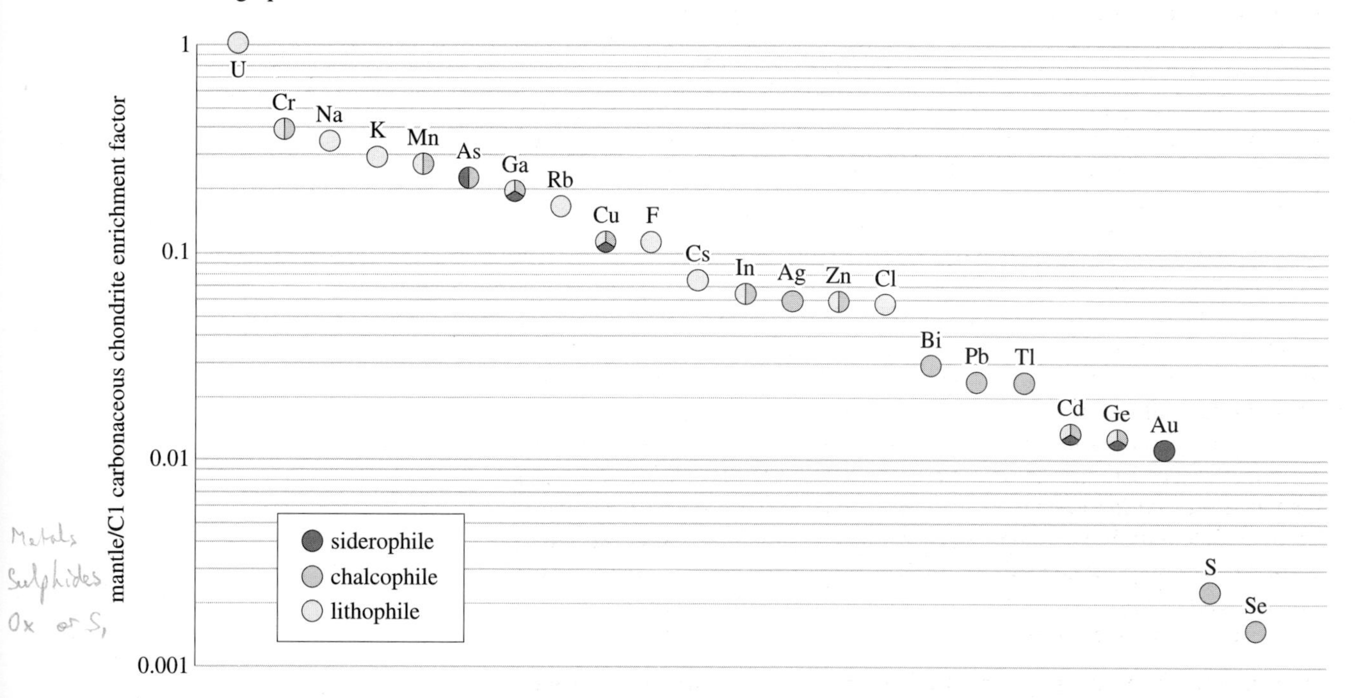

❑ Which two non-metallic elements exhibit the greatest degree of depletion?

■ Sulphur (S) and selenium (Se), which both have EFs approaching 0.001 (in other words, they are about a thousand times less abundant in the mantle than in chondrites).

❑ Which six metallic elements exhibit the greatest degree of depletion?

■ Bi, Pb, Tl, Cd, Ge and Au, with EFs approaching 0.01.

❑ Can you spot anything that the elements Bi, Pb, Tl, Cd, Ge, Se and S have in common? (Don't worry about Au: it's rather anomalous.)

■ All the metals (with the exception of Au, which is siderophile) are chalcophile — i.e. they have an affinity for sulphur. However, that's not what we had in mind. More importantly, if you check with Figure 1.5 you will see that all seven elements (i.e. except Au) are volatiles.

In fact, there is, with some exceptions, a fairly strong correlation between EF and volatility for minor and trace elements — and that notwithstanding that some elements are prone to migrate towards the crust and that some may or may not be prone to migrate towards the core. In other words, there is a strong hint here that the abundance of elements that accreted to form the Earth was strongly influenced by their volatility and that, generally, the more volatile an element is, the less likely it is to be found in the Earth in chondritic proportions.

SUMMARY OF SECTION 1.11

If the Earth is taken to be chondritic in composition, the four most abundant refractory elements in the planet should be oxygen (O), iron (Fe), silicon (Si) and magnesium (Mg). Experiments using shock waves, combined with the mixing of meteorite compositions to (a) give material having a mean atomic mass that agrees with that obtained from the shock-wave experiments and (b) provide the core : mantle mass ratio corresponding to that of the real Earth, also indicate that the four most abundant refractory elements in the Earth are O, Fe, Si and Mg. This is not a very good (partial) test of the chondritic Earth model (CEM), however, because the compositions obtained from meteorite mixing are not themselves chondritic. Thus while it is probably true that the Earth's four most abundant refractory elements are indeed O, Fe, Si and Mg, such a combination is consistent with both a chondritic Earth and certain non-chondritic compositions.

Geophysical data such as seismic velocities and density are insufficient to enable the compositions of the Earth's major layers to be specified, because the ranges of the values of such parameters overlap for a wide variety of common rocks, although the data are sometimes sufficient to exclude certain rock types in particular circumstances. Nevertheless, geophysical data combined with information from other sources (e.g. geochemical analysis of crustal rocks and the theoretical mixing of rock types) suggest that the Earth's upper continental crust has an average composition similar to that of the rock granodiorite and that the lower crust (which seismic velocities indicate to have a different composition) is granulite, having a composition broadly similar to that of metamorphosed basalt. Apart from its sedimentary veneer, the oceanic crust is largely basalt and gabbro (see Section 1.7.4).

Information from geophysical studies, from laboratory experiments, from modelling based on meteorite compositions, and from ophiolites, olivine nodules and kimberlite pipes, suggests that the mantle consists largely of the rock peridotite, the main constituent of which is the mineral olivine. Olivine undergoes phase changes at pressures and temperatures corresponding to depths in the Earth of about 400 km and 670 km, and the consequential jumps in density at those depths probably give rise to the seismic discontinuities there (see Section 1.7.5). As a result of the phase change at 670 km, the minerals in the lower mantle probably have a perovskite structure. The upper and lower mantles may also be chemically different, at least to the extent of there being a higher proportion of iron in the lower mantle.

The outer core appears to consist mainly of iron, possibly in association with a small percentage of nickel. However, the seismic velocities in the outer core suggest that it must also contain 5–15% of lighter elements. These could be oxygen, silicon, sulphur, carbon, hydrogen or potassium (and perhaps others), or a combination of two or more of these, although at present oxygen appears to be the main option favoured by geochemists. The composition of the solid inner core is even more uncertain, with opinions covering the complete range from iron with one or more lighter elements, through pure iron, to iron and nickel (i.e. iron plus a heavier element).

A comparison of the composition of the mantle, as represented by garnet peridotite, with the average composition of carbonaceous chondrites suggests that the Earth is indeed chondritic in respect of the major elements Si, Mg, Al and Ca as represented by their oxides. When the large amount of iron in the core is taken into account, the Earth also appears to be chondritic in iron. Testing the CEM in respect of minor elements is much more difficult, however, partly because such minor elements (by definition not very abundant) could have migrated to the crust and/or core, partly because they tend to be volatile, and partly because they often lie within crystal lattices of other minerals. Thus, although definitive proof is hard to come by, information from the Earth's individual layers appears to support the conclusion reached in Section 1.4.2, namely, that the Earth *is* chondritic, at least in respect of the most abundant refractory elements.

Objectives for Section 1.11

When you have completed this Section, you should be able to:

1.1 Recognize and use definitions and applications of each of the terms printed in the text in bold.

1.41 Outline the case in favour of the proposition that the four most abundant refractory elements in the Earth are oxygen (O), iron (Fe), silicon (Si) and magnesium (Mg).

1.42 Describe in general terms the compositions of the upper and lower crust, and list the types of evidence used to define them.

1.43 Summarize the sources of evidence for the composition of the mantle, and describe the probable nature of the mantle and its chief phase changes.

1.44 List briefly the evidence for supposing that the Earth's core consists mainly of iron, and discuss the options for the lighter element(s) in the outer core.

1.45 Summarize a method of testing the chondritic Earth model by comparing the proportions of major elements in the Earth (as represented by the composition of the mantle) with those in carbonaceous chondrites, and discuss the results obtained therefrom.

1.46 Perform simple calculations based on topics covered by Objectives 1.42–1.45, where appropriate.

Apart from Objective 1.1, to which they all relate, the four ITQs in this Section test the Objectives as follows: ITQ 41, Objectives 1.41 and 1.46; ITQs 42 and 43, Objectives 1.42 and 1.46; ITQ 44, Objectives 1.45 and 1.46.

You should now do the following SAQs, which test other aspects of the Objectives.

SAQS FOR SECTION 1.11

SAQ 50 (*Objectives 1.1, 1.41, 1.42, 1.43, 1.44, 1.45 and 1.46*)

State, giving reasons where appropriate, whether each of the following statements is true or false.

(a) Assuming the Earth to be chondritic, the fifth, sixth and seventh most abundant refractory elements in the planet in terms of number of atoms per 10^6 atoms of Si should be aluminium (Al), calcium (Ca) and nickel (Ni).

(b) The four most abundant refractory elements in the Earth are oxygen (O), iron (Fe), silicon (Si) and carbon (C).

(c) According to the experimental relationship derived from shock waves, the Earth's average seismic-wave velocity suggests that the Earth has a mean atomic mass of about 27.

(d) Using the Nafe–Drake curve it is possible to determine the density of near-surface rock formation from knowledge of the P-wave velocity within it, and hence specify uniquely the type of rock involved.

(e) A near-surface rock with a P-wave velocity of $6\,km\,s^{-1}$ is unlikely to be clay.

(f) The proportion (in mass per cent) of SiO_2 in the Earth's crust as a whole is probably in excess of 55%.

(g) Chemically identical upper and lower continental crust would be consistent with the seismic data from both parts of the crust.

(h) Where all parts of the sequence are present, the lowest layer in an ophiolite sequence is always mainly peridotite.

(i) The proportion (in mass per cent) of SiO_2 in the mantle is lower than in the crust, but the proportion of MgO is higher.

(j) More than 98% of the mantle can be accounted for by just six oxides.

(k) A case can be made for suggesting that the lower mantle is richer in iron than is the upper mantle.

(l) The seismic discontinuities at 400 km and 670 km are probably due to phase changes in granodiorite.

(m) It is clear that the crust and mantle could contain all the iron necessary for the Earth to be considered chondritic in iron.

(n) Seismic velocities in the fluid outer core are too low to be explained on the basis of a core of pure iron.

(o) FeO mixes readily with iron at core pressures and temperatures.

(p) Carbon (C) is relatively less depleted in the mantle, compared to carbonaceous chondrites, than is iodine (I).

(q) Potassium (K) appears to be depleted in the crust and mantle relative to carbonaceous chondrites.

(r) TiO_2 appears to be enriched in the Earth as compared to carbonaceous chondrites, whereas MnO is depleted.

SAQ 51 (*Objectives 1.41 and 1.46*)

Calculate the relative abundance (in terms of mass) of chromium (Cr) in the Earth compared to silicon, assuming the Earth to be chondritic and given that the atomic mass of Cr is 52.00. (As a first step, determine the abundance of Cr in terms of number of atoms to the nearest power of 10.)

SAQ 52 (*Objectives 1.42 and 1.46*)

The P-wave velocity in a near-surface rock formation is found to be about $8.5\,\mathrm{km\,s^{-1}}$. What is the most likely type of rock involved, given a choice limited to those in Table 1.5?

1.12 THE EARTH'S HEAT

In Section 1.3.3, we pointed out that as every geological process taking place on or within the Earth involves the transfer of energy, and as that energy is always ultimately expressed as heat, heat can be regarded as the Earth's most fundamental property. Unfortunately, however, it is also the most difficult to investigate, at least as far as the Earth's deep interior is concerned. The Earth's internal structure may be determined by seismic and gravity measurements made at the surface, and the planet's chemical constitution may be deduced (however imperfectly) by a combination of means, including the chemical analysis of near-surface rocks and comparisons with cosmic abundances. The flow of heat from the Earth may also be measured at the surface; but because heat is a much less tangible phenomenon than either physical structure or chemical composition, surface measurements are able to tell us less about its distribution at depth. More generally, heat generation and flow are simply difficult to quantify, as we again saw in Section 1.3.3, where we found it difficult, and often impossible, to put very precise figures on the amount of heat generated by various processes supposedly occurring during the Earth's formation.

One figure that is known fairly precisely, however, is that for the total amount of heat now escaping through the Earth's surface; and it is instructive to compare this internal heat with that from other sources. The primary energy sources responsible for Earth processes may be classified into two main groups:

A *external* sources: (i) solar energy
(ii) gravitational influence of the Moon and Sun

B *internal* sources: (i) Earth's internally generated heat
(ii) Earth's rotation and gravity field

ITQ 45

Use the list below to correlate some of the common processes operating on and within the Earth with one or more of the sources listed above. You should decide in each case which you think to be the most significant energy source(s).

(a) ocean tides

(b) atmospheric circulation

(c) biological activity

(d) ocean currents

(e) weathering, erosion, sedimentation

(f) volcanism

(g) metamorphism

(h) earthquakes

Looking at the answers to ITQ 45, you might be inclined to feel intuitively that the external sources of energy available to the Earth are of less importance than the internal sources, if only because earthquakes and volcanoes, for instance, seem to be so much more substantial phenomena than, say, tides, currents and winds. In fact, the reverse is true. The Sun supplies some 1.8×10^{17} W to the Earth, of which about 60% penetrates the atmosphere ($\sim 10^{17}$ W) to reach the surface. By contrast, the flow of heat from the Earth's interior (largely by conduction and convection) amounts to only some 4×10^{13} W.

❑ How many orders of magnitude difference are there between the amount of energy supplied to the Earth's surface by the Sun and that supplied from the Earth's interior?

■ The Sun supplies 10^{17} W to the surface, and 4×10^{13} W reaches the surface from the interior. The difference is more than three orders of magnitude. The flow of energy from the Earth's interior is some 2 500 times smaller than the available solar power, *at the Earth's surface*.

Below the continental surface, however, the picture alters dramatically. Daily temperature variations produce temperature changes of less than 1 K at a depth of 1 m below the land surface; and seasonal temperature variations are not registered below depths of about 20 metres at the outside. The reason why the effects of solar radiation are limited to the topmost few metres of the Earth is that rocks are poor conductors; in other words, they have low thermal conductivity, a term that we shall define more precisely later. The low thermal conductivity of rocks prevents the flow of heat from the Earth's interior (the geothermal heat loss) being 'swamped' by the solar energy received at the surface, except close to the surface itself. This means that below depths of a few metres it is possible to make accurate measurements of the heat flow from the Earth's interior. However, the low thermal conductivity of rocks also means the conduction of heat through the crust is very slow; a sudden temperature change at depths of even a few kilometres may take many years to be felt or registered at the surface. Accordingly, although we may be able to make quite accurate determinations of the geothermal heat flow at many points on (or just beneath) the Earth's surface, these determinations tell us nothing directly about the sources of heat and very little about how the heat at depth is distributed.

The many measurements that have now been made of the flow of heat from the Earth's interior indicate that the world average heat flow is about 0.08 W m^{-2}, or 80 mW m^{-2}, although this average conceals many variations.

ITQ 46

To help you appreciate the magnitude of this figure, try this exercise. A football field has the dimensions 100 m × 70 m. Assuming that it is characterized by a representative heat flow of 80 mW m^{-2}, how many 100-watt electric light bulbs could be continuously illuminated, in theory, by the heat evolved?

In this Section, we shall look at how heat flow from the Earth is measured and how the results may be interpreted. In doing so, we shall be introducing a number of quantities, which, for convenience, are summarized in Table 1.10 together with their units. The unit of heat energy is the joule (J), and the rate at which heat flows is expressed in joules per second (J s^{-1}). However, joules per second are also watts (W, a unit of power), and it is more usual in heat flow work to use watts (or, more conveniently, milliwatts, mW). Finally, a word of warning. The term 'heat flow' is used in two different ways, and you have to learn to distinguish between the two by the context. Very often it simply means 'the flow of heat' in general terms. In other instances, however, it means very specifically 'the rate at which heat flows through one square metre of the Earth's surface' (W m^{-2}). Thus, for example, when we say that the heat flow in region A is twice that of region B, we are referring to W m^{-2} (or mW m^{-2}). By the same token, the term 'heat generation' means 'the rate at which heat is generated in (depending on context) one cubic metre or one kilogram of rock'. The shorthand terms 'heat flow' and 'heat generation' are widely used to avoid having continually to write 'rate of'.

Table 1.10 Heat flow quantities and units

Quantity	Symbol	Unit
heat flow	q	$W\,m^{-2}$ ($J\,s^{-1}\,m^{-2}$)
temperature gradient	$\Delta T/\Delta z$	$K\,m^{-1}$ or $°C\,m^{-1}$ *
thermal conductivity	k	$W\,m^{-1}\,K^{-1}$ ($J\,s^{-1}\,m^{-1}\,K^{-1}$)
heat generation	A	$W\,m^{-3}$ ($J\,s^{-1}\,m^{-3}$) or $W\,kg^{-1}$ ($J\,s^{-1}\,kg^{-1}$)

1.12.1 Long-lived radioactive isotopes

❑ In Section 1.3.3, we discussed various sources of heat in the Earth. Can you recall what these were?

■ Accretion, compression, core formation, short-lived radioactive isotopes, long-lived radioactive isotopes, and tidal dissipation.

The context in which we were discussing these sources was the formation of the Earth, during which some of the sources were much more important than they are now. We may presume, for example, that very little, if any, heat is now being generated by accretion, compression and short-lived radioactive isotopes. There is a hypothesis — neither proven nor disproved — to the effect that the Earth's core has been growing throughout the planet's history, albeit latterly at a very slow rate; and to that extent heat could still be being generated by core formation. But if so, it is likely to be quite small, as, indeed, is the heat generated by tidal dissipation (see Section 1.3.3). That just leaves long-lived radioactive isotopes as the main contender for the production of heat in the Earth today. (Remember, though, that although long-lived radioactive isotopes may be a source of *new* heat in the Earth today, because of the low thermal conductivity of rocks some of the heat now escaping through the Earth's surface is heat produced as long ago as the time of the planet's origin.)

❑ Can you recall the four long-lived radioactive isotopes that are significant heat producers?

■ ^{40}K, ^{232}Th, ^{235}U and ^{238}U (see Section 1.3.3), being those radioactive isotopes that are relatively abundant and have half-lives comparable with the age of the Earth. (The half-life of a radioactive isotope, you should recall, is the time it takes half of any quantity of it to decay.)

But although long-lived radioactive isotopes must be a source of heat today, just how significant are they, especially in relation to the heat still present from the time of the Earth's formation? Could they, for example, account for most, or even all, of the heat now escaping from the Earth's surface?

For each of the heat-producing long-lived radioactive isotopes, it is possible to determine, from the known properties of the isotopes, the rate at which heat is generated. The results are shown in column 3 of Table 1.11. To assess the importance of radioactive heat generation, we need to know the concentrations of these elements in different Earth layers. Because of the problem of accessibility, however, such concentrations are unknown for the lower mantle and core, so we cannot easily make such an assessment, although we can make some intelligent guesses by starting with heat generation values for various rock types. The average concentrations of K, Th and U and their important radioactive isotopes in

* When plotting experimental data, Earth scientists often use °C. (A temperature change of 1 °C is, of course, the same as 1 K.)

three rock types are shown in Table 1.12. For a rock with given proportions of K, Th and U, it is relatively straightforward to combine the concentrations of heat-producing isotopes with the heat generation rates ($mW\,kg^{-1}$) in Table 1.11 to determine the heat generation in the rock (also measured in $mW\,kg^{-1}$).

Table 1.11 Half-lives and heat generation values for long-lived radioactive isotopes

	Half-life (10^9 years)	Heat generation, A ($mW\,kg^{-1}$)
^{40}K	1.30	2.8×10^{-2}
^{232}Th	13.9	2.6×10^{-2}
^{235}U	0.71	56×10^{-2}
^{238}U	4.50	9.6×10^{-2}

Table 1.12 Concentrations of main heat-producing isotopes (in parts per million) and heat generation rates of representative rock types

Rock type	K (total)*	^{40}K	^{232}Th (total)$^{+}$	U (total)‡	^{238}U	^{235}U	Heat generation ($10^{-8}\,mW\,kg^{-1}$)
granodiorite	35 000	3.5	18	4	3.97	0.03	
gabbro	9 600	0.96	3	0.8	0.794	0.006	18.45
peridotite	12	1.2×10^{-3}	0.06	0.01	0.01	7×10^{-5}	0.26

* At present, natural potassium contains 0.01% ^{40}K.
$^{+}$ At present, natural thorium can be considered to consist exclusively of ^{232}Th.
‡ At present, natural uranium contains 0.7% ^{235}U and 99.3% ^{238}U.

ITQ 47

As you should recall from Sections 1.7.4, 1.11.1 and 1.11.2, the most representative rock types of the upper continental crust, the oceanic crust and the mantle are, respectively, granodiorite, gabbro and peridotite. Table 1.12 lists concentrations of K, Th and U and their heat-producing isotopes in representative granodiorite, gabbro and peridotite. The heat generation values for the gabbro and peridotite are entered in the table. Use the heat generation values in Table 1.11, together with the concentrations of U, Th and K in Table 1.12, to calculate the heat generation value of granodiorite, and enter your answer to ITQ 47 in the appropriate place in the table. What do you deduce about the distribution of heat generation in the Earth from your answer?

Earlier in this Block, we argued that the Earth may be chondritic. But is such a model consistent with the heat generation data? In fact, the data in Table 1.12 may be used to comment on the chondritic Earth model (CEM).

ITQ 48

(a) Concentrations of ^{238}U, ^{235}U, ^{232}Th and ^{40}K in an average chondrite are, in ppm: ^{238}U, 0.01; ^{235}U, 7×10^{-5}; ^{232}Th, 0.04; ^{40}K, 0.10. Use these values to calculate the heat generation (in $mW\,kg^{-1}$) of chondrites (and hence the heat generation per kilogram of the chondritic Earth).

(b) Listed below are the masses and heat generation values for the different parts of the crust, mantle and core. Use these data to calculate the bulk radiogenic heat generation of the Earth (in $mW\,kg^{-1}$).

	Heat generation (10^{-8} mW kg^{-1})	Mass (kg)
upper continental crust	96	8×10^{21}
lower continental crust	40	8×10^{21}
oceanic crust	19	7×10^{21}
mantle	0.26	4×10^{24}
core	—*	2×10^{24}

* For the purpose of this ITQ we shall assume that the core does not contain any heat-producing isotopes. However, this may not be entirely valid (see Section 1.11.3).

(c) What do you deduce about the heat generation predicted by the CEM (part (a)) and that predicted from consideration of the heat generations of the component parts of the Earth (part (b))?

Taken at face value, therefore, the result obtained in ITQ 48 suggests that much, if not all, of the present heat production matches that to be expected on the basis of the CEM. However, this does not constitute proof of the CEM, not least because there is a weak link in the argument.

❑ Have you spotted what this weakness is?

■ You should have spotted that for the upper continental crust, the oceanic crust and the mantle, the heat generation figures given in ITQ 48(b) are those given in Table 1.12. Thus, for example, the heat generation in the upper continental crust is that of granodiorite. The weakness here, however, is that, although the upper continental crust appears to have the average composition of granodiorite, it is not homogeneous granodiorite, which is not, in any case, a rock of fixed composition (see Section 1.11.1). It doesn't necessarily follow, therefore, that the heat generation figure given in Table 1.12 precisely matches that of the upper continental crust. There is therefore some leeway; and as we pointed out in the answer to ITQ 48(c), it doesn't take much leeway to produce a perfect match between the bulk-Earth heat generation and the CEM heat generation. Similar arguments apply in respect of the oceanic crust and the mantle. The figure for the lower continental crust is even more uncertain, being based on a theoretical argument to the effect that the lower continental crust should have a lower concentration of radioactive isotopes than does the upper continental crust.

In fact, the figures in ITQ 48(b) are almost certainly wrong. As we saw in Sections 1.11.3 and 1.11.4, potassium, for example, appears to be depleted in the mantle and crust relative to chondritic abundances even though there is good reason for supposing the Earth's potassium to be preferentially concentrated in the crust. The figures in ITQ 48(b) are therefore probably too high. Moreover, the result obtained in ITQ 48(b), even if it were perfectly valid, does not mean that all the heat now observed at the Earth's surface is derived from radioactivity, because there is good reason for supposing that heat resulting from the Earth's formation still plays a significant part. One recent estimate suggests that 50–80% of the Earth's heat flow is of radiogenic origin, which means that heat from the Earth's formation might still account for up to 50% of observed heat flow. Finally, don't forget that, because of the low thermal conductivity of rock, there is a long time gap between the generation of heat deep in the Earth and its appearance at the surface. So even if present heat flow matched present radiogenic heat production perfectly, the coincidence would be practically meaningless, because present heat flow at the Earth's surface is largely the result of heat generated at various times in the *past*.

What, then, does the above exercise show? Simply this — that, on the basis of the evidence, *heat from radioactive decay is probably the most important source of heat being generated in today's Earth.* Certainly, there appears to be no need to search for any other to explain any of the Earth's thermal characteristics.

1.12.2 TEMPERATURE GRADIENTS WITHIN THE EARTH

One aim of studying the thermal behaviour of the Earth is to determine how temperature varies with depth and how such temperature variations reflect processes occurring in the Earth's interior. However, studies of this kind are based entirely on measurements made on, or within a few kilometres of, the Earth's surface during the last few decades.

Before we can use these measurements to tell us something about likely temperatures at greater depths, we need to quantify the increase of temperature with depth near the surface — but below the level at which daily or seasonal variations of temperature are felt. In other words, the **temperature gradient** (also called the thermal gradient, or **geothermal gradient**) must be determined. Figure 1.86 is a schematic profile of temperature versus depth. The most obvious feature of Figure 1.86 is that the profile is not a straight line; the temperature gradient changes with depth. Over the depth interval $z_2 - z_1$, the temperature gradient is given by

$$\frac{T_2 - T_1}{z_2 - z_1}$$

and over the depth interval $z_4 - z_3$, the gradient is given by

$$\frac{T_4 - T_3}{z_4 - z_3}.$$

❑ Which of these two gradients is the steeper — that is, which of them shows the greater increase of temperature with depth?

■ The temperature interval between T_1 and T_2 is the same as that between T_3 and T_4. But $z_4 - z_3$ is greater than $z_2 - z_1$. So the gradient near to the surface is steeper, because the temperature increases more rapidly with depth there.

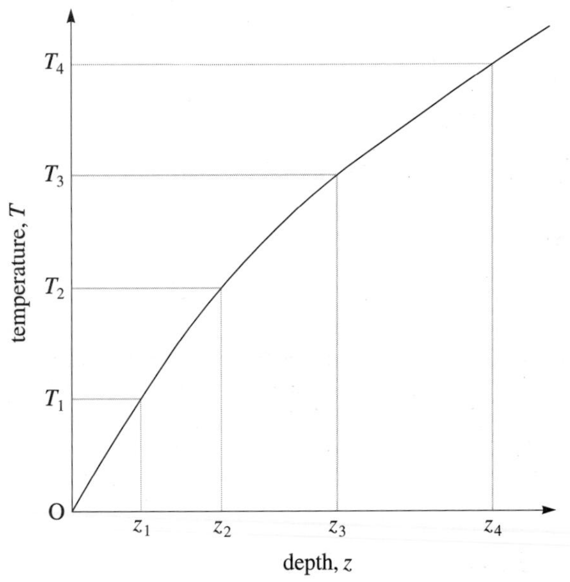

Figure 1.86 A plot of temperature (T) against depth (z) within the Earth, showing the determination of the temperature gradient ($\Delta T/\Delta z$).

The average temperature gradient near to the Earth's surface (say, within a few kilometres) is about 0.03 K m^{-1} (30 K km^{-1}), but values down to about 0.01 K m^{-1} are found in ancient continental crust and very high values are found in areas of active volcanism. At greater depths within the mantle, the average temperature gradient decreases to about $0.25\,\mathrm{K\,km^{-1}}$.

Once the temperature gradient has been measured, it can be used to determine the rate at which heat is moving upwards through a particular part of the Earth's crust. As the heat is generally moving upwards through solid rock, the principal mechanism of heat transfer must be conduction. The quantity of heat flowing by conduction through unit area ($1\,\mathrm{m^2}$ in this case) of solid rock in a given time (i.e. the rate of heat flow) is proportional to the temperature gradient. That is:

$$q \propto \frac{\Delta T}{\Delta z} \quad \text{or} \quad q = k\frac{\Delta T}{\Delta z} \qquad \text{(Equation 1.51)}$$

where q is the **heat flow**, ΔT is the temperature difference over the depth interval Δz, and k is the constant of proportionality, which is known as the thermal conductivity of the rock.

Thermal conductivity, k, is defined as the amount of heat conducted per second through an area of 1 square metre, when the temperature gradient is $1\,\mathrm{K\,m^{-1}}$ perpendicular to that area. The unit of thermal conductivity is therefore $\mathrm{W\,m^{-1}\,K^{-1}}$. Equation 1.51 can be used to calculate the heat flow for any part of the Earth's crust in which both the temperature gradient and the thermal conductivity of rocks can be measured.

ITQ 49

Figure 1.87 shows a hypothetical and schematic segmented temperature profile through a portion of continental crust. Thermal conductivities of the rocks in each layer are given.

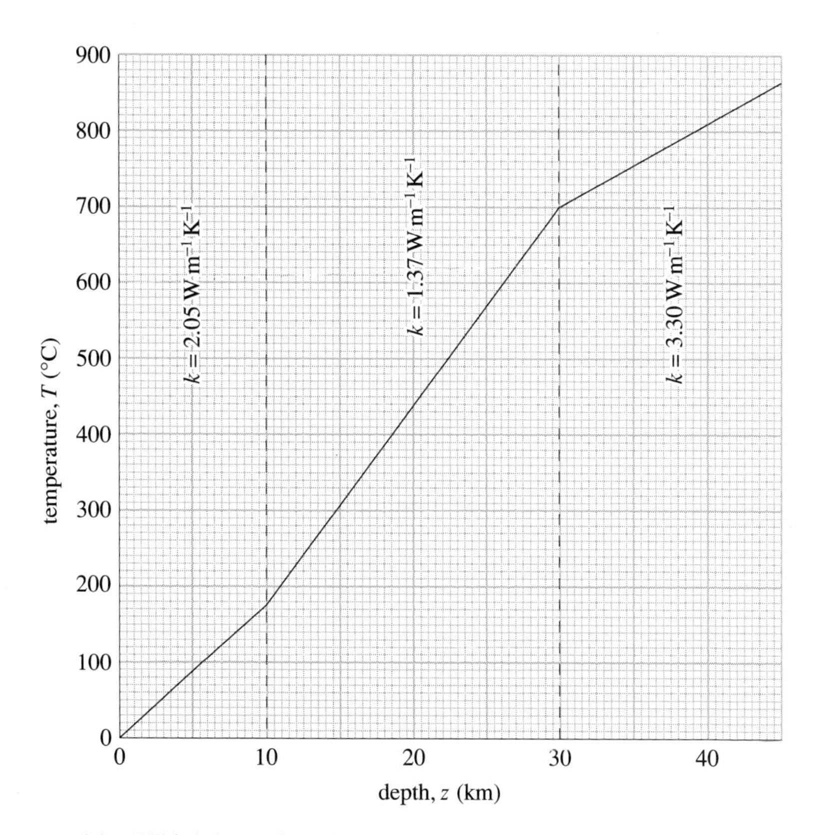

Figure 1.87 A plot of temperature (T) against depth (z) within the Earth through three layers of contrasting thermal conductivity (k).

(a) Which layer has the steepest (highest) temperature gradient?

(b) Is the heat flow through the bottom layer greater or less than that through the top layer?

You should now listen to audiovision sequence AV 04, 'Measurement of Heat Flow', which explains the methods by which temperature gradient, thermal conductivity and hence heat flow may be measured. (This lasts about 15 minutes.)

The next step is to examine how the heat-flow data may be interpreted. As the interpretation of heat-flow results is intimately bound up with plate-tectonic processes, however, we first need to describe briefly what those processes are. We shall be looking at plate tectonics in much more detail in Block 2; so what follows here in Section 1.12.3 is the briefest of summaries.

1.12.3 PLATE TECTONICS: A BRIEF SUMMARY

A diagram illustrating the basic concepts of plate tectonics is shown in Figure 1.88; the locations and types of boundary between the major lithospheric plates are shown in Figure 1.89. The basic concept of plate tectonics is that the rigid outer shell of the Earth, or lithosphere, is divided by a network of boundaries into separate blocks, which are termed lithospheric plates. The boundaries between plates are of three types:

(a) constructive plate margins, or actively spreading oceanic ridges, where two plates are moving apart, so permitting the upwelling and solidification of magma to form new lithosphere;

(b) destructive plate margins, which occur at ocean trenches, and mark places where two plates are converging so that one plate sinks below the other and is eventually reabsorbed into the mantle, or 'destroyed'; and

(c) conservative plate margins — faults where two plates slide past each other, so that lithosphere is neither created nor destroyed. In this case, the direction of relative motion of the two plates is parallel to the fault. Conservative plate margins occur within both oceanic and continental lithosphere, but the commonest conservative plate margins are oceanic transform faults.

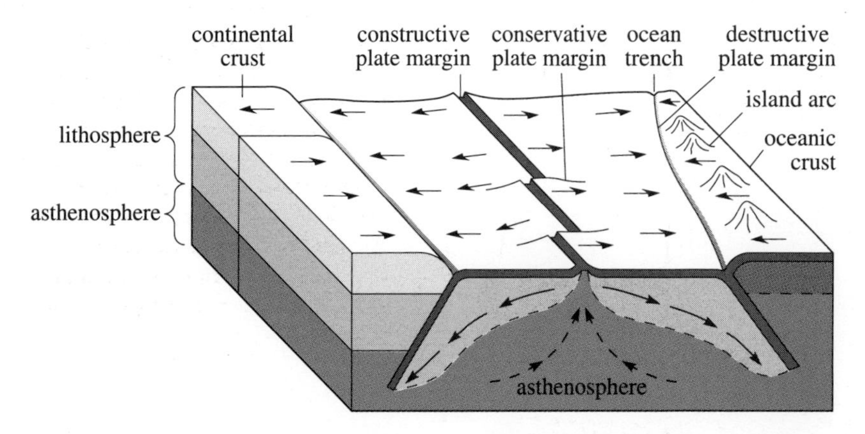

Figure 1.88 Diagram showing the basic concept of plate tectonics. Plates of rigid lithosphere (which include the oceanic or continental crusts and uppermost mantle) overlie a layer of relatively low strength called the asthenosphere. Mantle material rises below constructive plate margins (oceanic ridges) and plate material descends into the mantle at destructive plate margins (ocean trenches). At conservative plate margins, plates slide past each other, being neither created nor destroyed.

The different types of plate margin defined above were originally distinguished on the basis of their seismicity. Constructive and conservative margins are characterized by narrow zones of shallow-focus earthquakes. Destructive margins have shallow-, intermediate- and deep-focus earthquakes, which define a seismic zone with the earthquake foci lying along an inclined plane, or Wadati–Benioff zone, which reaches the surface at an ocean trench (see Section 1.5.2). Plate margins are illustrated further in Figure 1.90.

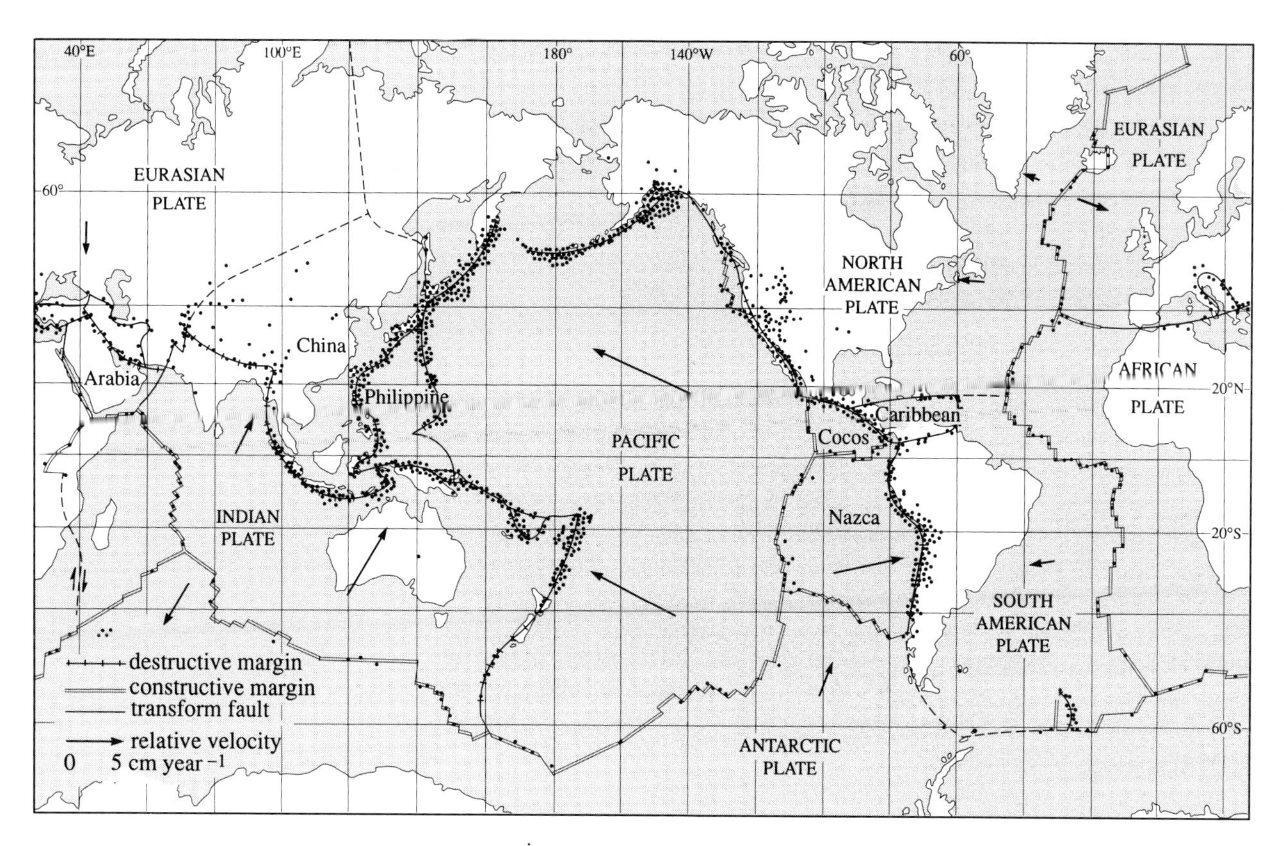

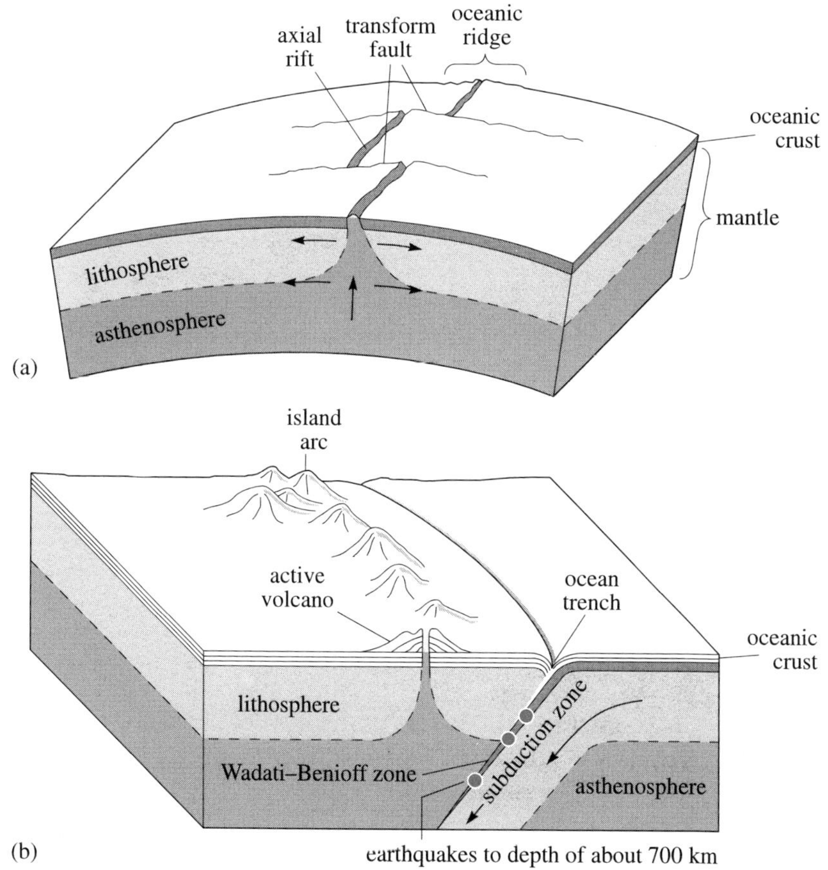

Figure 1.89 *(above)* The world pattern of plates, oceanic ridges, ocean trenches and transform faults in relation to deep-focus earthquake epicentres (indicated by dots). Tentative positions of plate margins are indicated by dashed lines. There are seven major plates — African, North American, South American, Antarctic, Eurasian, Indian and Pacific — and six minor — Arabia, China, Cocos, Nazca, Philippine and Caribbean; there are also several smaller ones. The length and direction of the arrows indicate the relative velocities of the plates at present. The African Plate is assumed to be stationary. (The arrow shown in the key corresponds to a relative velocity of 5 cm a^{-1}.)

Figure 1.90 *(left)* Geological features and processes characteristic of (a) a constructive plate margin and (b) a destructive plate margin.

One other aspect of plate tectonics, useful to know at this stage, is that as oceanic lithosphere spreads away from oceanic ridges it increases in age. So if spreading rates are known, as they often are, it is possible to deduce the age of the lithosphere at any distance from a ridge. As a result, when considering heat flow through the oceanic lithosphere, it is usual, as in the next Section, to plot heat-flow values not against distance from a ridge but against the age of the lithosphere.

As we said at the outset, this is the simplest of plate-tectonic summaries, presented here for the purpose of making sense of the heat-flow results that follow. If you are familiar with basic plate-tectonic processes already, you will probably have learned nothing new. However, if you are not, and you find this summary unsatisfactory, remember that we shall be examining plate tectonics in much more detail in Block 2.

1.12.4 OCEANIC HEAT FLOW

Mean values for heat flow through the floors of the major oceans are summarized in Table 1.13 and in Figure 1.91. All oceans show a similar distribution of heat flow. For lithosphere up to about 35 Ma old, the heat flow is rather variable and generally high, particularly near spreading oceanic ridges. For lithosphere older than about 65 Ma, the mean heat flow, $\overline{q}$, is relatively constant. The increase of heat flow values close to oceanic ridges is confirmed by a more detailed plot of heat flow against age of the lithosphere (Figure 1.92). The lowest heat flow values of all come from ocean trenches (Figure 1.93), where most values are similar to those of ocean basins, but many values are less than 30 mW m^{-2}. In summary, therefore, heat flow through the oceanic lithosphere decreases generally with increasing age.

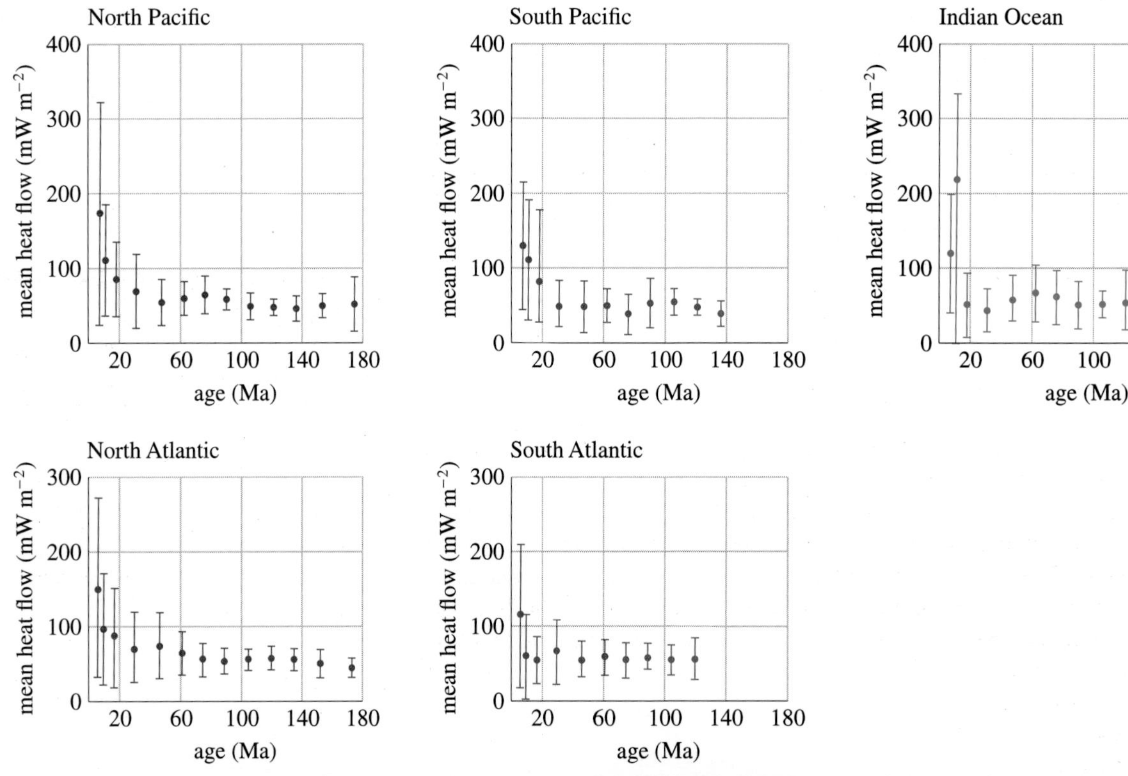

Figure 1.91 Mean heat flow ($\overline{q}$, dots) and standard deviation (σ, vertical lines) plotted against the age of the oceanic lithosphere for five major oceanic basins.

S11

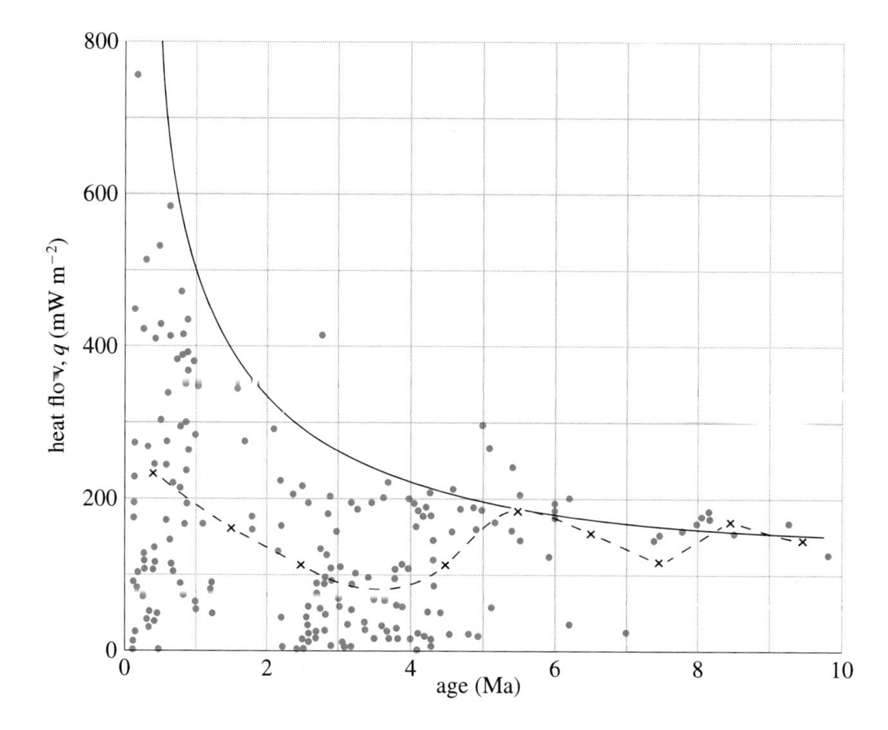

Figure 1.92 Heat flow values plotted against age of oceanic lithosphere near the Galapagos spreading centre. Only values for locations where the oceanic lithosphere has a well-defined age are plotted. Circles represent heat flow values. Crosses represent means for 1 Ma intervals, and are connected by a dashed curve. The solid curve is the heat flow expected from theoretical models of the formation of oceanic lithosphere which are discussed in the text.

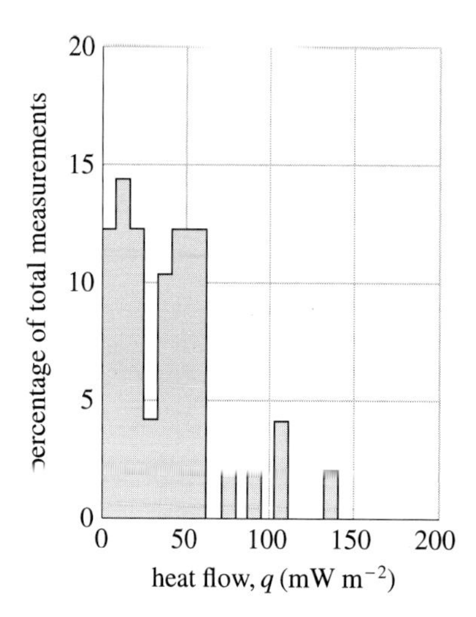

Figure 1.93 Histogram of heat-flow values from sites in Pacific Ocean trenches.

Table 1.13 Mean values of oceanic heat flow corresponding to different oceanic lithosphere of known age. N is the number of measurements, $\bar{q}$ is the mean heat flow and σ is the standard deviation

Age (Ma)	0–4	4–9	9–20	20–35	35–52	52–65	65–80	80–95	95–110	110–125	125–140	140–160	>160
N	506	444	470	304	252	265	277	204	193	162	178	66	26
$\bar{q}$ (mW m^{-2})	149	117	71	60	57	62	57	54	54	55	49	49	50
σ (mW m^{-2})	126	180	52	38	35	31	28	20	17	21	16	16	25

As newly formed lithosphere moves away from an oceanic ridge, it gradually cools, contracts and increases in density. Its surface therefore subsides, and the overlying ocean water gets deeper. In fact, the depth to the ocean floor increases from 2–3 km at oceanic ridges to 5–6 km for abyssal plains. The question then arises: is the cooling and contraction of the lithosphere *the* major control on the depth of the ocean floor, because, if it is, we might expect there to be a simple relationship between depth and the age of the lithosphere.

ITQ 50

(a) Figure 1.94 is a graph of ocean floor depth against the *square root* of the age of oceanic lithosphere for the North Pacific and North Atlantic oceans. Is the correlation between depth and age a positive or negative one in the two oceans? (Note that depth increases downwards on the left-hand axis.)

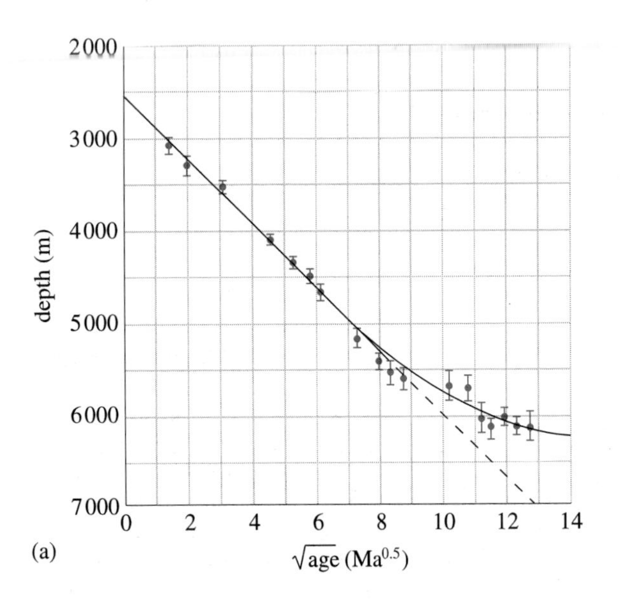

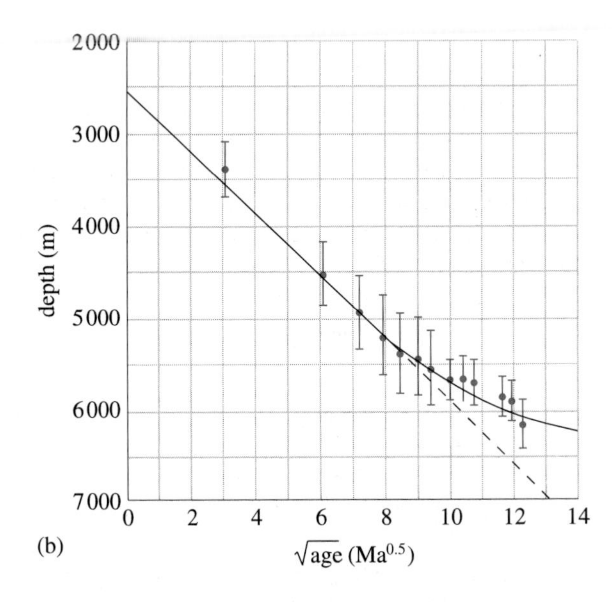

Figure 1.94 Mean ocean depths (and standard deviations) plotted against the square root of the age of the lithosphere for (a) the North Pacific and (b) the North Atlantic.

(b) At what age does the relationship depart from the linear correlation?

(c) Does this departure imply that older oceanic lithosphere is warmer or cooler than would be predicted from the simple linear relationship?

For all the oceans taken together, the relationship between oceanic depth (d, metres) and lithospheric age (t, Ma) up to about 70 Ma is

$$d = 2500 + 350\sqrt{t}\,. \qquad \text{(Equation 1.52)}$$

Geophysicists have also been able to devise a more complicated formula for lithosphere older than 70 Ma, taking into account the extra warming. When the two formulas are plotted together they show the depth to the ocean floor to be as in Figure 1.95, which appears to explain the oceanic depth in most regions quite adequately.

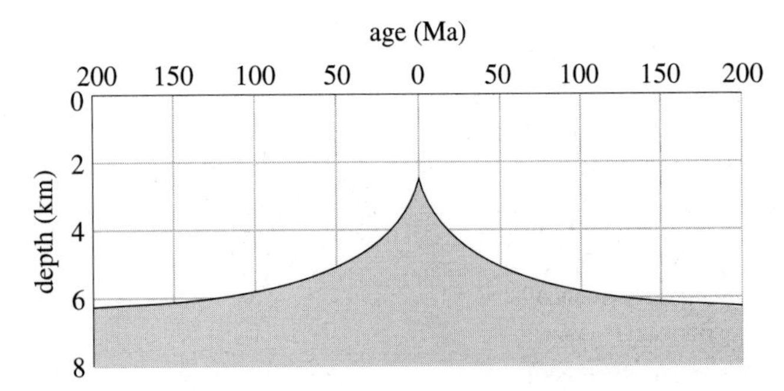

Figure 1.95 The relationship between oceanic depth and lithospheric age as plotted from Equation 1.52 and the more complicated formula beyond about 70 Ma.

But what of the base of the lithosphere, and hence its thickness, as it moves away from an oceanic ridge? Here, of course, no direct observations are possible, so we must rely on models, of which we shall consider two.

(a) Plate model

The simplest model was developed in the wake of the discovery of plate tectonics, and is illustrated in Figure 1.96a. The basis of this model is that lithosphere with a *constant* plate thickness is produced at the oceanic ridge and that the temperature below the lithospheric plate corresponds to the temperature of formation. Because the model postulates a plate of uniform thickness, it is referred to as the **plate model.**

(b) Boundary-layer model

A second, more sophisticated, model assumes that the oceanic lithosphere does not have a constant thickness, but that after formation it cools and *thickens* as it moves away from the spreading centre (oceanic ridge). This model refers to a layer, the lithosphere, that forms the upper boundary of the mantle, and which cools and thickens as it moves away from the spreading ridge. This is therefore referred to as the **boundary-layer model** (see Figure 1.96b).

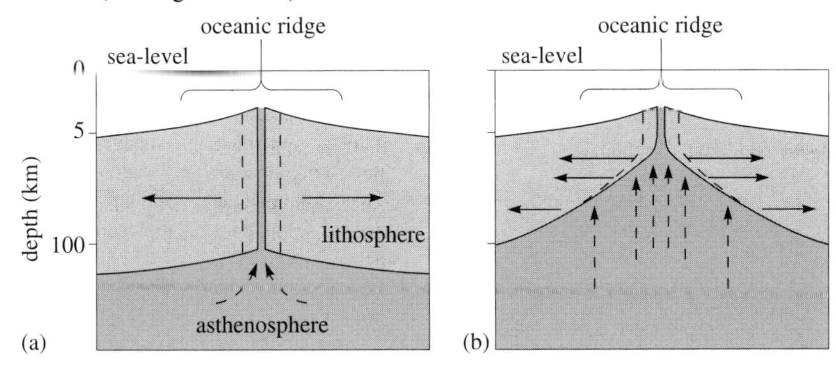

Figure 1.96 Schematic sections through oceanic lithosphere formed at an oceanic ridge according to two models. Note that, in order to indicate the ocean-floor topography, the scale from 0–5 km is larger than that from 5–100 km depth. The dashed lines in both diagrams indicate new lithospheric material added in a given time interval.
(a) Plate model, in which oceanic lithospheric thickness remains constant as the lithosphere moves away from the ridge.
(b) Boundary-layer model, in which oceanic lithosphere thickens as it moves away from the ridge.

In both models, the lithosphere cools by conduction and contracts as it moves away from the spreading ridge. Consequently, its density increases and it subsides into the mantle. The result of these processes is that the top of the oceanic lithosphere subsides to a uniform depth, corresponding to the depth of the abyssal plain*. The two models also both predict that heat flow decreases away from the oceanic ridge — i.e. with increasing age of the oceanic lithosphere. The predicted heat flow for the two models is shown plotted against lithospheric age in Figure 1.97. The average measured heat flow through oceanic lithosphere of known age is also shown.

❑ Where is the agreement between theory and practice least good?

■ The predicted heat flow near the ridge crest (younger lithosphere) is much higher than the measured heat flow. For older lithosphere, particularly that older than 50 Ma, the agreement between both models and the measured values is very good. So we cannot select either model on the basis of Figure 1.97. Although the models are not, of course, mutually exclusive, the boundary-layer model is more plausible and is also consistent with one version of the seismic evidence.

* You may now be puzzled, asking yourself how, if the oceanic lithosphere cools and contracts as it spreads away from a ridge, it is possible or reasonable to envisage a lithosphere which maintains a constant thickness with age (as in the plate model) or which even increases in thickness with age (as in the boundary-layer model). The answer is that the contraction is too small to be relevant to the models, which depend on other, more significant phenomena. Observations (see Figure 1.94) show that during the first 60 Ma contraction causes the surface of the oceanic lithosphere to subside by about 2.7 km. Bearing in mind that the average depth of the oceans is 4.5 km, this amount of contraction-induced subsidence is highly significant as far as ocean depth is concerned; indeed, as we have seen, the subsidence is the major control on the depth of the ocean floor. However, it is much less significant as far as the thickness of the lithosphere is concerned; for a lithosphere which is, say, 125 km thick (see Figure 1.96) a contraction of 2.7 km is only about 2%. In the context of the plate model, this can be ignored; it's not large enough seriously to affect the assumption of constant plate thickness. The boundary-layer model, on the other hand, assumes that the lithosphere increases in thickness with age as a result of the gradual cooling and solidification of the upper part of the underlying asthenosphere, and that this increase is far in excess of any decrease in thickness caused by contraction due to cooling.

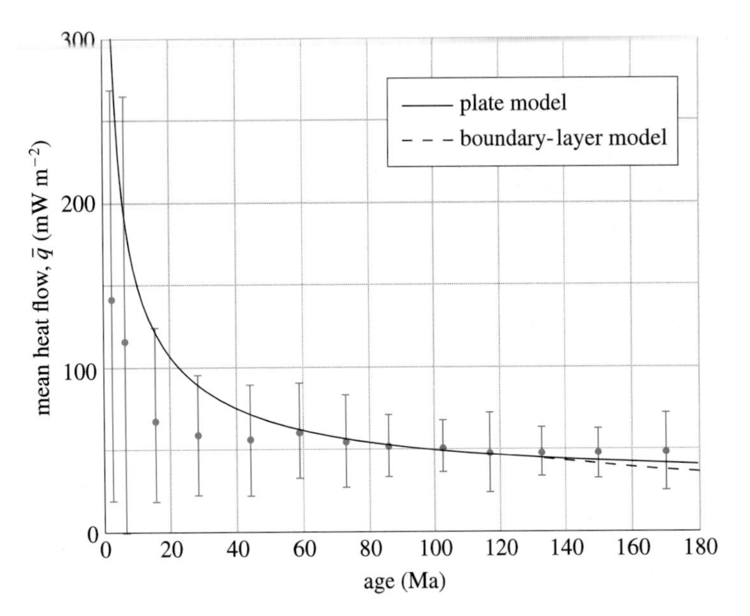

Figure 1.97 Mean heat flows averaged for all oceans (filled circles) and standard deviations as a function of age. Also shown are the expected heat flows (curves) from the plate and boundary-layer models. The curves are coincident for ages of less than about 130 Ma.

You will recall that in Section 1.7.5 we saw that there were two versions of the seismic evidence for the thickness of the oceanic lithosphere, one taking no account of the anisotropy of the mantle and one making allowances for it. In Figure 1.98, we show how both the plate model and the boundary-layer model compare with the seismic data that take no account of anisotropy. This comparison leaves no doubt at all that the plate model is the least satisfactory. However, it also leaves us with another problem.

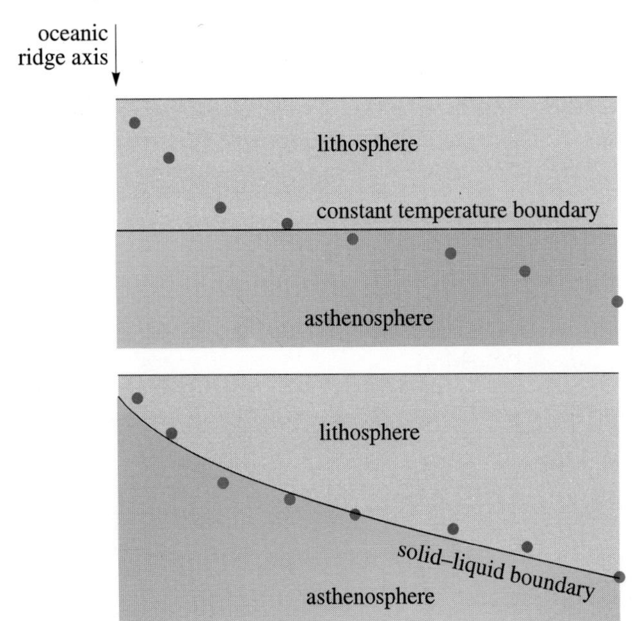

Figure 1.98 A comparison of the plate and boundary-layer models with seismic data (circles) for the thickness of the lithosphere. (These are the seismic data that do not take the anisotropy of the mantle into account.) Note that the scale of these diagrams is such that the subsidence of the surface of the lithosphere does not show up.

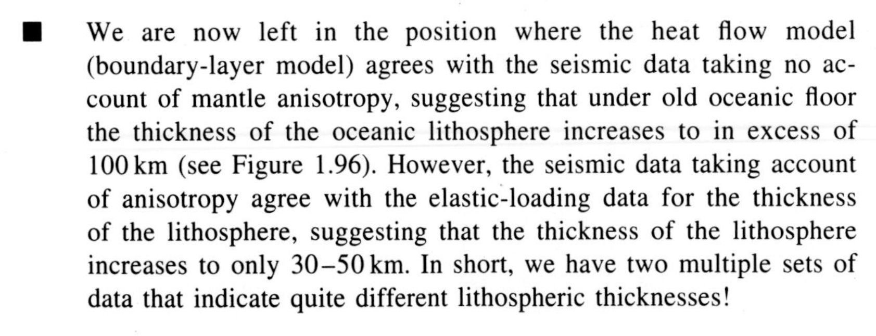

❑ Have you spotted what it is? (Think of the thickness of the lithosphere, with reference to Section 1.7.5.)

■ We are now left in the position where the heat flow model (boundary-layer model) agrees with the seismic data taking no account of mantle anisotropy, suggesting that under old oceanic floor the thickness of the oceanic lithosphere increases to in excess of 100 km (see Figure 1.96). However, the seismic data taking account of anisotropy agree with the elastic-loading data for the thickness of the lithosphere, suggesting that the thickness of the lithosphere increases to only 30–50 km. In short, we have two multiple sets of data that indicate quite different lithospheric thicknesses!

We shall look anew at the question of the thickness of the lithosphere in Section 1.13. In the mean time, we return to the specific question of heat flow.

The discrepancy between both the plate and boundary-layer models and the measured heat flow in young oceanic lithosphere (<50 Ma) is explained like this: the models are based on the assumption that heat is lost from the oceanic lithosphere solely by conduction. The measured heat flow is much *lower* than it would be if this were the only mechanism of heat transfer. The implication is that there is another mechanism of heat transfer, a more efficient one, which removes the heat *faster* than is possible by conduction. The formation of oceanic lithosphere at ridge crests involves the contact between hot rocks and cold seawater. The rocks crack and fracture as they cool, allowing seawater to penetrate them, down to depths of at least a few kilometres. The interaction between hot rocks and cold seawater has the effect of setting up a very powerful system of convection currents of water within the oceanic lithosphere which act to remove heat from the rocks much more quickly than could be achieved by conduction alone. That is why the heat flow measured in oceanic lithosphere near ridge crests is *lower* than predicted. Only part of the heat is being lost by conduction through the rocks; the rest is escaping as heated water. Dramatic confirmation of these convecting systems has come from underwater exploration near ridge crests, where in some places hot plumes erupt from the ocean floor at temperatures in excess of 620 K. (At the very high pressures on the ocean floor, this temperature is below the boiling temperature of water.) The circulation and heating of seawater in hot rocks at oceanic ridges is known as **hydrothermal circulation.**

Thermal models of cooling oceanic lithosphere less than 120 Ma old are characterized by a simple relationship between heat flow and age, namely,

$$q \propto \frac{1}{\sqrt{t}} \quad \text{or} \quad q \propto t^{-0.5}.$$

Expressed in words, this means that heat flow (q) is inversely proportional to the square root of the age (t) of the oceanic lithosphere. Writing the relationship as an equation, we have

$$q = Ct^{-0.5} \qquad \text{(Equation 1.53)}$$

where C is a constant. By taking logarithms of Equation 1.53, we can convert it into a more readily usable form, namely,

$$\log q = \log C + \log t^{-0.5} \quad \text{or} \quad \log q = \log C - 0.5 \log t. \qquad \text{(Equation 1.54)}$$

We now have a linear relationship between $\log q$ and $\log t$, with a gradient of -0.5.

Figure 1.99 is such a plot of $\log q$ against $\log t$ for the North Pacific. It shows a clear negative correlation, with a slope of -0.5, for oceanic lithosphere from about 3 Ma to about 120 Ma in age. For ages over 120 Ma, there is an increasing discrepancy between the model and the $q \propto t^{-0.5}$ relationship. In the range 3–120 Ma, the relationship between q and t shown in Figure 1.99 can be represented by the empirical relationship

$$q = \frac{473}{t^{0.5}} \text{ mW m}^{-2}. \qquad \text{(Equation 1.55)}$$

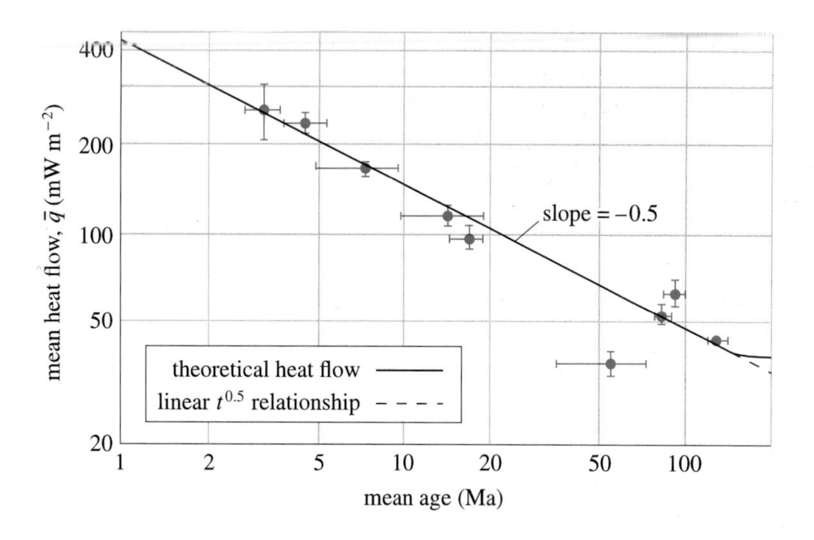

Figure 1.99 Graph of mean heat flow values against mean age for the North Pacific. Note that both scales are logarithmic. The theoretical curve (i.e. that derived from the models used to obtain Equations 1.53 and 1.54) has a slope of −0.5 up to about 120 Ma.

ITQ 51

Figure 1.99 shows a relative scarcity of heat flow measurements on oceanic lithosphere of age 30–40 Ma. What heat flow would you predict for oceanic lithosphere 36 Ma in age?

Finally, what of the special circumstances at subduction zones, where oceanic lithosphere descends into the mantle? In Figure 1.93, you saw that low values of oceanic heat flow are found in ocean trenches. Figure 1.100 illustrates very elegantly the contrast in heat flow between ocean ridges, ocean basins and ocean trenches, as well as showing the depth of these features below sea-level. Note that there is a sharp increase in the heat flow, and an abrupt decrease of depth, on the Japan side of the trench.

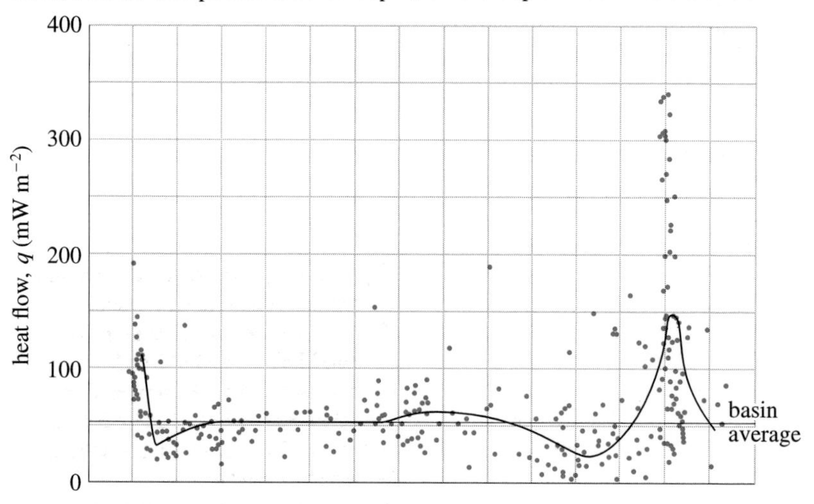

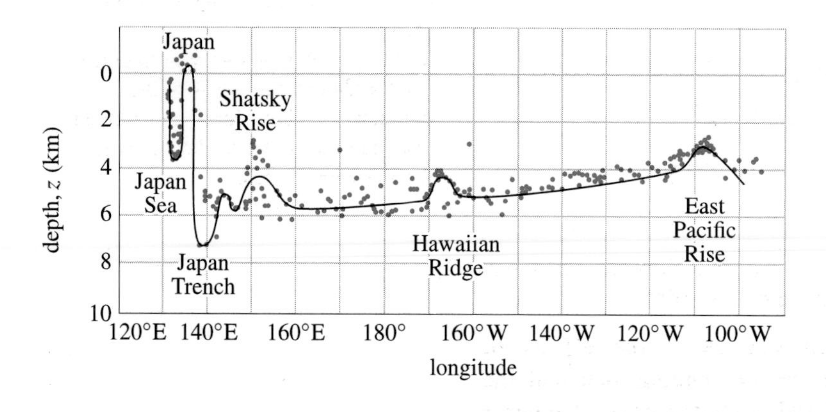

Figure 1.100 Profile of heat flow (top) and site elevation (bottom) from the Sea of Japan, across the Pacific, to the East Pacific Rise. Oceanic lithosphere is being formed at the East Pacific Rise and is being subducted below Japan. The solid lines are the approximate mean values drawn by eye through the points.

In brief, as the oceanic plate is subducted, it is heated partly by friction along its upper surface and partly by conduction from the hot mantle into which it is descending. There is also dehydration of the minerals formed by metamorphism of oceanic lithosphere (as we shall see in Block 4), and as this is an endothermic process (see Section 1.4), it has a cooling effect on the upper part of the slab. Because rocks have very low thermal conductivity, the main bulk of the slab will take a very long time to heat up by conduction and reach thermal equilibrium with the mantle into which it is descending. It is relatively easy to calculate the time involved, and it amounts to something of the order of several hundred Ma. When you compare that time period with the period of about 10 Ma that it takes for a point on the lithospheric slab to reach a depth of 500 km (assuming a reasonable subduction rate of $5\,\text{cm}\,\text{a}^{-1}$), it is clear that *the major thermal effect at a subduction zone is the descent of a relatively cold oceanic plate with low thermal conductivity into hotter mantle.* This effect has been modelled for varying rates of subduction, varying amounts of frictional heating and with different estimates of cooling by dehydration for different times following the start of subduction. Figure 1.101 illustrates one such model, describing the descent of an oceanic plate below a young island arc. The crucial feature in this and all such models is the way in which the temperature contours descend into the mantle more-or-less parallel to the subducted slab. Also shown is the surface heat flow above the subduction zone for 20 Ma intervals after initiation of subduction. As we have already seen (Figure 1.93), heat flow in the trench itself is very low.

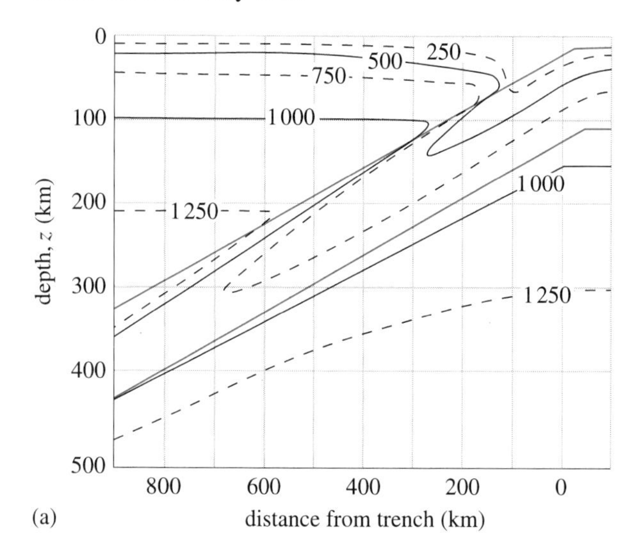

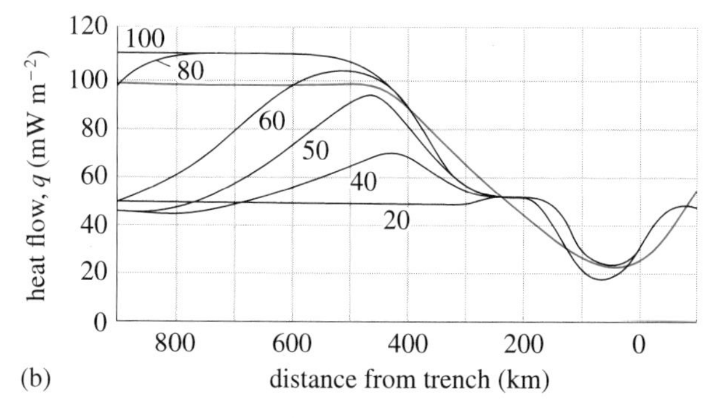

Figure 1.101 (a) Model temperature distribution in a descending slab, 100 Ma after the start of subduction. The solid lines are 500 °C and 1 000 °C isotherms. The broken lines show the 250 °C, 750 °C and 1 250 °C isotherms. The coloured lines outline the descending plate.
(b) Surface heat flow resulting from descent of slab according to model shown
in (a). Heat flow is shown at intervals from 20 Ma to 100 Ma after the initiation of subduction. The coloured line shows the observed heat flow for the Japanese islands.

❑ But how would you explain the increase in heat flow on the Japan side of the trench, so clearly shown in Figure 1.100?

■ The answer is quite simple when you recall the effects of subduction on the *upper* parts of the descending slab and the immediately overlying mantle. The basaltic rocks forming the upper part of the descending slab are first metamorphosed, and either melt or release a watery fluid into the overlying mantle. This causes melting in the overlying mantle wedge and the net result is the formation of magmas. Some of these intrude the island arc and some are erupted from volcanoes. The increase in heat flow in island arcs is therefore a result of intrusion and eruption of igneous rocks (as we shall see in more detail in Block 4).

1.12.5 CONTINENTAL HEAT FLOW

As you should have appreciated from AV 04, measuring heat flow on land is more difficult than on the ocean floor. Chiefly for this reason, continental heat flow measurements are both less numerous and less evenly distributed than in oceanic areas. The measurements are also more difficult to interpret, because the continents have had a much longer and more complex geological history than the oceans. The ages of some continental regions go back to 3 800 Ma, whereas oceanic crust and lithosphere are nowhere much older than about 200 Ma. Moreover, age determinations on continents and in oceans tells us fundamentally different things. The age of a piece of oceanic lithosphere represents the time of its formation at a spreading ridge. The 'age' of a piece of continental lithosphere often simply indicates the last time that it experienced an episode of magmatism and/or metamorphism. In some cases, that time may also be the time of formation of that particular piece of continental crust, but sometimes it is not.

The average heat flow is plotted against age for four continents in Figure 1.102, and estimates of mean heat flow for areas of different age in all continents are shown in Table 1.14. The mean values range from 41 to 76 mW m^{-2}.

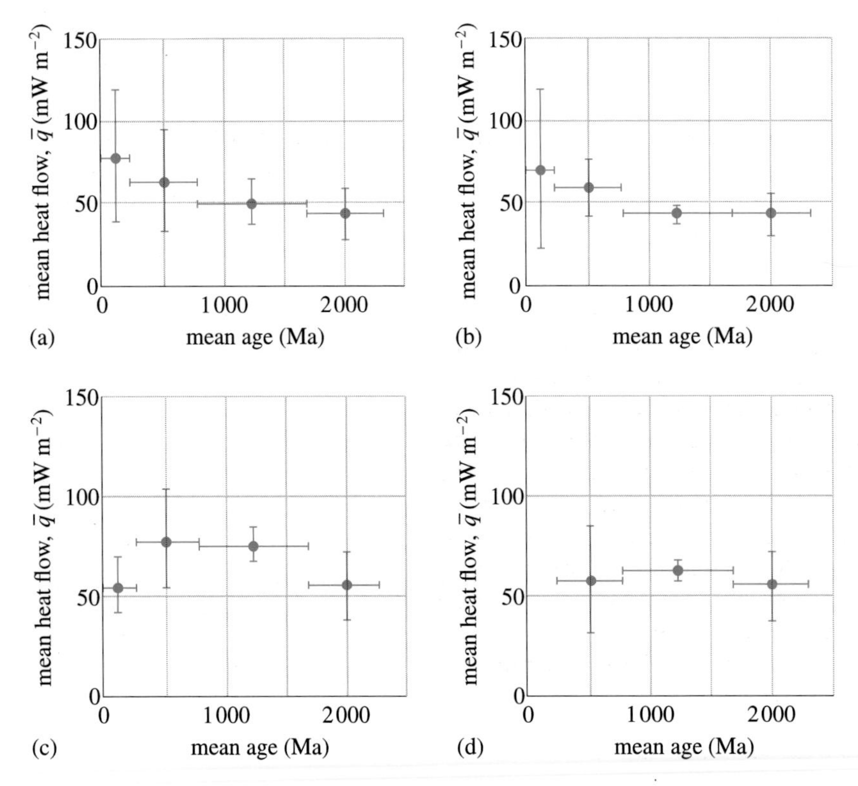

Figure 1.102 Mean heat flow values ($\bar{q}$) and standard deviations (σ) plotted against mean age for (a) North America, (b) Eurasia, (c) Australia and (d) Africa. The plotted points have been arranged for mean ages within four age groups: > 1 700 Ma, 1 700–800 Ma, 800–250 Ma and 250–0 Ma. Note: The standard deviations for heat flow are the vertical bars. The horizontal bars simply express the age ranges given above.

❑ Can you see any correlation in Figure 1.102 between the heat flow and age of provinces? (Look particularly at the graphs for North America and Eurasia (Figures 1.102a and b), which are based on the most data.)

■ The data in Figure 1.102 and Table 1.14 show that there is a clear relationship between heat flow and age in some large-scale geological provinces. This relationship is brought out much more clearly by Figure 1.103, which is a graph showing heat flow as a function of age for all the continents together.

Table 1.14 Mean values of continental heat flow corresponding to continental lithosphere of known age. N is the number of measurements, $\bar{q}$ is the mean heat flow and σ is the standard deviation. Two sets of data are shown calculated for different age ranges. Set (a) are the data plotted in Figure 1.103.

(a)

Age (Ma)	0–250	250–800	800–1 700	> 1 700
N	398	500	138	375
$\bar{q}$ (mW m^{-2})	76	63	50	46
σ (mW m^{-2})	53	21	10	16

(b)

Age (Ma)	0–70	70–230	230–400	400–570	570–1 500	1 500–2 500	> 2 500
N	587	85	514	88	265	78	136
$\bar{q}$ (mW m^{-2})	71	73	61	52	54	51	41
σ (mW m^{-2})	37	29	18	17	20	21	11

ITQ 52

Compare Figure 1.97, which may be taken as reasonably representative of ocean basins generally, with Figure 1.103. You should observe (a) an obvious and important similarity, and (b) a less obvious but equally important difference, between the two. Can you identify them?

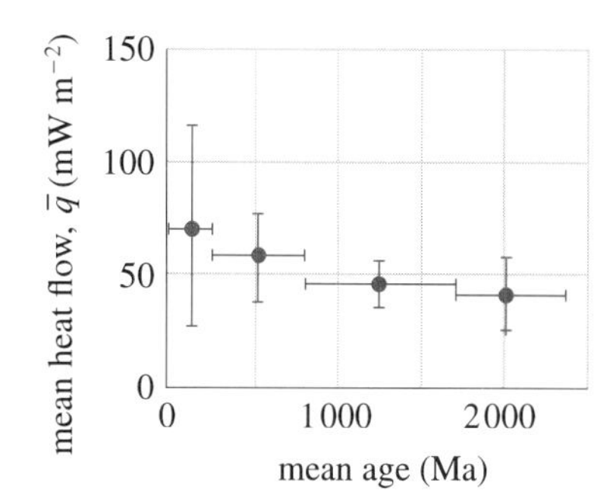

Figure 1.103 Mean heat flow values ($\bar{q}$) and standard deviations (σ) plotted against mean age for all continents. Points are plotted as in Figure 1.102.

The answer to the first part of ITQ 52 should not be a surprise. After all, the time of formation of oceanic lithosphere represents a single massive input of heat, after which the lithosphere is left to cool down as it moves away from the site of formation (the oceanic ridge). The latest episode of regional metamorphism/magmatism in a segment of continental lithosphere likewise represents a single (albeit more prolonged) massive input of heat, after which the lithosphere is, so to speak, 'left alone', and is thereafter unaffected by subsequent disturbances.

The answer to the second part is less obvious. The thermal conductivities of oceanic and continental crustal rocks are generally similar (of the order of a few W m^{-1} K^{-1}). The heat flows are also generally comparable, so the temperature gradients cannot be very different either (Equation 1.51). For oceanic areas, the patterns of heat flow can be satisfactorily modelled by assuming simply that heat is lost entirely by conduction. Such models work well, except for the immediate vicinity of ocean ridges where, as explained in Section 1.12.4, we know that hydrothermal convection is important, and for lithosphere older than about 70 Ma where some additional input of heat is required. However, the fact that heat flow declines so slowly with increasing age in continental lithosphere must surely imply that there is an additional source of heat in continental lithosphere, which is not available in the ocean.

❑ Can you suggest what that additional source of heat might be?

■ You should recall from ITQ 47 that the heat-producing isotopes — ^{40}K, ^{232}Th, ^{235}U and ^{238}U — are enriched in continental crust relative to oceanic crust, and that these would therefore provide an additional source of heat. However, given the wide variety of rock types in the upper crust, and hence the uneven distribution of heat-producing isotopes, measured continental heat flow values vary widely even within relatively small areas (a few 100 km^2). These variations began to be understood in the late 1960s, when rocks sampled near sites of heat flow measurements were analysed for their content of radioactive elements (K, Th and U) and the resulting heat generation values calculated, as we outlined in Section 1.12.1.

Figure 1.104 is a graph of heat flow against heat generation for part of the northeastern USA. The straight line is a graphical expression of the linear relationship

$$q = bA + q^* \quad \text{(Equation 1.56)}$$

where q is the measured heat flow ($mW\,m^{-2}$), A is the heat generation ($\mu W\,m^{-3}$) and b and q^* are constants, of which b has units of depth (m) and q^* has units of heat flow.

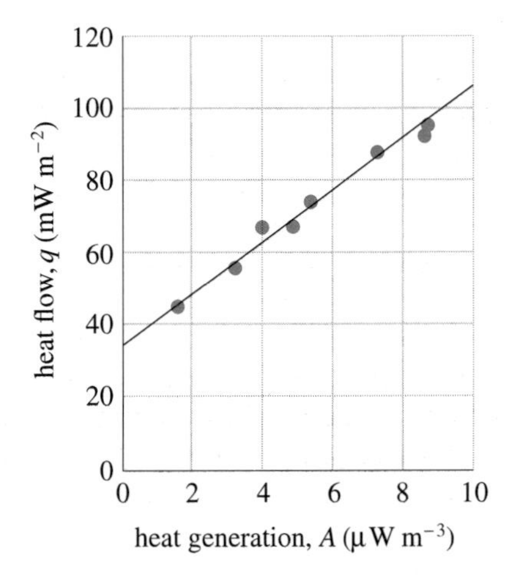

Figure 1.104 Plot of heat flow (q) against heat generation (A) for part of the northeastern USA.

ITQ 53

(a) Which quantity in Equation 1.56 is represented by the gradient of Figure 1.104?

(b) Which quantity in Equation 1.56 is represented by the intercept on the heat flow axis on Figure 1.104?

(c) Determine both of the quantities in your answers to (a) and (b) from Figure 1.104.

A simple model that explains the linear relationship in Figure 1.104 and Equation 1.56 is that the heat generation due to radioactive decay comes from a layer of uniform thickness (b), and that there is no radioactive heat generation from rocks below this depth. This might seem intuitively unlikely, but it is the *simplest* model that explains the relationships summarized in Equation 1.56. In this simple model, b represents the thickness of an upper crustal layer with a uniform concentration of the heat-producing elements, K, Th and U, whereas q^* represents the heat flow coming from the underlying lower crust and mantle. The quantity q^* is referred to as the **reduced heat flow**.

Using this approach, it is possible to identify distinctive areas of continental crust, each characterized by a linear relationship between q and A, but with different slopes (b) and intercepts (q^*). Such areas, which are known as **heat-flow provinces**, have been identified in many parts of North America, Australia and Europe (including England and Wales). They are usually areas of regional extent (e.g. the Canadian Shield) which have experienced a common history.

The simple model works very well, but it is a bit unrealistic, because there is unlikely to be an abrupt cut-off of heat-producing isotopes at some specific depth that varies from place to place (as indicated by the varying slopes obtained for the q–A graphs in different provinces); and even if there were, it is unlikely that the isotopes would have uniform concentration within it. In fact, it has been shown that the linear relationship portrayed in Figure 1.104 and Equation 1.56 can also be satisfied by a model in which the concentration of *heat-producing isotopes decreases exponentially* with depth throughout the crust. This idea is illustrated in Figure 1.105 in schematic form, along with the simple model. Most models of the lower crust predict a lower concentration of heat-producing isotopes such as ^{40}K, as compared with the upper crust. It is not clear, however, whether the actual distribution of such isotopes through a given section of crust approximates more closely to the simple model (the black line in Figure 1.105) or the exponential model (the coloured line in Figure 1.105).

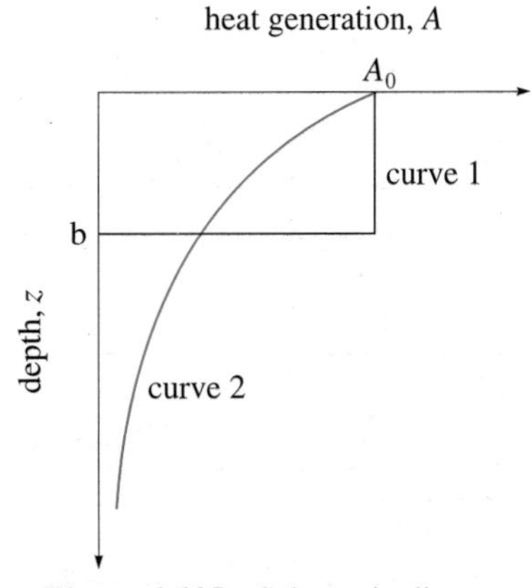

Figure 1.105 Schematic diagram of two different heat-generation distributions consistent with Equation 1.56. In the simple model (curve 1), heat generation (A) is constant at A_0 down to depth b and zero thereafter. In the exponential model (curve 2), heat generation varies exponentially with depth (z), being A_0 at the surface.

Whatever the explanation of the relationship in Figure 1.105 and Equation 1.56 may be (that is, regardless of whether the vertical distribution of heat-producing elements approximates the exponential curve or the rectangle in Figure 1.105, or neither), the relationship characterizes many areas of continental crust. The reduced heat flow, q^*, represents — at least in part — the amount of heat entering the upper crust from the lower crust and mantle. By studying q^*, we should be able to 'see through' the upper crust and find out about the pattern of lower crustal and mantle heat flow.

The relationships in Equation 1.56 and Figure 1.105 can be explained in two ways:

(i) either q^* is everywhere constant and q depends only on variations in the quantity bA;

(ii) or q depends on variations in both bA and q^*.

We have already seen that the first of these is most unlikely, because of our earlier statement that different heat flow provinces give different slopes (b) and intercepts (q^*) when A is plotted against q as in Figure 1.104. So the second alternative is obviously the one to pursue. The relationship between q and q^* is examined in Figure 1.106, which is a plot of q^* against mean heat flow, $\overline{q}$, for different heat flow provinces in various parts of the world. It shows a strong positive correlation between mean heat flow, $\overline{q}$, and reduced heat flow, q^*. The straight line plot passes through the origin; that is, when $\overline{q} = 0$, $q^* = 0$ (as you might expect). Figure 1.106 provides evidence that a more or less *constant proportion* of the total heat flow is provided by reduced heat flow.

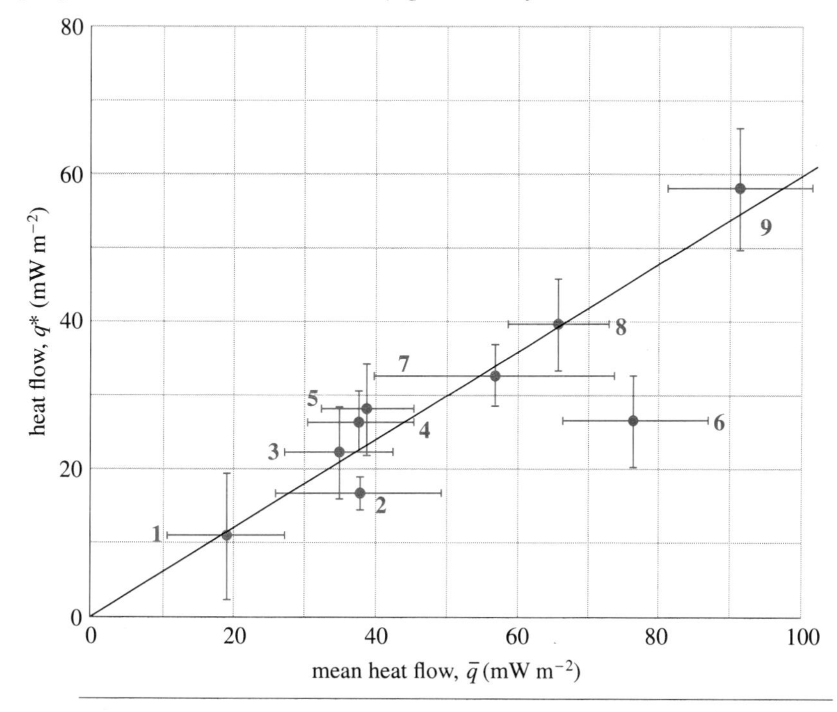

Figure 1.106 Graph of reduced heat flow (q^*) against mean heat flow ($\overline{q}$) for various heat flow provinces. Horizontal bars represent standard deviation about the mean heat flow; vertical bars represent uncertainty of estimate of reduced heat flow. Regions represented: 1, Niger; 2, Baltic Shield; 3, Sierra Nevada; 4, western Australia; 5, Canadian Shield; 6, eastern USA; 7, Zambia; 8, central Australia;
9, Basin and Range Province, USA.

ITQ 54

Use Figure 1.106 to estimate what proportion of the total flow is contributed by the reduced heat flow. In other words, find the slope of the straight line.

The answer to ITQ 54 indicates that, within a continental region, some 60% of the heat flow comes from lower crustal or mantle sources. The remaining 40% comes from upper crustal sources — that is, mainly heat-producing isotopes. We can express the relationship between heat flow and reduced heat flow illustrated in Figure 1.106 by the simple empirical equation

$$q^* \sim 0.6\,\overline{q}. \qquad \text{(Equation 1.57)}$$

Therefore continental heat flow appears to be derived from two components. There is a radiogenic component produced within the upper crust and a 'background' component rising from the mantle and lower crust. These are, of course, in addition to the injection of heat during the latest magmatic/metamorphic event.

Figure 1.107 illustrates the results of this discussion. It shows that continental lithosphere up to about 1 500 Ma old has a heat flow made up of three components:

(1) the slowly decaying radiogenic component of the upper crust (area I on Figure 1.107);

(2) the more rapidly decaying 'heat of formation' (area II on Figure 1.107);

(3) the constant deep background rising from the lower crust and underlying mantle (area III on Figure 1.107).

In terms of our heat flow symbols, q and q^*, the mean heat flow is $\overline{q}$, made up of all three components (where present) on Figure 1.107. The reduced heat flow is q^*, which is made up of components II and III (where both are present). Clearly, in continental lithosphere older than 1 500 Ma, the 'heat of formation' (component II on Figure 1.107) no longer has any effective contribution, and so the heat flow is due entirely to a combination of radiogenic heating and the *deep* background rising from the mantle below the lithosphere. Heat flow in younger lithosphere is made up of all three components.

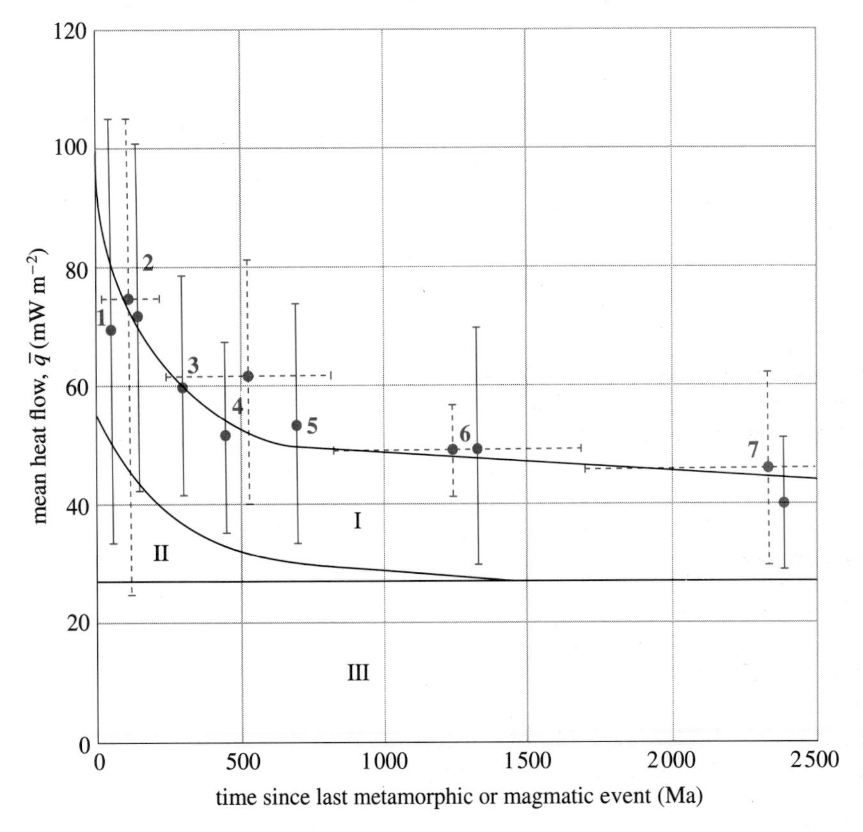

Figure 1.107 Decrease of continental heat flow with age and its three principal components: component I is radiogenic heat from the crust, component II is heat from a transient thermal perturbation associated with metamorphism and magmatism, and component III is background heat flow from deeper sources. The upper line is a visually fitted curve through the circles with the solid error bars. The circles with the dashed error bars are plotted from the data set (a) in Table 1.14 and the circles with the solid error bars are plotted from data set (b). 1 denotes Cenozoic; 2, Mesozoic; 3, late Palaeozoic; 4, early Palaeozoic; 5, late Precambrian; 6, middle Precambrian; and 7, early Precambrian. Points are plotted at the mean age of the respective age ranges. The vertical bars represent the standard deviation of mean heat flow; the horizontal bars represent the ranges over which the measurements are averaged. Note that component I decreases slowly from 0 to 2 500 Ma; component II decreases from 0 to 1 500 Ma; component III is shown as constant over the time.

1.12.6 GLOBAL HEAT FLOW

There are many thousands of heat flow measurements currently available. Unfortunately, however, these are distributed very unevenly over the Earth, which makes it very difficult to produce a map showing how heat flow varies over the whole surface of the planet.

There are two different possible ways of solving the problem of unevenly distributed data. Firstly, and ideally, heat flow measurements could be made in areas where measurements are sparse, but this will obviously take many years to do! However, an alternative ingenious solution has been devised as follows. The surface of the Earth was divided into 5° latitude × 5° longitude regions. Where data were available, the mean heat flow was calculated for each region. Where heat flow data were absent, as in 68% of regions, geological maps were used to determine the proportion of each region which had a given age, and hence calculate the heat flow. By using the relationship shown in Figures 1.97 and 1.103, the heat flow for each region was predicted. This therefore allowed a set of heat flow values to be deduced, which was then used to produce the first global heat flow map in 1975. A revised version of this map is shown as Plate 1.11. This map is based on 7 200 measurements, together with new predicted values for unknown 5° × 5° regions and some adjustment of oceanic-ridge heat flow to account for loss of heat by hydrothermal circulation through the oceanic lithosphere (Section 1.12.4). Before we comment on global heat flow, you should use Plate 1.11 to answer ITQ 55, which will summarize the regional heat flow variations.

ITQ 55

By comparison of Plate 1.11 with Figure 1.89, which shows the locations and types of boundary between the major lithospheric plates, estimate the range of heat flow values for geological provinces listed below.

Geological province	Heat flow range ($mW\,m^{-2}$)
(a) oceanic ridges	
(b) old ocean floor (for example, in the western Pacific Ocean, southeast of Japan)	
(c) ancient continental crust (for example, the Canadian Shield in central Canada)	
(d) island arcs and continental destructive margins	

From maps such as Plate 1.11, geophysicists have estimated that the mean heat flow through continental lithosphere is $60\,mW\,m^{-2}$ and through oceanic lithosphere is $95\,mW\,m^{-2}$; the global mean is $81\,mW\,m^{-2}$. The importance of oceanic crust in determining global heat flow is emphasized by these data; half the global heat loss is from Cenozoic oceanic lithosphere (<65 Ma in age), representing only 31% of the Earth's surface and formed in less than 2% of geological time!

What this emphasizes, of course, is the important role in the Earth of convective heat transfer (see Section 1.3.3). A high proportion of the heat now observed at the Earth's surface is delivered there by convective motions — largely by the rise of magma at oceanic ridges to form new oceanic lithosphere, which subsequently cools. In Section 1.14, therefore, we shall look in more detail at the important phenomenon of convection in the Earth.

SUMMARY OF SECTION 1.12

The rate at which the Earth's surface receives heat from the Sun is about 2 500 times greater than that at which it receives heat from the planet's interior; but because of the low thermal conductivity of rocks, the influence of daily and seasonal variations in solar heat is limited to the upper few metres (exceptionally up to perhaps 20 metres) of the continental crust. Once below that depth, it is possible to measure the flow of heat from the Earth's interior free of interference from external disturbances. (Incidentally, the oceanic crust remains unaffected by solar heat because it is shielded by water.) The average flow of heat from the Earth's interior is about $80\,\mathrm{mW\,m^{-2}}$, although there are large departures from this mean in particular regions.

Calculations based on (a) the (accurately) known decay characteristics of the four heat-producing radioactive isotopes ^{40}K, ^{232}Th, ^{235}U and ^{238}U, and (b) the proportions of those isotopes in representative rock types (granodiorite, gabbro and peridotite, standing proxy for, respectively, the upper continental crust, the oceanic crust and the mantle) suggest that the material of the upper continental crust generates heat at the highest rate (but see SAQ 55). Comparison of the heat generated by a typical chondrite with an estimate of the heat generated in the bulk Earth suggests that, if the Earth is chondritic, heat now being generated by the four heat-producing isotopes could account for at least 80% of the heat flow currently observed, and possibly all of it (given the uncertainly in the estimate of the heat generation in the bulk Earth). However, there are reasons for supposing that, because of the low thermal conductivities of the Earth's rocks and hence the long time it takes for heat to travel through them, current heat flow must also include a component (possibly up to 50%) of heat generated at the time of the Earth's formation. In any event, heat from the decay of long-lived radioactive isotopes is almost certainly the most important source of *current* heat generation.

Temperature gradients in the upper few kilometres of the Earth can be as low as $0.01\,\mathrm{K\,m^{-1}}$ in ancient continental crust, but the average is about $0.03\,\mathrm{K\,m^{-1}}$. The measured temperature gradients at any point in the upper crust (but below the influence of solar variations) can be combined with independent knowledge of the thermal conductivities of the rocks in the vicinity to enable heat flow to be determined.

Heat flow through the oceanic lithosphere generally decreases with increasing age of the lithosphere, with generally high and rather variable values near active oceanic ridges and generally very low values in the vicinity of ocean trenches. As newly formed lithosphere moves away from an oceanic ridge it cools and contracts, and so its surface subsides. In fact, the contraction resulting from cooling is the main control on the depth (d) of the ocean floor, which is related to the age (t) of the corresponding lithosphere by a relationship of the form $d = a + b\sqrt{t}$, where a and b are constants. As a and b can be determined from observations, the age of the lithosphere can be predicted from the depth to its surface, and vice versa, at least up to a lithospheric age of about 70 Ma, at which point the simple relationship breaks down.

Spreading oceanic lithosphere can be modelled either as a layer of constant thickness (the plate model) or as a layer that increases in thickness with age (the boundary-layer model). Unfortunately, the heat-flow data cannot be used to determine which model is the most realistic because both models predict a similar variation of heat flow with the age of the lithosphere. The observed heat flow differs from that predicted by either model at lithospheric ages of less than about 50 Ma (because the conductive heat flow is reduced by hydrothermal circulation in the vicinity of oceanic ridges). In so far as the seismic and elastic-loading data

(Figure 1.66) suggest that the thickness of the oceanic lithosphere increases with age, they would seem to support the boundary-layer model. There is still a problem, however, because the boundary-layer model predicts that the thickness of the lithosphere rises to over 100 km, which conflicts with the elastic-loading and some seismic data (see Section 1.13). Up to about 120 Ma, both thermal models of cooling oceanic lithosphere predict that heat flow varies as $1/\sqrt{\text{age of lithosphere}}$. At subduction zones, where a relatively cold slab is descending into a hotter mantle, the temperature contours are depressed into the mantle despite frictional heating of the slab.

Heat flow through continental lithosphere also decreases with the age of the lithosphere, albeit over a longer period and from a lower maximum. This suggests that the continental lithosphere or crust has a source of heat not available to the oceanic lithosphere. This source is evidently the high proportion in the continental crust of heat-producing radioactive isotopes, which also (because of non-uniform distribution) give rise to a wide variation of measured heat-flow values even within local areas. Notwithstanding these variations, however, distinct provinces of the continental crust can be recognized within each of which there is a relationship between heat flow (q) and heat generation (A) of the form $q = bA + q^*$, where b and q^* are constants within each province. q^* can be interpreted as the heat entering the upper crust from the lower crust and mantle. Analysis of the data from many heat-flow provinces shows that, although q^* varies from province to province, it is almost always a constant proportion (0.6) of the mean heat flow in any province. In other words, within any given province, 60% of the heat flow comes from sources in the lower crust and mantle and 40% comes from sources in the upper crust.

The mean heat flow through continental lithosphere is about 60 mW m^{-2} and that through oceanic lithosphere is about 95 mW m^{-2}.

OBJECTIVES FOR SECTION 1.12

When you have completed this Section, you should be able to:

1.1 Recognize and use definitions and applications of each of the terms printed in the text in bold.

1.47 Recognize the differences between sources of heat internal and external to the Earth.

1.48 Summarize the heat-producing role of long-lived radioactive isotopes in the present Earth.

1.49 Explain the basic principles of heat-flow determination in terms of temperature gradient and thermal conductivity.

1.50 Summarize the main features of oceanic heat flow and of thermal models for the spreading oceanic lithosphere.

1.51 Summarize the main features of continental heat flow and of models consistent with the linear relationship between heat flow and heat generation in continental heat-flow provinces.

1.52 Quote global heat-flow averages and relate deviations from the averages to specific types of geological environment.

1.53 Perform simple calculations based on the topics covered in Objectives 1.48–1.52.

Apart from Objective 1.1, to which they all relate, the 11 ITQs in this Section test the Objectives as follows: ITQ 45, Objective 1.47; ITQ 46, Objectives 1.52 and 1.53; ITQ 47, Objectives 1.48 and 1.53; ITQ 48, Objectives 1.48 and 1.53; ITQ 49, Objectives 1.49 and 1.53; ITQ 50,

Objective 1.50; ITQ 51, Objectives 1.50 and 1.53; ITQ 52, Objective 1.51; ITQ 53, Objectives 1.51 and 1.53; ITQ 54, Objectives 1.51 and 1.53; ITQ 55, Objective 1.52.

You should now do the following SAQs, which test other aspects of the Objectives.

SAQs FOR SECTION 1.12

SAQ 53 (*Objectives 1.1, 1.47, 1.48, 1.49, 1.50, 1.51, 1.52 and 1.53*)

State, giving reasons where appropriate, whether each of the following statements is true or false.

(a) The Earth's surface receives much more heat from the Sun in any given period of time than it does from the planet's interior.

(b) Assuming that heat generation is measured on a volume basis (i.e. per cubic metre), the ratio of heat flow to heat generation has the unit of length.

(c) A kilogram of gabbro generates more radiogenic heat in a given time than does a kilogram of granodiorite.

(d) The basic heat-flow equation (Equation 1.51) applies only to heat transfer by conduction.

(e) If the temperature gradient is $27.3\,K\,km^{-1}$ in rocks having a thermal conductivity of $1.50\,W\,m^{-1}\,K^{-1}$, the heat flow is $40.1\,W\,m^{-2}$.

(f) The depth of the ocean floor where the oceanic lithosphere is 49 Ma old is 4 950 m.

(g) The depth of the ocean floor where the oceanic lithosphere is 144 Ma is 6 700 m.

(h) In the boundary-layer model, the thickness of oceanic lithosphere is assumed to be independent of age.

(i) The heat flows predicted by the plate and boundary-layer models agree best with observation in the age range 0–50 Ma.

(j) Where the oceanic lithosphere is 100 Ma old, the predicted heat flow is $47.3\,mW\,m^{-2}$.

(k) Continental heat flow generally decreases with the age of lithosphere, but less rapidly than does oceanic heat flow and from a lower maximum.

(l) If, in a continental heat-flow province, the heat flow is $20\,mW\,m^{-2}$ where the heat generation is $4\,\mu W\,m^{-3}$ and the heat flow is $30\,mW\,m^{-2}$ where the heat generation is $8\,\mu W\,m^{-3}$, the values of b and q^* in the province are, respectively, 5 km and $10\,mW\,m^{-2}$.

(m) Mean heat flow through continental lithosphere is lower than that through oceanic lithosphere.

SAQ 54 (*Objectives 1.48 and 1.53*)

Determine the rate of heat generation per kilogram in gabbro from ^{40}K alone.

SAQ 55 (*Objectives 1.48 and 1.53*)

In ITQ 47 and Table 1.12, we saw that 1 kg of granodiorite (representing the upper continental crust) generates far more heat from radioactive isotopes than does 1 kg of peridotite (representing the mantle). However, in the answer to ITQ 47, we also pointed out that the mantle is much more massive than the upper continental crust. So assuming that granodiorite is representative of the upper continental crust and that peridotite is representative of the mantle, which actually generates the more heat, the upper continental crust or the mantle?

SAQ 56 (*Objectives 1.49 and 1.53*)

In a divided bar apparatus for measurement of thermal conductivity, the brass rods are each 0.05 m long and the rock sample placed between them is 0.005 m thick. The extreme ends of the brass rod are maintained at 27 °C and 7 °C, respectively, and the measured temperature at the inner end of the cooler brass rod is 8 °C. What is the thermal conductivity of the rock sample, given that the thermal conductivity of brass is $113\,W\,m^{-1}\,K^{-1}$?

SAQ 57 (*Objectives 1.49 and 1.53*)

In a measurement of the thermal conductivity of a sediment by the needle-probe method, the needle was 60 mm long and 36 J of heat was applied to it each minute. After 10 s, the measured temperature was 30.3 °C. The temperature thereafter rose linearly to 34.9 °C at 1 000 s. Calculate the thermal conductivity of the sediment. (Note: If you have a reasonably good scientific calculator, it should enable you to determine ln values directly. If not, you will need to know that ln 1 000 = 6.9 and ln 10 = 2.3).

SAQ 58 (*Objectives 1.51 and 1.53*)

Within a particular continental region, there are found to be two distinct heat-flow provinces (X and Y). In province X, q was observed to be $36\,mW\,m^{-2}$ and $96\,mW\,m^{-2}$, where A was determined to be $1\,\mu W\,m^{-3}$ and $5\,\mu W\,m^{-3}$, respectively. In province Y, q was observed to be $50\,mW\,m^{-2}$ and $80\,mW\,m^{-2}$, where A was determined to be $1.7\,\mu W\,m^{-3}$ and $7.7\,\mu W\,m^{-3}$, respectively. What are the b values for the two provinces?

SAQ 59 (*Objectives 1.51 and 1.53*)

What are the q^* values for the two provinces in SAQ 58?

SAQ 60 (*Objectives 1.51 and 1.53*)

The average heat flow in province Y in SAQ 58 is $59.3\,mW\,m^{-2}$. What proportion of the total heat flow in the province is *not* generated by the decay of radiogenic isotopes in the upper crust?

SAQ 61 (*Objective 1.51*)

Which three of the following conclusions can be drawn with reasonable certainty from the data in SAQs 58–60 if the crust concerned is more than 2 000 Ma old?

(a) The observed average surface heat flow within province Y can be accounted for entirely by radiogenic heat production in the upper crustal layer.

(b) The observed average surface heat flow within province Y can be accounted for entirely by heat rising from the lower crust and mantle.

(c) The proportion of the heat flow observed at the surface that is not generated by radioactive decay in the upper crustal layer is about 117% of the average proportion for the Earth's continents as a whole.

(d) The proportion of the heat flow observed at the surface that is not generated by radioactive decay in the upper crustal layer is about 83% of the average proportion for the Earth's continents as a whole.

(e) The continental crust concerned has probably been subjected to a magmatic and/or metamorphic episode during the past few hundred million years.

(f) The continental crust concerned cannot have been subjected to a magmatic and/or metamorphic episode during the past few hundred million years.

(g) Assuming that no magmatic and/or metamorphic episodes affect province Y over the next 500 Ma, the average heat flow in the province at the end of that time will be down to 40–50 $\mathrm{mW\,m^{-2}}$.

(h) Assuming that no magmatic and/or metamorphic episodes affect province Y over the next 500 Ma, the average heat flow in the province at the end of that time will be up to 70–80 $\mathrm{mW\,m^{-2}}$.

1.13 A NOTE ON THE THICKNESS OF THE LITHOSPHERE

If, before starting this Course, you ever read a popular book on the Earth sciences, you almost certainly came across a statement of the form 'the lithosphere is about x km thick', where x is 50, 75, 100, 125, or whatever. Such a statement, if not actually nonsense, borders on oversimplification. For one thing, there is no single 'thickness of the lithosphere'. The lithosphere is almost certainly thicker (and more complicated) in continental regions than in oceanic ones, and the oceanic lithosphere probably increases in thickness with age. If we are to adopt a standard for defining the thickness of the lithosphere at all, therefore, it would probably be best (simplest) to opt for mature oceanic lithosphere, which means oceanic lithosphere about, or rather less than, 200 Ma old. The obvious alternative — to consider the average lithospheric thickness — would not be very helpful, for it would include, at one extreme, the thick lithosphere beneath continental mountain ranges and, at the other, the thin lithosphere at spreading oceanic ridges.

Even if mature oceanic lithosphere be adopted as 'standard', however, there is still a major difficulty.

ITQ 56

Make a list of the evidence adduced thus far in this Block for the thickness of the lithosphere (bearing in mind that we are now defining 'standard' lithosphere as mature oceanic lithosphere).

So why are the data so disparate? In fact, a more detailed examination of the situation suggests that this may be the wrong question. It would probably be more realistic to ask: why shouldn't they be? The problem lies in the multiplicity of definitions of lithosphere.

In very general terms, the lithosphere can be defined (as Joseph Barrell first defined it in 1914) as the mechanically strong (elastic) outermost layer of the Earth, now thought to comprise the crust and uppermost mantle. In equally general terms, the asthenosphere can then be defined as the underlying, weaker, deformable (plastic) layer within the mantle. Having made these simple definitions, however, an examination of the various properties thought likely to characterize the two layers reveals different concepts of thickness. In short, it becomes clear that the lithosphere, and hence its thickness, can be defined in a number of different ways.

The elastic lithosphere

The **elastic lithosphere** is defined as the layer that flexes when loaded but returns to its original shape when the load is removed. In other words, it behaves elastically. For example, when a volcano or mountain belt builds up, the lithosphere bends downwards under the weight, but bends back into shape when the load is eroded away. By contrast, the underlying asthenosphere can flow. When the lithosphere bends downwards it displaces part of the asthenosphere, and when it returns to its original shape asthenospheric material flows back into the space thus re-created. But just as when a stick is poked into a pond and then removed, the material that flows back into the space is not the same material that was displaced in the first place. The overall configuration of the asthenosphere remains the same, but its detailed arrangement has been changed for good.

In practice, no Earth material is perfectly elastic and so it is usual to set numerical limits on what the term 'elastic' means. In the case of the lithosphere, the layer is usually said to behave elastically if it can regain its shape after having been subjected to loading stresses in excess of $10^8\,N\,m^{-2}$ for over 1 Ma. Volcanoes (known as **seamounts**) on the ocean floor come into this category; and from observations of the flexure of oceanic lithosphere due to seamounts, it is possible to deduce that the elastic thickness of mature oceanic lithosphere is 30–40 km (see (4) in the answer to ITQ 56). There is a slight complication with the concept of elastic lithosphere in that the flexure of the lithosphere depends on both stress and time in such a way, that counter-intuitively perhaps, a small load present for a short time (e.g. a glacier) will make the elastic lithosphere appear thicker than will a large load acting for a long time (e.g. a seamount). The difference can be considerable (tens of kilometres); but as in this Course we are concerned mainly with large stresses acting over geological time-scales, the thickness of the elastic lithosphere can be taken as 30–40 km.

The seismic lithosphere

The **seismic lithosphere** is defined as the high(er)-velocity 'lid' overlying the low(er)-velocity zone (LVZ). Modern seismic-velocity models of the Earth (see (1) in the answer to ITQ 56) place the top of the LVZ, and hence the base of the lithosphere according to our definition of asthenosphere (see Section 1.7.5), at 40–60 km. Other types of seismic evidence, if taken at face value (see (2) in the answer to ITQ 56), indicate a much thicker lithosphere of about 125 km, but when corrected on the basis of mantle anisotropy (see (2) in the answer to ITQ 56) suggest a lithospheric thickness of more like 60 km. The seismic data taken as a whole thus indicate the seismic lithosphere to be thicker than the elastic lithosphere, but only marginally so.

The seismogenic lithosphere

The **seismogenic lithosphere** is defined as the layer in which earthquakes can take place. Earthquakes are the result of the rupture of brittle material and cannot occur where the material is capable of ductile flow. It follows, therefore, that the depth of the deepest earthquakes should mark the boundary between the brittle lithosphere and the ductile asthenosphere. You might suppose from this that, as earthquakes take place down to depths of about 700 km, that is where the lithosphere–asthenosphere boundary lies. However, such deep earthquakes are characteristic of the extraordinary circumstances of plate margins, whereas we are here concerned with mature oceanic lithosphere. In the latter environment the maximum depth of well-located earthquakes is 45–50 km.

The thermal lithosphere

The **thermal lithosphere** is defined as the layer in which heat is transferred largely by conduction, the underlying asthenosphere being the layer in which heat is transferred largely by convection. At first sight, this appears to conform well with the general concepts of lithosphere and asthenosphere; the solid lithosphere conducts, as solids tend to do, and the partially molten asthenosphere below convects, as partially molten material might be expected to do. However, things cannot be that simple. The thermal models of the spreading oceanic lithosphere (Figures 1.96 and 1.98) were designed on the basis of a conducting lithosphere and a convecting asthenosphere, and they suggest a lithospheric thickness of at least 75 km and possibly as much as 125 km (see (5) in the answer to ITQ 56). But the LVZ, interpreted as a partially molten zone, begins at a depth of only 40–60 km (Section 1.7.5). So if the thermal models are right, and unless the LVZ in not partially molten after all, the thermal lithosphere must include part of the partially molten zone, which implies that in the upper part of the partially molten zone heat is transferred largely by conduction.

General conclusions

The observations of lithosphere thickness are summarized in Table 1.15. That the original, uncorrected seismic-wave dispersion data and the later thermal models agree in indicating mature oceanic lithosphere to be about 125 km thick appears to be a coincidence — and perhaps an unfortunate coincidence, for it seems to have misled some Earth scientists during the 1980s into supposing that mature oceanic lithosphere is indeed 125 km thick. In fact, as Table 1.15 makes clear, the weight of observational evidence is in favour of much thinner mature oceanic lithosphere. The elastic, seismic (plus seismic-wave dispersion) and seismogenic data are indeed in remarkable agreement, bearing in mind that they are based on several different definitions of lithosphere which would not necessarily be expected to lead to identical concepts of thickness, that the transition from lithosphere to asthenosphere is unlikely to be very sharp anyway (i.e. it might be expected to occur over a depth range of 10 km or more) and that the data, as all data, are subject to observational error.

Table 1.15 The lithospheres and their mature oceanic thicknesses.

Type	Thickness (km)
elastic	30–40
seismic	40–60*
seismogenic	45–50
thermal	75–125

* Supported by corrected seismic-wave dispersion data, which give a thickness of 60 km.

The obvious way of interpreting the data in Table 1.15 would be to say that mature oceanic lithosphere is 30–60 km thick; but, though realistic, that is unsatisfyingly imprecise. We might speculate, however, on where, within the range, the 'true' thickness lies. You will recall that, when discussing the elastic lithosphere, we pointed out that data from low-stress, short-period phenomena make the lithosphere appear to be thicker than do data from high-stress, long-period phenomena. Seismic waves are low-stress, short-period phenomena; and on that basis it is possible to speculate that seismic data might be overestimating lithospheric thickness. If that be so, it would imply that the true thickness of mature oceanic lithosphere lies towards the lower end of the range quoted above. 40 km would therefore be a reasonable guess, but it's only a guess. It's interesting to note, however, that if there indeed be such a thing as a 'true' lithospheric thickness it appears to be close to the thickness of the elastic lithosphere, the definition of which comes closest to the classic definition of lithosphere as a rocky or strong outer shell.

You may be wondering where all this leaves the thermal models, which demand a much thicker lithosphere. Certainly, the thermal lithosphere is the odd data set out in Table 1.15. It's impossible to say much about that at this stage, except to report that at least one distinguished geophysicist (Don Anderson of the California Institute of Technology) is on record as claiming that the thermal lithosphere, being by definition thermally controlled, should not be expected to be equivalent to the elastic lithosphere, which is a mechanical concept. The thickness of the thermal lithosphere might be proportional to that of the elastic lithosphere but it should not be equated with it. For that reason, the thermal lithosphere should not be called 'lithosphere' at all, but simply the **conduction layer**.

Finally, there is one other lithosphere of considerable interest, namely, the **plate-tectonic lithosphere**, which refers to the thickness of the layer that translates coherently during the major horizontal motions of plate-tectonic processes. In other words, how thick are the plates in plate tectonics? Clearly, this thickness will range from 'thick' where continents are concerned to 'thin' in the vicinity of spreading oceanic ridges; but as far as mature oceanic lithosphere is concerned, it is generally assumed that the plate-tectonic lithosphere is more or less equal in thickness to the elastic lithosphere (i.e. about 40 km). This is, of course, the thickness derived from the observation of phenomena that apply to the lithosphere stresses similar to those of plate-tectonic processes themselves and over similar time-scales.

OBJECTIVES FOR SECTION 1.13

When you have completed this Section, you should be able to:

1.1 Recognize and use definitions and applications of each of the terms printed in the text in bold.

1.54 Discuss the problems of defining the lithosphere and its thickness, and compare the observational data from different sources

ITQ 56 relates to Objectives 1.1 and 1.54. There are no SAQs for this Section.

1.14 Mantle convection

The Earth's mantle is a discrete layer separated by seismic discontinuities from the crust and outer core. Although it represents 80% of the Earth's volume, it is essentially hidden from our view. Nevertheless, as you have already seen, we can make many deductions about its chemical and physical properties by making inferences from geochemical and geophysical data. In this Section, we shall take a closer look at the physical properties of the mantle and consider the role of convection (that is, the upwards transfer of heat by rising plumes of hot buoyant material and the sinking of cooler denser material) in plate tectonics.

We need to find a mechanism to account for the horizontal movements of the lithospheric plates. Most geoscientists agree that plate tectonics is related to mantle convection, but to see why the two are related, we need to introduce some new concepts and answer several questions. For example, how does a solid mantle convect? Is all the mantle involved in convection? And why are the plates not more regular in their shape and their movements if convection drives them?

1.14.1 What is convection?

Mention of convection usually conjures a mental image of hot soup in a pan. It has been suggested that this could be a model for the Earth, with the lithospheric plates being carried along like the skin. But convection within the Earth's mantle is within a spherical shell, unlike soup in a pan. Although some heat in the mantle is likely to come from the core, heat is also derived from radioactive isotopes in the mantle itself; that is, there is **internal heating**, in contrast to soup heated only by an external source. Also the physical properties of mantle peridotite are different from those of most soups! Therefore before attempting to explain convection inside the Earth, we must look at the properties of convecting fluids in more detail.

All other things being equal, the density of the mantle will increase with depth owing to the pressure caused by the overlying material. The temperature will also increase with depth because of this pressure increase.

❑ Do you remember what this natural change of temperature with pressure is called?

■ It is the adiabatic temperature gradient, which you met in Section 1.10.

As long as we know or can estimate pressure, we can calculate the temperature at any depth. The value will be approximate though, because pressure is not the only source of heat in the mantle, as we saw in Section 1.12. If the temperature gradient is exceeded as a result of heat sources within and below the mantle, the gradient becomes greater than the adiabatic gradient and is then called super-adiabatic.

In order for convection to occur, hot material must underlie cooler material, whether it is tomato soup or mantle peridotite. There is one other necessary condition to be satisfied before convection will begin. The hotter and cooler materials do not simply change places by leaking through each other. During convection, columns of relatively hot material (and therefore lower density) rise up, and columns of relatively cool and therefore more dense material sink.

❑ Can you think now what other condition must be met for convection to occur?

■ There must be some lateral variation in density within the material to accommodate the uprising and sinking convection columns.

So convection will occur in a fluid provided that the lower material is hotter than the overlying material and there are lateral density variations. For a very deep fluid, the lower parts will be hotter simply as a result of compression. This natural adiabatic temperature gradient must be exceeded before convection can commence.

As an example, imagine a tank of fluid heated only from below. At first the fluid will be stable, following the adiabatic gradient, and convection does not occur. Increased heating of the lower layers reduces their density and increases their buoyancy, until a point is reached at which any slight lateral density irregularity will allow lighter material to rise. The first convective pattern developed in this way is characterized by regular convection cells, roughly as wide as the depth of the fluid and with hexagonal plan, so enabling the individual cells to fit together like the cells of a honeycomb. Hot fluid ascends like a fountain in the centre of each cell and descends at the edges (Figure 1.108). This is a stable configuration and it is found that stirring a fluid convecting in this way cannot disrupt the patterns for long. The cells of the hexagonal convective pattern form a group of **Bénard cells**, so called after the French physicist H. Bénard, who first published experimental evidence for such convection in 1906.

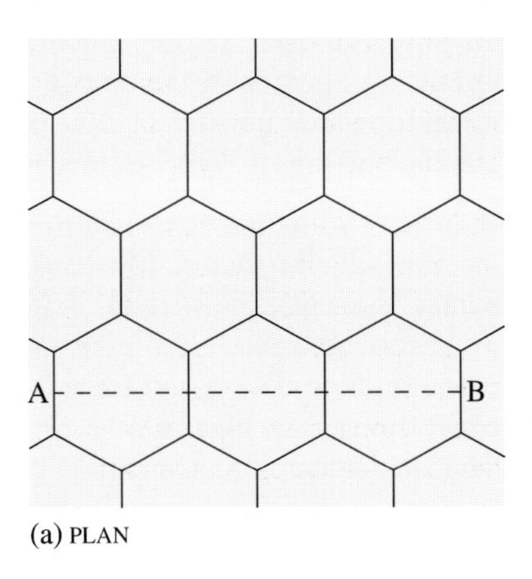

(a) PLAN

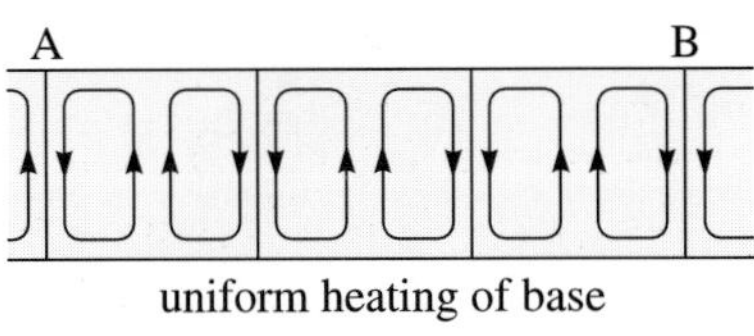

(b) SECTION AB

Figure 1.108 Diagram to illustrate the honeycomb pattern of Bénard convection cells (a) as seen from above (plan view) and (b) in cross-section (from the side). The arrows indicate the direction of movement of the fluid.

Following the establishment of Bénard cells in a uniformly heated fluid, further heating causes the convective patterns to become unstable. Increased heating increases the vigour of the system, until eventually the regularity of the cells starts to break down. They start to change in position and size and the hexagonal pattern is replaced by irregular convection with columns and blobs of lighter material rising from the base at random.

Bénard's experiments showed that a fluid passes through three types of behaviour when subjected to uniform heating. Firstly, before convection begins, heat is transferred only by conduction; secondly, stable Bénard cells are established; and thirdly, irregular, or **turbulent convection** develops. Although Bénard's work was published in 1906, it was not until 1916 that his results were explained quantitatively by the British physicist, Lord Rayleigh. He showed that the onset of convection depends not on any one or two properties taken singly but on the combined properties of the fluid. Lord Rayleigh combined the fluid properties together as a quantity now called the **Rayleigh number**, Ra, which can be used to predict the conditions under which convection will be initiated.

1.14.2 The Rayleigh number

It can be shown mathematically that convection will only occur if the Rayleigh number (Ra) exceeds a critical value $\mathrm{Ra}_{\mathrm{cr}}$. The Rayleigh number is defined as follows:

$$\mathrm{Ra} = \frac{\alpha \Delta T g d^3 \rho}{\kappa \eta}. \quad \text{(Equation 1.58)}$$

The terms in this expression are as follows, with their units in parentheses:

α: volume coefficient of thermal expansion (K^{-1}; see note (a) below)

ΔT: temperature difference (K; see note (b) below)

g: gravitational acceleration ($10\,\mathrm{m\,s}^{-2}$; see note (c) below)

d: height of the fluid layer (m; see note (d) below)

κ: thermal diffusivity ($\mathrm{m}^2\,\mathrm{s}^{-1}$; see note (e) below)

η: dynamic viscosity (Pa s*; see note (f) below)

ρ: density (kg m^{-3})

There are several important points to note about this expression.

(a) The **volume coefficient of thermal expansion**, or cubic expansivity, α, of a substance is the increase in volume per unit increase in temperature of that substance, expressed as a fraction of the original volume. The unit of cubic expansivity is therefore K^{-1}. For example, for a unit volume of rock for which $\alpha = 2 \times 10^{-5}$ K^{-1} (appropriate for the Earth's mantle), a temperature increase of 1 K will cause a volume increase of 2×10^{-5} of the original volume.

(b) You should recall from Section 1.10 that compression causes an increase in temperature with increasing depth. This is known as the adiabatic temperature gradient, and it must be exceeded for convection to occur. As the adiabatic temperature increase in passing through a pan of cold soup is tiny (since the depth is small), the application of only a very small amount of heat will exceed the gradient. Although exceeding the adiabatic temperature gradient might be easy in a pan of soup, the situation is less clear within the Earth's mantle. As the mantle is much denser than soup and has a much greater vertical extent, the adiabatic temperature gradient, about 0.25 K km^{-1}, is substantial, and there is a significant adiabatic temperature increase of about 750 K in passing from the top to the bottom of the mantle simply due to compression. For convection to occur, this gradient must be exceeded somewhere in the mantle, and this is reflected in the ΔT term in the Rayleigh number. Thus for a given depth,

$$\Delta T = T(\text{actual}) - T(\text{adiabatic}).$$

ΔT, the temperature *difference* between the actual temperature and the adiabatic temperature, is measured in K.

(c) Gravitational acceleration, g, is almost constant throughout the mantle, at 10 m s^{-2} (see Section 1.10 and Figure 1.80).

(d) The term d is the height of the fluid layer, but the units are m^3 because in Equation 1.58 we use d^3.

(e) **Thermal diffusivity**, κ, is given by the thermal conductivity (k) of the fluid divided by its density (ρ) and specific heat capacity (c)

$$\kappa = \frac{k}{\rho c}$$

where

k: thermal conductivity (W m^{-1} K^{-1}; see Section 1.12.2)

ρ: density (kg m^{-3})

c: specific heat capacity (J kg^{-1} K^{-1}) [The specific heat capacity of a substance is the amount of heat (measured in J) needed to raise the temperature of 1 kg of the substance by 1 K, so the units of c are J kg^{-1} K^{-1}.]

The unit of thermal diffusivity is therefore m^2 s^{-1}.

(f) **Dynamic viscosity**, η, is a measure of the ease with which layers of a fluid can move relative to one another. For example, the low dynamic viscosity of water means that a barge can move along a canal quite easily

* Pa stands for pascal, a unit of pressure or stress. Thus far in this Block, we have used the unit N m^{-2} for pressure or stress, but we could just as easily have used Pa. 1 Pa = 1 N m^{-2} and as N = kg m s^{-2}, N m^{-2} = kg m^{-1} s^{-2}. Therefore Pa s, the unit of viscosity, is kg m^{-1} s^{-1}.

since the water in contact with the barge is able to move separately from the water in contact with the bottom of the canal. In other words, water has a low shear modulus — which is why it does not transmit seismic S-waves. If you now imagine a canal full of treacle or tar, then because it does not flow as easily as water, the barge would not move so easily and seismic S-waves could be transmitted. This ease of relative movement between layers of a fluid is directly related to the dynamic viscosity.

Put mathematically, dynamic viscosity is:

$$\eta = \frac{\text{shear stress}}{\text{strain rate}}.$$

This equation is reminiscent of the shear modulus used for calculating density from seismic wave velocity. Strain rate is simply the rate of change of strain with time.

Before we begin to discuss the viscosity of the mantle in detail, it is worth placing it in perspective. Table 1.16 lists the dynamic viscosities of some common liquids, including melts of common rocks. You will perhaps be surprised to see that the dynamic viscosity of porridge is very similar to that of basalt, but much lower than that of andesite and rhyolite.

Table 1.16 Viscosities of some common substances at room temperatures and molten lavas at their melting temperatures.

Substance	Density, ρ (kg m^{-3})	Dynamic viscosity, η, at atmospheric pressure (Pa s)
water	$1.0. \times 10^3$	1.0×10^{-3}
mercury	1.4×10^4	1.6×10^{-3}
porridge	1.1×10^3	1.1×10^2
basalt lava (1 150 °C)	2.7×10^3	2.7×10^2
andesite lava (1 000 °C)	2.5×10^3	2.0×10^3
rhyolite lava (850 °C)	2.4×10^3	2.4×10^5

Dynamic viscosity can be measured in a variety of ways, one of them being to measure the rate of descent of spheres of known size and density within the fluid. This technique is obviously impossible for the mantle since we can't watch spheres sinking through it, and so indirect methods must be used to estimate mantle viscosity.

All the physical quantities involved in the Rayleigh number, together with their units, are listed in Table 1.17.

The quantities in the numerator (above the line on the right-hand side of Equation 1.58) of the Rayleigh number are those that encourage convection when they have a large value, whereas those in the denominator (below the line) tend to inhibit convection when they have a large value. A large volume coefficient of expansion, α, and temperature difference, ΔT, encourage convection because the decrease in density with increasing temperature provides the driving force for convection. The depth (d) of the fluid layer is important because the taller the column, the greater is the difference in pressure at the base of adjacent hot and cold columns. Gravity, g, is relevant because it pulls down less on a lighter fluid, allowing it to rise. High values of both dynamic viscosity, η, and thermal diffusivity, κ, inhibit convection. The effect of viscosity is fairly obvious (heated water convects more vigorously than heated treacle!). The thermal diffusivity is a measure of heat loss by conduction; high heat loss from the uprising cells by conduction will diminish their potential for driving convection.

Table 1.17 Summary of physical quantities, symbols and units involved in the Rayleigh number, Ra.

Physical quantity	**Unit**
volume coefficient of expansion, α	K^{-1}
temperature difference, ΔT	K
gravitational acceleration, g	$m\ s^{-2}$
thickness of fluid layer, d	m
thermal diffusivity, κ	$m^2\ s^{-1}$
dynamic viscosity, η	$kg\ m^{-1}\ s^{-1}$
density of fluid, ρ	$kg\ m^{-3}$

The Rayleigh number is therefore a measure of the factors that tend to promote convection, acting against the factors that tend to inhibit convection. It is the most important parameter in determining whether or not mantle convection will occur. As we said earlier, the Rayleigh number must exceed a critical value, Ra_{cr}, before convection can occur. The exact value of Ra_{cr} depends on several factors not yet considered — for example, the three-dimensional geometry of the convecting cells and the proportion of heat derived from below compared with that generated internally. Calculations have shown that for liquids in a spherical shell, convection is initiated when $Ra_{cr} \sim 10^3$ and that convection becomes more vigorous and turbulent as Ra increases.

ITQ 57

What are the units of the Rayleigh number?

In summary, the theory of fluid dynamics tells us that when the Rayleigh number, which depends on the physical properties of a substance, exceeds its critical value then convection will occur. We shall now consider how convection actually occurs and whether it does so in stable Bénard cells or more chaotically. Firstly, though, we need to put numbers into Equation 1.58. We cannot directly measure any of the values in the mantle, so we have to estimate them. There isn't space here to describe how we obtain these estimates for all the terms in the equation, so we shall only discuss η, the dynamic viscosity.

1.14.3 Evidence for movement within the mantle

The Earth's lithosphere 'rebounds' when vertical stresses applied to it are released; this is the principal of isostasy that we introduced in Section 1.9. An example of such stress is the loading of continental areas by large ice sheets during the last glaciation. If the lithosphere is depressed by a large mass of ice, then the mantle material below that lithosphere — the asthenosphere — has to move out sideways to make room for it. Conversely, when the load is removed, the lithosphere slowly rises to its former level, and this is accomplished by a return flow of mantle below the depressed area. The faster the rate of flow, the lower the viscosity must be. So if the rate of such flow can be determined, then we are well on the way to obtaining a measure of the dynamic viscosity of the upper mantle.

1.14.4 Estimating the dynamic viscosity of the mantle

Figure 1.109 is a map showing the uplift of the area of Norway, Sweden and Finland — together called Fennoscandia — since the most recent glaciation (about 20 000 years ago). Evidence for this uplift comes from features such as raised beaches (ancient beaches now lifted up above sea-level) and other geological records of progressively changing tidal level, as the sea-level has fallen relative to the land. Repeated precise levelling surveys reveal changes in relative height across the region, and show that although global sea-level has risen, Scandinavia has risen more. The maximum uplift relative to sea-level is nearly 300 m, and has taken place in the last 10^4 years, which represents an average rate of $3\,cm\,a^{-1}$, or $9.5 \times 10^{-10}\,m\,s^{-1}$, but is much slower at the present time.

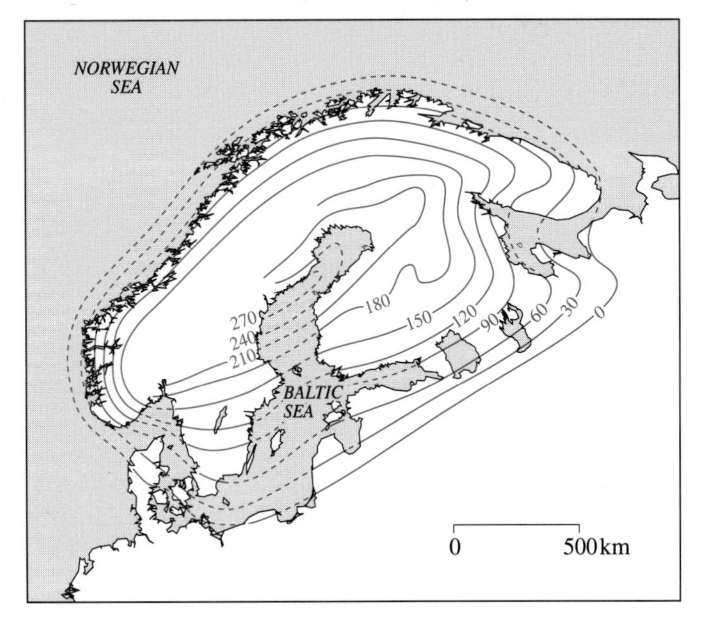

Figure 1.109 Post-glacial uplift of Fennoscandia as determined from the elevation of the highest tide mark of the sea. Contours are in metres of uplift. The dashed contours show estimated uplift in areas largely covered by sea.

This uplift was produced in response to the removal of the mass of ice that formerly covered the area in Figure 1.109. We need to know the size of the mass, because that will tell us the magnitude of the force that was responsible for deforming the asthenosphere below by 'squeezing' it out sideways. We also need to know the thickness of the asthenosphere, and to have some idea how far the uplift has gone, and how far it has to go. This can be determined from a study of regional gravity anomalies.

Gravity measurements for Fennoscandia are characterized by small negative Bouguer and free-air anomalies.

❑ From your study of gravity anomalies earlier in this Block (Section 1.8.3), what does this indicate?

■ The negative anomalies indicate that Fennoscandia is not yet in a state of isostatic equilibrium; the small negative values indicate a mass deficiency — and that the area is therefore still rising.

However, it can be shown from the small magnitude of the gravity anomalies that only about 30 m of uplift remains to be accomplished. Thus Fennoscandia has nearly completed a cycle of depression and uplift, the magnitude and time-scale of which are known quite well. It would seem that the load was applied over a period of some 100 000 years (the duration of the last glaciation), and then released as the ice melted over some 10 000 years. The total vertical displacement of the crust is estimated to have been about 300–350 m, caused by an ice sheet up to 2.5 km thick and covering an area of about $9.5 \times 10^{11}\,m^2$.

In order to use these uplift data to estimate dynamic viscosity at depth, we have to consider exactly where and how the flow occurs. This has been determined on the basis of models which have a three-layer mantle structure. The top layer consists of rigid mantle lithosphere with very high viscosity and subject only to elastic and brittle deformation. This overlies the asthenosphere, a region of lower viscosity mantle that behaves plastically and deforms under stress but does not go back to its original shape unless forced to. This plastic layer in turn overlies an elastic lower mantle of higher, but unknown, viscosity. Two models have been proposed to account for the flow of mantle material. These are termed the deep-flow and channel-flow models (Figure 1.110).

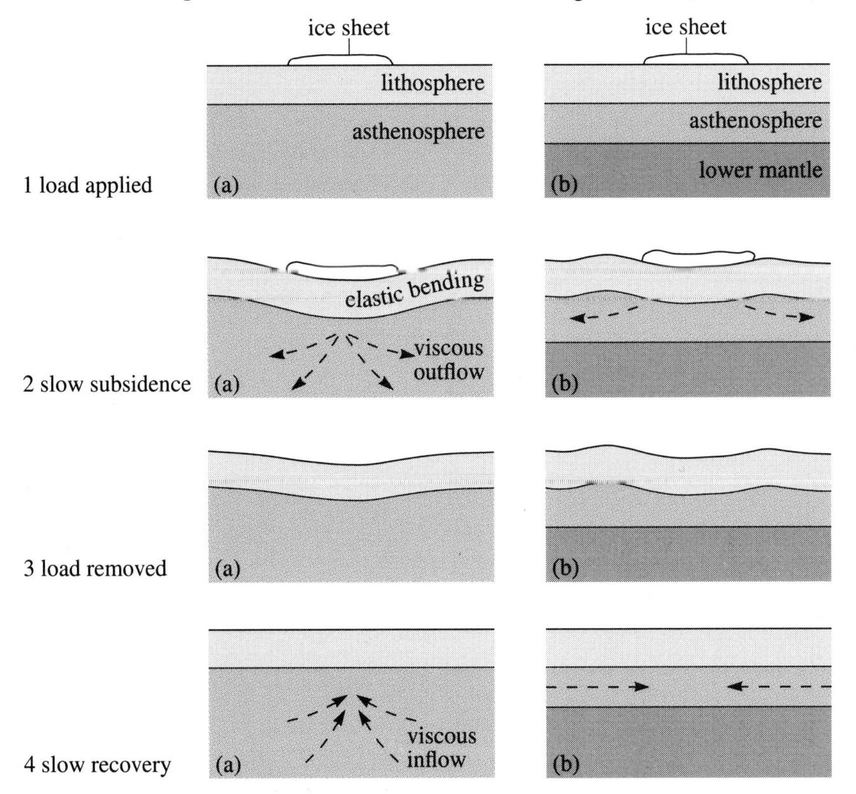

Figure 1.110 The viscous response of the asthenosphere to the application and removal of a load such as an ice sheet on the surface of the elastic lithosphere.
(a) deep-flow model; (b) channel-flow model. Numbers 1–4 show succeeding stages of glaciation, deglaciation and isostatic recovery.

According to the **deep-flow model**, the mantle flow caused by glacial loading occurs over a wide depth range, to a depth equivalent to at least the radius of the ice sheet (Figure 1.110a). By contrast, in the **channel-flow model** (Figure 1.110b), mantle flow associated with ice loading is restricted to a relatively shallow depth range (thickness less than 200 km) — that is, much less than that involved in the deep-flow model. It is therefore very important to determine which model applies to the mantle deformation below an ice sheet.

If the deep-flow model is valid, then the viscosities determined will apply to a relatively deep section of the mantle (up to half the horizontal extent of the ice sheet). If channel flow is the more appropriate model, then viscosities determined from deformation around ice sheets will apply only to a relatively small channel within the upper mantle.

You may also have noted on Figure 1.110 that in the channel-flow model there is a pronounced *bulge* in the crust around the depression caused by the ice sheet. This would result from flexing of the lithospheric plate and mantle material being squeezed sideways and upwards from below the centre of the ice sheet because the underlying mantle is assumed to be rigid. In contrast, according to the deep-flow model the mantle material is squeezed out over a much greater depth range, so that any bulge is absent or small in size. *The occurrence of a pronounced peripheral bulge is therefore the most important geological difference between the deep- and channel-flow models* and provides the key to deciding which of these two models is applicable to the Fennoscandian ice sheet.

Computer models for the behaviour of Fennoscandia as a result of ice loading according to the two models are shown in Figures 1.111 and 1.112. Figure 1.111 shows the isostatic adjustment due to deep flow that best fits the uplift data following the removal of an ice sheet of radius 550 km. This model shows the ground profile around the model ice sheet during glacial unloading from 10 000 years ago to the present, and indicates that up to 40 m of post-glacial uplift remains. There are two important distinctive features. These are firstly that during glacial unloading, there is a relatively *small* amount of uplift of the peripheral bulge up to 20 m, and secondly that this is followed by slower peripheral sinking (Figure 1.111b).

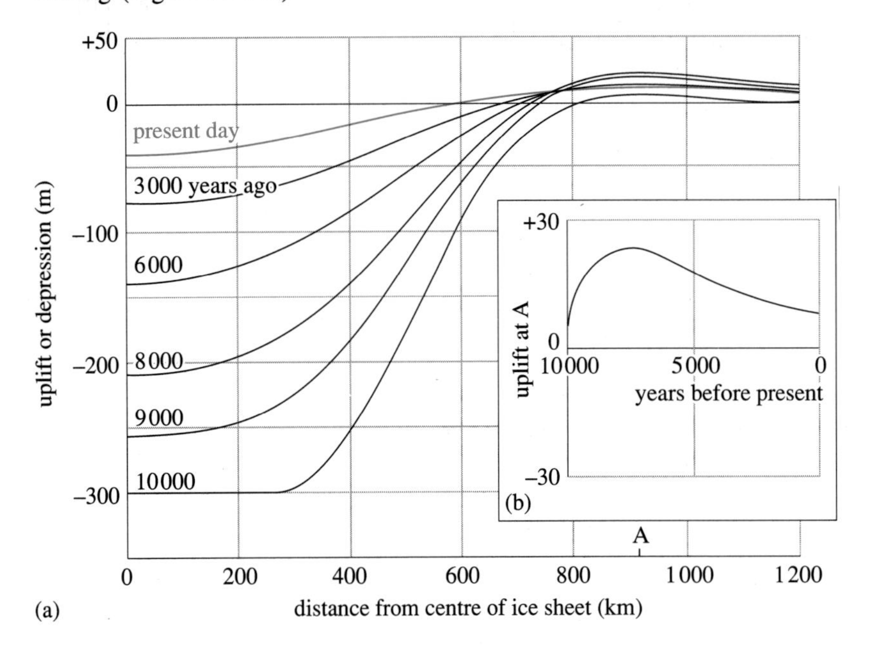

Figure 1.111 Deep flow mantle model.
(a) Plots of relative elevation against distance from the centre of the Fennoscandian ice-sheet, for the present day and various times up to 10 000 years ago, when deglaciation commenced. Thus isostatic adjustment of ground level after removal of the ice sheet is indicated. The model was calculated by assuming that an ice sheet of radius 550 km and thickness 1 100 m was present for 20 000 years and was then removed 10 000 years ago. The deformation associated with the depression and uplift cycle occurred within the *mantle below the lithosphere,* as shown schematically in Figure 1.110a.
(b) The uplift and depression plotted against age at a single location, A, near to the margin of the ice sheet. This shows that the area surrounding the ice sheet experiences rapid uplift to about 20 m 8 000 years ago, following sinking from about 8 000 years ago to the present day.

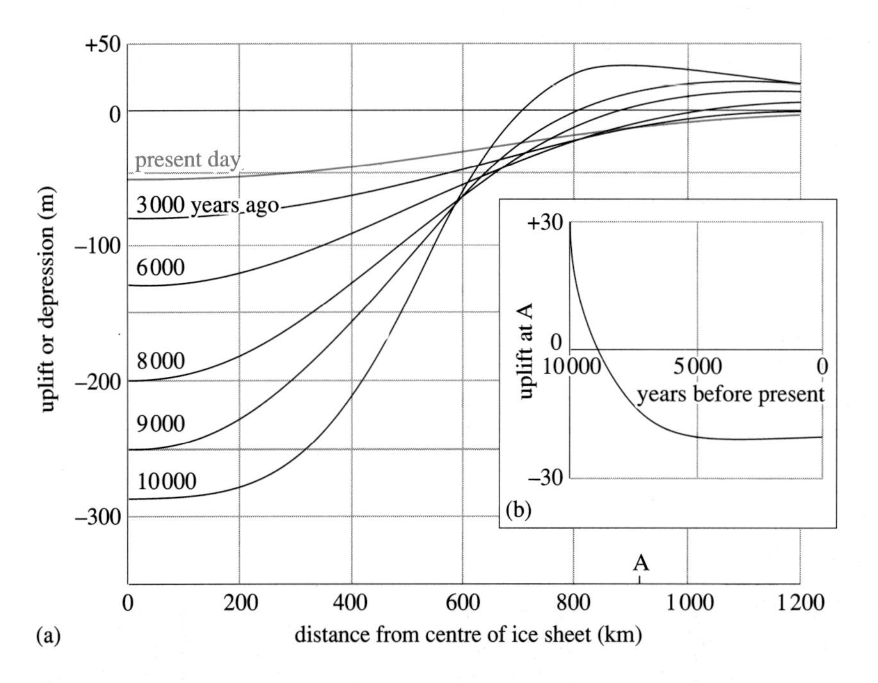

Figure 1.112 Channel-flow mantle model.
(a) Plot of relative elevation against distance from the centre of the Fennoscandian ice sheet, for the present day and various times up to 10 000 years ago, when deglaciation commenced. The model was calculated using the same parameters as used in Figure 1.111a, but assuming that the deformation associated with the depression and uplift cycle was restricted to *a channel underlain by rigid mantle* as shown schematically in Figure 1.110b.
(b) The uplift and depression plotted against time at a single location, A, near to the margin of the ice-sheet. This shows that at A a large bulge of about 30 m is rapidly produced, but quickly decays by about 50 m (to about – 20 m) by very rapid post-glacial sinking. The area around A has shown uplift from about 3 000 years ago to the present day.

Figure 1.112a shows the isostatic adjustment to the unloading of a similar ice-load if the flow is restricted to a shallow channel. As in Figure 1.111a, this shows the ground profile around the same model ice sheet during glacial unloading from 10 000 years ago to the present, and suggests that about 50 m of uplift still remains. In contrast to the deep-flow model, Figure 1.112a shows that according to the channel-flow model, a much *larger* peripheral uplift, about 30 m, is produced as a result of the 'squeezing out' of channel material. The bulge quickly disappears during unloading, and the peripheral regions are shown to have been *sinking* for the 10 000 years since the melting of the ice sheet commenced (Figure 1.112b). From Figure 1.112b, we can see that the total decay of the bulge since formation is 50 m.

ITQ 58

Look carefully at Figures 1.111 and 1.112 and answer the following questions.

(a) Briefly describe the size and shape of the peripheral bulge formed around the ice sheet according to the deep-flow and channel-flow models.

(b) After deglaciation, what happened to the peripheral bulge around the ice sheet, according to the deep-flow and channel-flow models?

(c) From (b), how do the contours showing zero uplift migrate after glaciation (in relation to the ice sheet), according to the deep-flow and channel-flow models?

From the answer to ITQ 58, you will realize that the *geological* difference between the deep- and channel-flow models should tell us which model applies to the Fennoscandian ice-sheet. *The lack of evidence for the past existence of a large bulge around the Fennoscandian ice sheet is inconsistent with the channel-flow model.* Therefore the uplift behaviour of the crust can be used to determine the viscosity of the mantle down to a depth similar to the radius of the ice sheet (the region of inflow and outflow) — that, is to about 550 km depth.

Dynamic viscosity is the ratio of shear stress to strain rate. Since stress is force (F) per unit area (A), and strain rate is the rate of uplift divided by the total amount of depression (or total uplift, to reach equilibrium), we can write an equation for viscosity:

$$\eta = \frac{F/A}{\text{rate of uplift/vertical distance affected}}.$$

The force on the mantle is equal to the mass of ice on top of it (m) multiplied by the acceleration due to gravity (g):

$$F = mg.$$

We can estimate m by first estimating the total volume of the ice cap when it was at its maximum and multiplying this by its density (900 kg m^{-3}). The area, A, of the ice sheet was about 9.5×10^{11} m^2, and its thickness was about 1 100 m on average, so we estimate its mass to be about 9.4×10^{17} kg.

$$\text{So } F \sim 9.4 \times 10^{17}\,\text{kg} \times 10\,\text{m s}^{-2}$$
$$= 9.4 \times 10^{18}\,\text{kg m s}^{-2} \text{ or } 9.4 \times 10^{18}\,\text{N}.$$

The rate of uplift we know is about 9.5×10^{-10} m s^{-1}, and although the total depression due to the ice was about 330 m, the depth of mantle affected is up to half the diameter of the ice cap, so about 550 km or 5.5×10^5 m. So viscosity η is:

$$\eta = \frac{9.4 \times 10^{18} / 9.5 \times 10^{11}}{9.5 \times 10^{-10} / 5.5 \times 10^5} \sim 10^{21}\,\text{Pa s}.$$

This rather crude calculation doesn't allow us to be any more precise but you should be able to see the way in which these sorts of calculations are done.

This value for mantle viscosity was estimated using data valid down to about 550 km.

❑ If we want to find out about viscosities deeper in the mantle how should we go about it?

■ We need to use the same procedure, but with a bigger ice sheet.

The North American ice sheet, which was centred on Canada, had a diameter of 4 000 km, and the loading and uplift cycle affected large areas of the USA and Canada.

The behaviour of this area has been modelled as described already for Fennoscandia. Firstly, there is no evidence for the formation of large peripheral bulges, predicted for the channel-flow model (cf. Figure 1.112b). If such bulges had existed, then the rates of peripheral subsidence would have been similar to those of the post-glacial central uplift (about 160 m). Such subsidence has not been observed. Secondly, and more important, is the uplift history of the east coast of the USA, which represents the peripheral region of the North American ice sheet. (Figure 1.113).

ITQ 59

Study Figure 1.113, which shows the peripheral uplift/subsidence history for the east coast of the USA during the last 13 000 years (deglaciation began about 18 000 years ago).

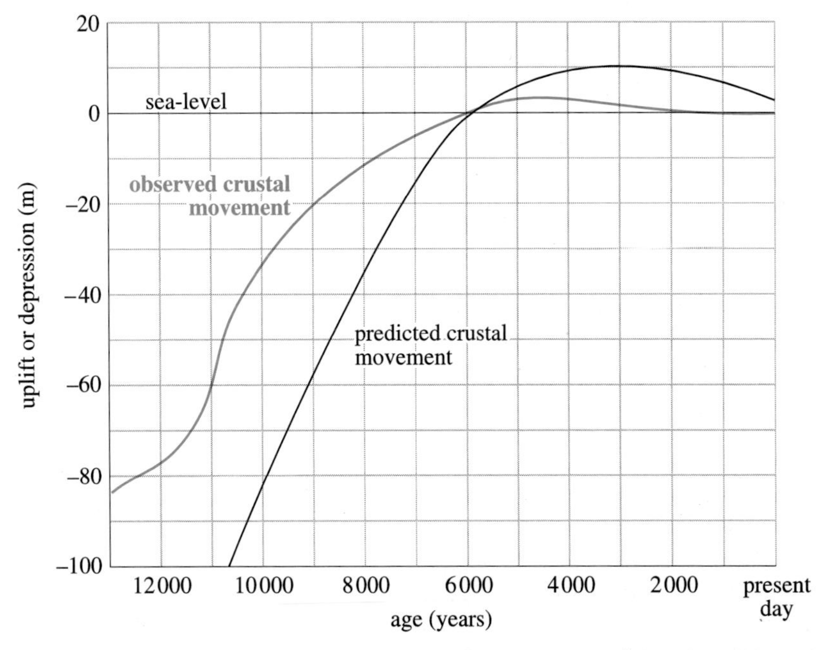

Figure 1.113 Plot of the relative elevation against age for Boston, Massachusetts. The coloured line shows the observed crustal movement. Similar behaviour characterizes many locations on the east coast of the USA. The calculated uplift behaviour for a model with a *uniform* 10^{21} Pa s mantle is shown by the black line.

(a) Briefly describe the pattern of observed peripheral uplift and subsidence.

(b) Compare this pattern with the patterns of uplift and subsidence characterizing the deep-flow and channel-flow models shown for Fennoscandia (Figures 1.111 and 1.112). Which of these models is better able to account for the uplift/subsidence behaviour of the east coast of the USA?

Mantle viscosity is estimated in just the same way as it was for the Scandinavian ice sheet, but this time, as the ice sheet was larger, deeper mantle was affected and so our estimate is for viscosity down to about 2 000 km. To within an order of magnitude, we end up with the same result as before. So, in conclusion, viscosity of the mantle down to at least 2 000 km is 10^{21} Pa s to within an order of magnitude.

ITQ 60

Using the values given below, evaluate Ra. What can you deduce from this by comparing your value with Ra_{cr}?

α, volume coefficient of thermal expansion = $2 \times 10^{-5}\,K^{-1}$
g, gravitational acceleration = $10\,m\,s^{-2}$
d, depth of mantle = 2 900 km ~ $3 \times 10^{6}\,m$
κ, thermal diffusivity = $10^{-6}\,m^{2}\,s^{-1}$
η, dynamic viscosity = $10^{21}\,Pa\,s = 10^{21}\,kg\,m^{-1}\,s^{-1}$
ρ, density = $3\,300\,kg\,m^{-3}$

The value for ΔT, the temperature difference between the actual and the adiabatic gradient, is not well known for any depth in the mantle, but it is unlikely to be much less than 1 K, or much more than 10^{2} K. Calculate Ra for both these values of ΔT.

1.14.5 Convection in a solid mantle

You should recall from Section 1.7 that seismic S-waves are transmitted throughout the Earth except in the outer core. The outer core cannot sustain shear movements because it is fluid, and the modulus of rigidity for fluids is zero or at most very low. The mantle transmits S-waves and so must be solid, but we have just seen in the previous Section that the mantle can flow. So how does the mantle manage to behave as a solid for seismic wave transmission and as a fluid when subjected to heavy loads? It may surprise you to learn that there are several everyday analogies of this type of behaviour. Imagine a wooden book shelf. If you karate chop it, it may break in two or it may remain perfectly intact depending on your technique. The important point is that it responds rigidly and does not deform. This karate chop is rather like the effect of a seismic wave in the mantle — it is very short-lived. Now consider the effect of books on the shelf. The force may be smaller, but it acts for a long time and eventually the shelf will bow downwards. So the shelf has deformed — or flowed in response to the force. The shelf is undoubtedly a solid, but if a force is exerted over a long period of time it behaves rather like a very viscous fluid. Another example is a steel bar. If you try to karate chop it, it will certainly behave as a solid. You can deform it, though, by exerting a force for long enough — and you can speed the process up by heating the bar. The same is true for a lead bar, and for rocks. Think of the rocks we see at the Earth's surface. They are brittle and solid, but if they were buried hundreds or thousands of kilometres beneath the surface and heated up, they would by analogy with our wood and steel examples behave like a viscous fluid. Over short time-scales, they would respond rigidly, passing seismic S-waves, but over long time-scales, they could flow and even convect. The process occurs in all crystalline solids because all crystals have imperfections within the lattice structure. These imperfections are weaknesses in the crystal and they tend to 'give' or move when stress is applied. This type of movement is termed **creep** (the solid-state creep of Section 1.7.5).

In summary, our best estimates of the Rayleigh number (Ra) for the mantle indicate that the mantle does indeed convect. It even convects unsteadily because Ra far exceeds Ra_{cr} even for modest estimates of the temperature in excess of the adiabatic gradient. Convection within the mantle is not inconsistent with the mantle being solid. Over short time-scales, the mantle behaves as a solid — it transmits seismic S-waves. Over longer time-scales, the crystalline structure of the mantle enables it to creep and thus move — it transfers heat upwards by convection. The rate of movement is slow — comparable with movements of the lithospheric plates above — and is likely to be of the order of $10^{-9}\,m\,s^{-1}$ (a few centimetres per year or the rate at which your finger nails grow).

1.14.6 MODELS OF CONVECTION IN THE MANTLE

Several geophysical and geochemical observations suggest that the convective flow in the upper and lower mantle could be separate systems. For example, there is a jump in seismic P-wave velocity at 670 km depth (Section 1.7.5) which most likely corresponds to a mineral phase change. Chemical evidence for a separation between the upper and lower mantle comes from subtle differences in the composition of basalt magma erupted from shallow depths (oceanic ridges) and from deeper sources (oceanic islands, such as Hawaii). If the whole mantle convected as one system, chemical heterogeneities would be destroyed. It seems that the shallow (or upper mantle) sources are depleted — or are short of some key components compared with the lower mantle sources. Also there is a remarkable uniformity within each of the two types of basalt magma worldwide, implying that the upper and lower mantle may be separate homogeneous regions. We will consider how these chemical differences can be observed later in the Course (Block 3).

These observations leave us with a problem; if the upper and lower mantle convect separately, how is it that the lithospheric plates are so large? For plate dimensions of several thousands of kilometres, the down-going parts of the convection cells should also be of the order of thousands of kilometres according to our simple model (Figure 1.108) This implies **whole-mantle convection** with convection cells extending from the core–mantle boundary to the base of the lithosphere. We know that convection does not occur in stable Bénard cells, and that the horizontal extent of cells is up to thousands of kilometres in the upper mantle, and that the upper and lower mantle are distinct. So we seem to have evidence that the whole mantle convects, and that the upper and lower mantle are separate.

Mathematical, computer, and laboratory models have been developed to solve this paradox. The difficulty with laboratory models is in finding materials that behave like the mantle and overriding lithosphere but operate at lower temperatures and pressures and over shorter time-scales. Syrup and oil or glycerine are popular ingredients since they move rather faster than the much more viscous mantle; there is no use having an experiment that takes several million years to complete!

Computer modelling has shown that convection may occur on two scales within the upper mantle. On one scale, convection cells are about as tall as they are wide. On the other scale, cells are much wider — comparable with the dimensions of the lithospheric plates — but the height is the same as for the smaller cells. If constant temperatures at the boundaries (the base and top of the upper mantle) and a constant heat flow across the upper boundary are assumed, then the computer prediction gives the temperatures and fluid flow lines for these conditions shown in Figure 1.114. It shows that in addition to cells that are as tall as they are wide, convection cells that are much wider than they are tall are also stable. These cells are said to have a high aspect ratio — the aspect ratio is the width of the cells divided by their height. The model also predicts that hot 'plumes' of material will rise towards the surface then spread out parallel to the surface. Whether this is a reasonable model for upper mantle convection or not is a matter for discussion. It does account for the apparently well-mixed nature of the upper mantle and for the relatively large dimensions of the lithospheric plates compared with the thickness of the convecting medium. However, it is only a model, and an obvious simplification is that it was assumed that the upper mantle is heated only from below. We don't know what percentage of the upper mantle's heat supply is internally generated by long-lived radioactive isotope decay, but this will certainly have an effect on the pattern of convection. A computer prediction of the temperatures and fluid flow lines assuming that all heat is generated within the upper mantle is shown in Figure 1.115. An important contrast between this model and the previous one is that the temperature is constant across the surface and increases quite uniformly with depth. This means that there are no rising plumes of hot material. Again high-aspect-ratio cells are stable and so are smaller more uniform cells. The two models, shown in Figure 1.114 and 1.115 are 'end members' or extremes. The truth may lie somewhere between the two.

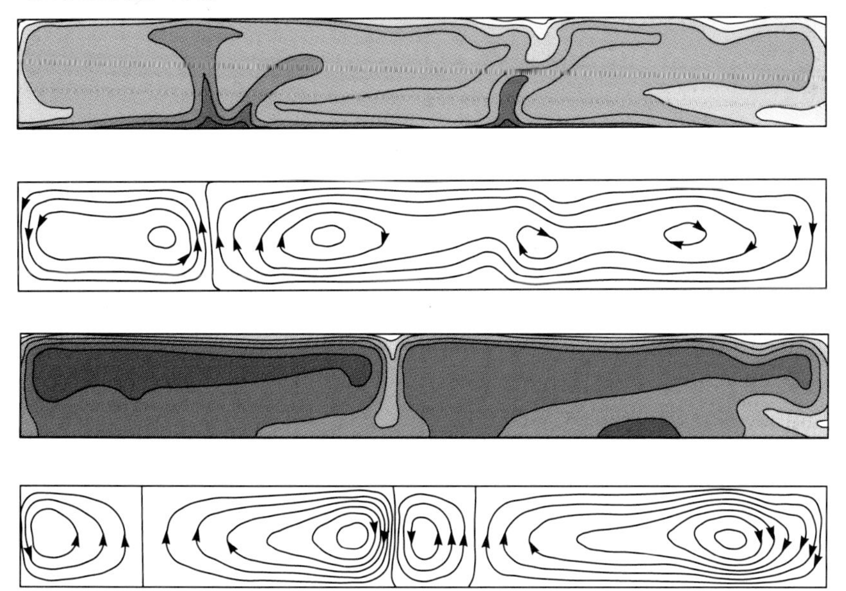

Figure 1.114 Temperature (upper) and fluid flow lines (lower) for computer models of convection in the upper mantle. Higher temperatures are indicated by darker shading. A Rayleigh number of 2.4×10^5 has been assumed. For this model, heat flow across the upper and lower boundaries are assumed to be constant.

Figure 1.115 This is the same as Figure 1.114 except that it is assumed that all the heat is supplied from within the mantle. Again heat flow at the upper boundary is constant.

A model that combines the attributes of both end members, and again would produce good mixing in the mantle, is shown in Figure 1.116. We cannot prove whether this illustrates upper mantle convection realistically or not, but it is a useful model and it does fit the observations. It is undoubtedly a simplification, but it is constrained by our indirect geophysical and geochemical measurements and its main purpose is to help us to visualize a working process that we can never observe at first hand.

Similar models have been developed for the lower mantle. Geophysical and geochemical evidence suggest that the upper and lower mantle support separate systems and that there is very little communication between the two. However, there is evidence for some communication; subducted lithospheric slabs can sometimes penetrate the lower mantle, and hot-spot volcanism — which we shall come to in Block 3 — is believed to originate in the lower mantle. Thus mantle convection is neither one layer (whole mantle) nor strictly two layers (completely separate upper and lower mantle), but one and a half layers.

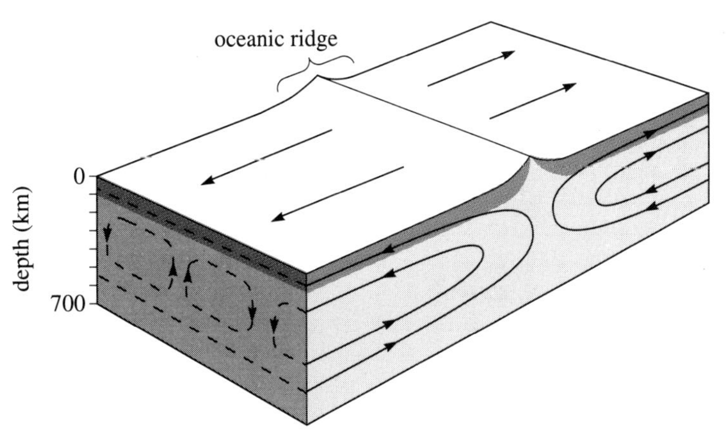

Figure 1.116 Two-scale mantle convection model. The large-scale convection (solid lines) corresponds to the horizontal movement of the plates themselves. This is accompanied by small-scale convection beneath the plates (shown as dashed lines). Both scales involve the mantle down to about 700 km. If the plates are moving apart fast enough, by 10 centimetres or more a year, the small-scale convection may be transformed into longitudinal cylinders (shown as cross-section parallel to the ocean ridge), with axes parallel to the directions in which the plates are moving. The coloured area represents the lithosphere.

1.14.7 Tomography, the mantle and the geoid

Lateral as well as vertical density variations in the mantle can be detected using seismic tomography (described in Section 1.7.6). Combining tomography with observations of the geoid (Section 1.8.2), we can build up a detailed picture of the three-dimensional structure of the mantle. A detailed discussion of results for the whole Earth is extremely complicated and well beyond the scope of this Course, but we can consider a few relatively simple examples.

Just as ocean waves approaching a beach are disrupted by the sea-bed when the depth of water is approximately the same as the wavelength (distance between the crests), so seismic surface waves are affected by features beneath them. The longer the wavelength, the deeper the waves can 'see'. The waves travel more rapidly or more slowly depending on the seismic velocity of regions beneath the surface at depths more or less equal to the seismic wavelength.

Rayleigh waves (which are not related to the Rayleigh number except in name) are most sensitive to heterogeneities in the mantle at depths of 100–600 km and so they 'see' rather deeper than Love waves (which are most sensitive at depths of 200–300 km). The slowest regions for Rayleigh waves include western North America, northeast Africa, the central Indian Ocean, the northeast Atlantic and the Tasman Sea–New Zealand–Campbell Plateau region. The fastest regions are the western Pacific, New Guinea–western Australia–east Indian Ocean–west Africa, northern Europe and the South Atlantic (Figure 1.117). You will use the Smithsonian Map in more detail with Block 2, but a glance at it now should lead you to some important conclusions based on the surface-wave tomography.

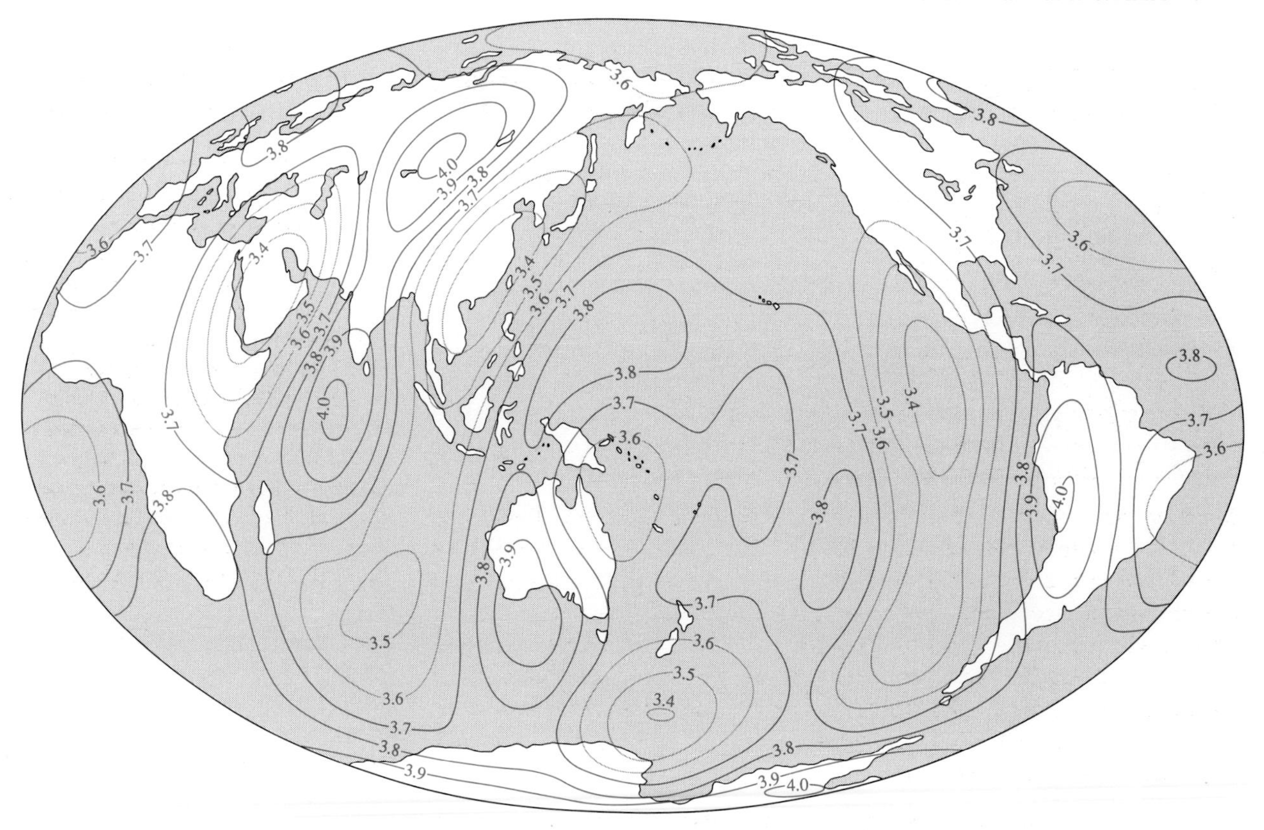

Figure 1.117 This map shows contours of Rayleigh wave velocity superimposed on outlines of coastlines. The contour interval is $0.1\,km\,s^{-1}$.

ITQ 61

Can you see any broad correlations between Rayleigh-wave velocity and the earthquake and volcanic regions marked on the Smithsonian Map?

There are many similarities between the results of Rayleigh and Love waves (Figures 1.117 and 1.118) and some differences related to the different properties of the waves that need not concern us here.

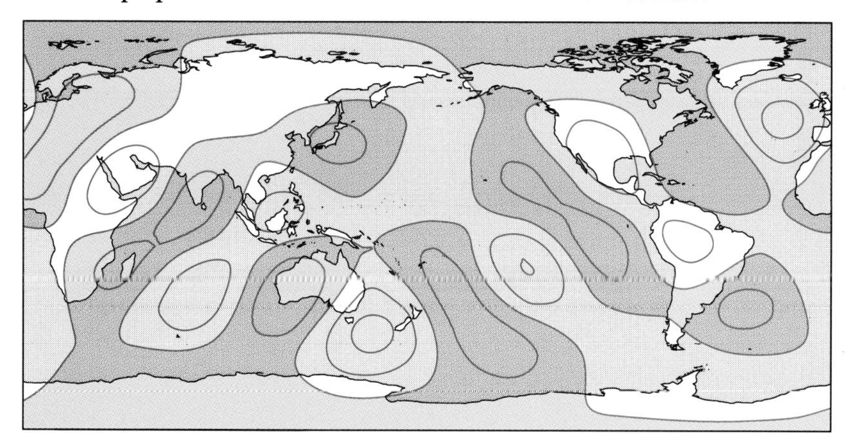

Figure 1.118 Love-wave velocities for the upper few hundred kilometres of the mantle. Orange shaded areas represent fast regions, while non-orange areas are slow. Most old continental lithosphere is underlain by fast material, while oceanic and active tectonic regions are underlain by slow material.

You should recall from Section 1.8 that the geoid is the surface on which sea-level lies. If we assume that there is a linear relationship between geoid anomaly height and seismic velocity, then a 'computed' or 'predicted' velocity contour map can be drawn. This computed map (Figure 1.119) is similar to the Love-wave map (Figure 1.118). The regions where the maps agree indicate that the geoid anomalies and the changes in wave velocity have the same cause, and are therefore at depths of 200–300 km. There is a very poor correlation between Rayleigh-wave speed and the geoid, which implies that those geoid anomalies that cannot be accounted for by Love-wave anomalies cannot be accounted for by Rayleigh-wave anomalies either and must be caused by much deeper features. To illustrate what these deeper features might be, we will consider the example of Iceland in a moment.

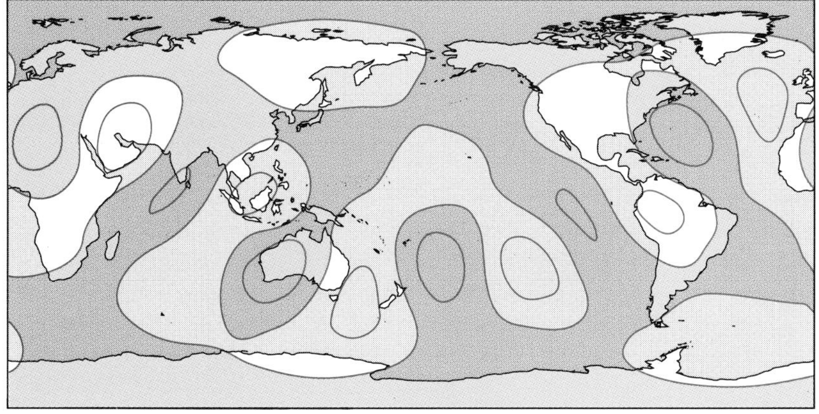

Figure 1.119 Assuming a linear correlation between geoid height and Love-wave velocity, the greater the geoid height the greater the Love-wave velocity should be and the two should increase or decrease evenly with each other. The Love-wave velocity may therefore be predicted from observed geoid height anomalies. This predicted velocity map compares well with the measured velocity (Figure 1.118). From the wavelength (width) of the geoid anomalies used to predict the Love-wave velocities, we know that they represent features in the upper mantle, so the fast and slow regions seen occur in the top few hundred kilometres of the mantle.

Maps of surface-wave velocity provide the best and most direct evidence we have of lateral heterogeneity in the mantle. In general, the shorter period waves (i.e. short wavelength), which sample only the crust and shallow mantle, correlate well with surface tectonics. The longer period waves, which penetrate into the transition zone (400–600 km), correlate less well with surface tectonics. There are not yet enough seismic stations worldwide to improve the horizontal resolution of this technique to better than about 2 000 km. Nevertheless, some very interesting and important results have been found.

Body-wave data are also used to construct a three-dimensional seismic model for the mantle in terms of P-wave and S-wave velocities. The results are easiest to see on a two-dimensional profile though (Figure 1.120). Orange regions are seismically fast, representing solid colder material, while grey regions are slow owing to the presence of partial melt. Features to note here are (i) for the western Pacific, low velocities at shallow depth (0–150 km) give way to faster regions below and (ii) the eastern Pacific is seismically slow at all depths. The Atlantic is fast below 400 km.

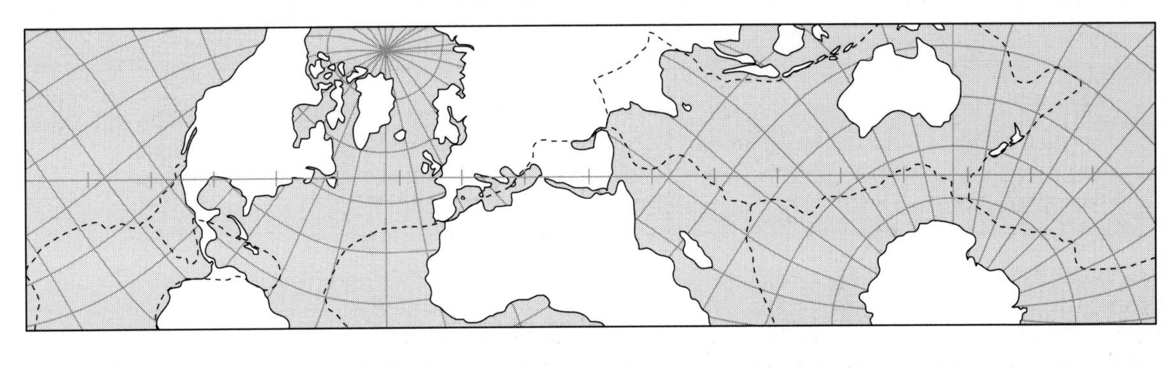

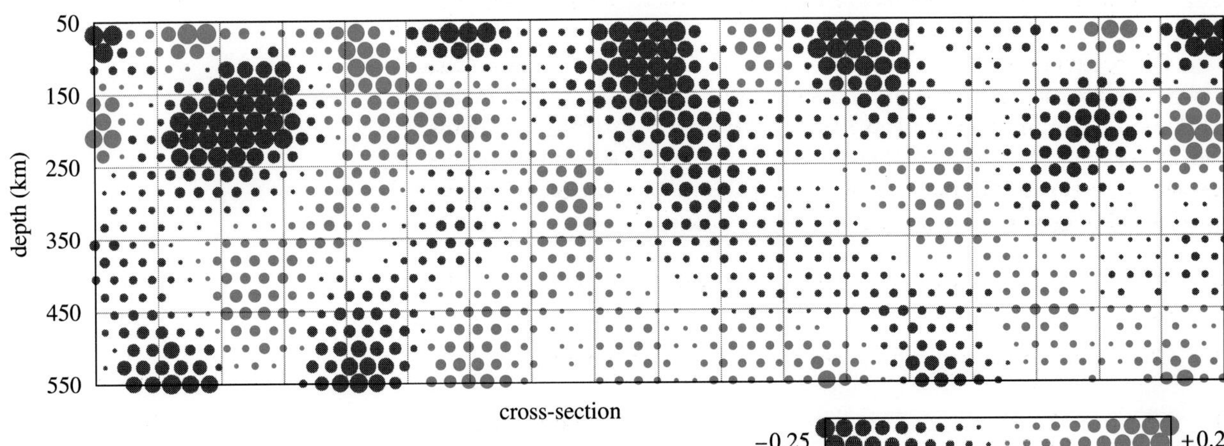

Figure 1.120 Seismic velocity in the upper mantle along the cross-section shown above. Velocity variations are given as fractions of the average, ranging from 25% slower to 25% faster than average. Note the fast region beneath the eastern USA and the slow region beneath the northern Mediterranean and the Indian Ocean; the slow region beneath the eastern Pacific is rather deeper. Note also the migration of fast and slow regions with depth. There is no anomaly — no particularly fast or slow material — beneath Britain.

The velocity of seismic waves through olivine, the major constituent of the mantle, depends on the alignment of the crystals. If olivine crystals are subjected to stress and begin to creep, they develop shear features and align themselves parallel to the stress. Seismic velocity is greater along this axis than it is perpendicular to it. Imagine a fast-flowing river carrying sticks, reeds, or even tree trunks. These objects align themselves parallel to the applied stress — the direction of flow, provided that the flow rate is not the same right across the river. In just the same way, crystals align themselves so that they offer least resistance to any applied stress. A dependence of seismic velocity on direction is called **velocity anisotropy**. This is different from inhomogeneity or heterogeneity, which in the mantle refer to changes in chemical composition or phase. Any flow within the mantle will tend to align the olivine crystals parallel to the direction of flow. We can make use of this fact to investigate whether flow is vertical or horizontal by looking at the relative velocities of **polarized seismic waves**. These are waves that are aligned in a particular direction — in this case either horizontally or vertically in the mantle. The results for both S-waves and P-waves waves do indeed show that in the upper mantle there is horizontal flow beneath continental lithosphere and vertical flow beneath oceanic lithosphere and subduction zones.

The movement of the lithospheric plates, which we shall return to in Block 2, is a consequence of these lateral heterogeneities. Here, we simply want to emphasize that the mantle is certainly not homogeneous but that there are discrete regions of partial melt extending up to the base of the lithosphere and identifiable as regions of horizontal and vertical flow.

1.14.8 The three-dimensional structure beneath Iceland

As an illustration of tomography and the geoid, we shall take a closer look at an area that has been studied intensively.

ITQ 62

Is Iceland characterized by a geoid anomaly, and if so, what does it mean? (Look at Figure 1.68 and revise Section 1.8.1 if you are unsure of this.)

The geoid surface is uplifted over a broad region in the Iceland area by about 50 m. This means that there is an overall excess of mass below this region. At first sight, this might be just what you would expect in an active volcanic zone. But in fact, excess mass would tend to sink, not rise, so we have a problem. Let us see whether seismic tomography can help. The behaviour of P-waves and S-waves in the crust and upper mantle is quite similar, so for simplicity we shall consider only P-waves. Figure 1.121 is a contour map of P-wave velocity for four discrete depths beneath Iceland. The velocities have been normalized (subtracted from the average velocity) so that the contours become P-wave velocity anomalies, rather like Bouguer gravity anomalies (Section 1.9).

❑ What happens to the region of anomalously low velocity as the depth increases?

■ The region bends towards the southwest and divides into two.

❑ What is the likely physical cause of the low-velocity regions beneath Iceland?

■ By analogy with the LVZ in the upper mantle (Section 1.7.5) this region probably represents a zone of partially molten material.

ITQ 63

What sort of geoid anomaly would you expect to see over a vertically extensive but laterally confined partial melt?

The three-dimensional picture deduced from seismic P-wave data (Figure 1.122) reveals a set of nested low-velocity regions. At shallow depths, the linear feature is probably related to the mid-Atlantic ridge that runs through Iceland (more about this in Block 2), but it is clear at greater depth that the low-velocity region becomes more confined and pipe-like.

The fact that there is a broad geoid high across Iceland tells us two things. Firstly, it is a positive anomaly and therefore not related to the low-velocity region described above because that would produce a geoid low. Secondly, the anomaly is broad, of long wavelength (width), indicating that the source of the anomaly is very deep. Although it is speculative at present, it is thought that this geoid high and others that do not correlate with more shallow density anomalies are due to mountainous features of the order of 10 km high on the core–mantle boundary. A mountain of core material (which has a much higher density than the mantle) penetrating into the mantle should produce an overall mass increase in the region which would be seen at the surface as a geoid high. Such a mountain would also cause a thermal perturbation or plume within the mantle. The plume would consist of hot buoyant material rising from the lower mantle and could even produce a hot spot when it reaches the lithosphere. We shall return to this subject in Block 2. Although these features do not necessarily lead directly to active volcanic regions at the Earth's surface, there is thought to be some link between thermal anomalies at the core–mantle boundary and volcanism. These features on the core–mantle boundary are related to heat flow from the core to the mantle and as such may be transitory structures. Indeed, their existence and nature are the subject of active research at present.

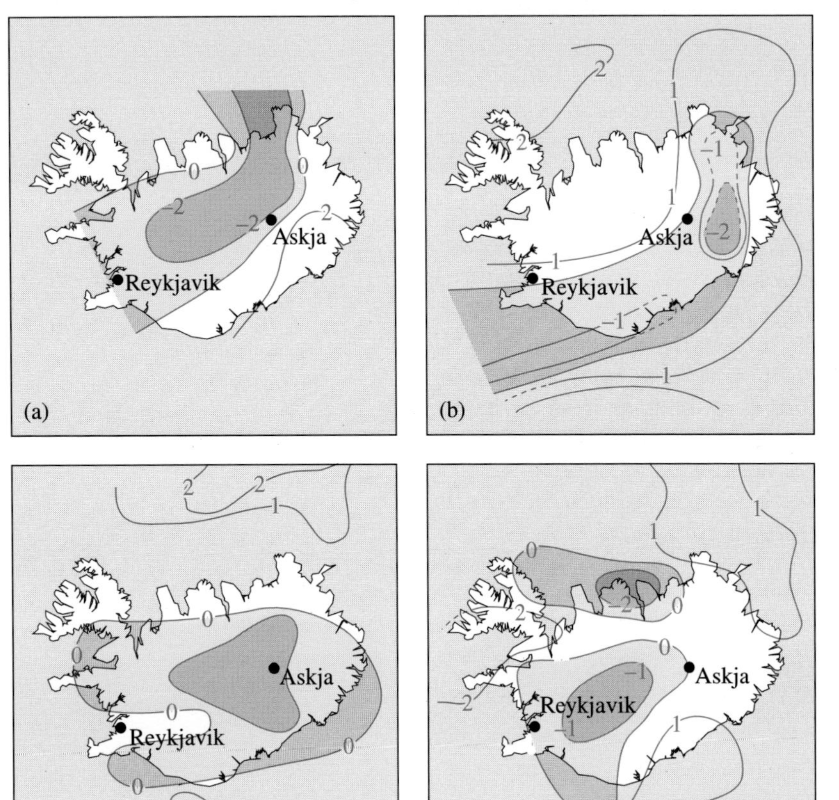

Figure 1.121 P-wave velocity variations beneath Iceland for depths of (a) 0–75 km; (b) 75–175 km; (c) 175–275 km; and (d) 275–375 km. Colour-shaded regions indicate increasingly low velocity, and contour values indicate percentage change in velocity.

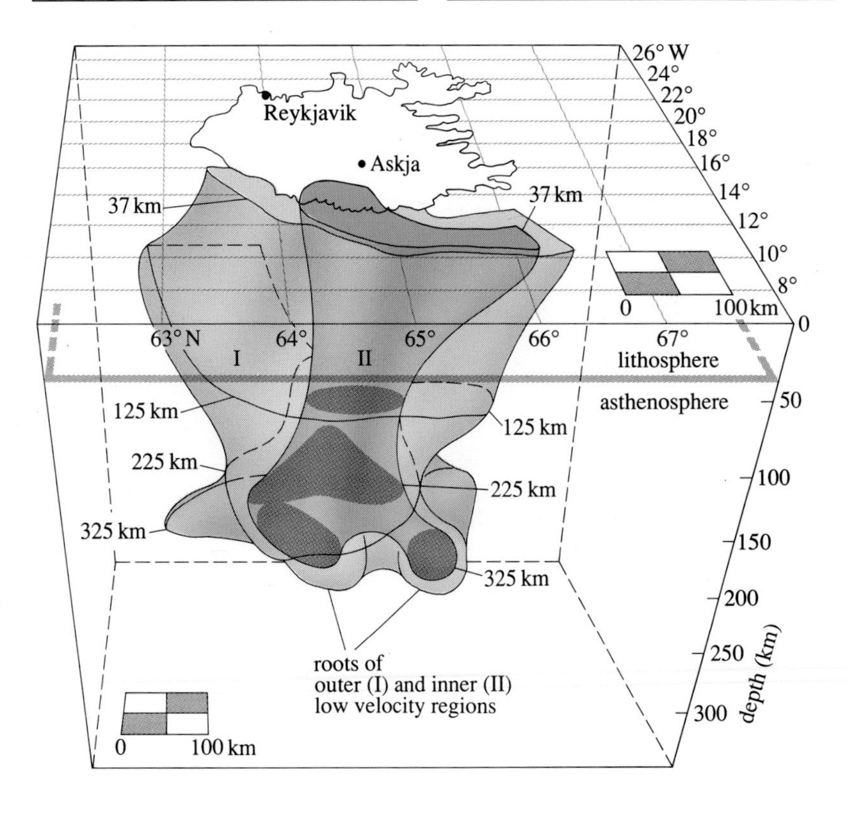

Figure 1.122 Three-dimensional structure of the Icelandic magma system looking westward away from Europe. The structure consists of two nested low-velocity regions. In region I, P-wave velocities are as much as 2% slower than average; and in region II, P-wave velocities are much more than 2% slower. The low-velocity region is not uniform beneath Iceland and rises up from a columnar plume-like shape at 325–225 km to an elongated shape at shallower depths. This reflects the complex structure of Iceland being formed from a combination of a deep-seated hot spot and shallow oceanic ridge spreading. More details of these processes will be given in Block 2.

SUMMARY OF SECTION 1.14

We know from the passage of seismic S-waves that the mantle is solid. However, by considering isostatic rebound after the retreat of ice caps in Scandinavia and North America, we can estimate a value for the dynamic viscosity of the mantle. When this value is combined with other realistic values for the physical properties of the mantle, the Rayleigh number is found to exceed the critical value by at least an order of magnitude (ITQ 60). Thus the mantle convects turbulently, but it moves very slowly, just a few centimetres per year.

Theoretical models developed on computers and laboratory models made with glycerine and other substances can simulate mantle convection. The mantle does not simply convect as a whole, but is divided into an upper and a lower part and these convect independently, so there are two layers. However, there is some limited interaction between these two layers. We refer then to mantle convection as a $1\frac{1}{2}$-layer process because it is partly 1-layer and partly 2-layer.

OBJECTIVES FOR SECTION 1.14

When you have completed this Section, you should be able to:

1.1 Recognize and use the definitions and applications of the terms printed in the text in bold.

1.55 Understand the importance of the Rayleigh number in convection.

1.56 Discuss the role of the various parameters that promote and inhibit mantle convection and describe how dynamic viscosity may be estimated.

1.57 Show how deformation within the mantle can be explained by either channel-flow or deep-flow models.

1.58 Discuss the relationship with depth for geoid anomalies, Love and Rayleigh waves.

1.59 Interpret a tomographic diagram showing slow and fast seismic regions in terms of hot and cold material and relate it to geoid anomalies.

Apart from Objective 1.1, to which they all relate, the seven ITQs in the Section test the Objectives as follows: ITQ 57, Objective 1.55; ITQs 58 and 59, Objective 1.57; ITQ 60, Objective 1.56; ITQs 61 and 62, Objective 1.58; ITQ 63, Objective 1.59.

You should now do the following SAQs, which test other aspects of the Objectives.

SAQS FOR SECTION 1.14

SAQ 62 (*Objectives 1.55 and 1.56*)

The Rayleigh number is defined in terms of the variables listed below. For each of these, indicate whether increase of the variable promotes or inhibits convection by ticking the appropriate column.

Variable	Promotes convection	Inhibits convection
(a) volume coefficient of thermal expansion (α)		
(b) temperature difference (ΔT)		
(c) gravitational acceleration (g)		
(d) depth of fluid layer (d)		
(e) thermal diffusivity (κ)		
(f) dynamic viscosity (η)		

SAQ 63 (*Objective 1.57*)

(a) How do the channel-flow and deep-flow models for mantle deformation associated with the growth of large ice sheets differ?

(b) What is the most important predicted distinction between the channel-flow and deep-flow models for the deformation behaviour of deglaciated areas?

SAQ 64 (*Objective 1.55*)

Given a uniform dynamic viscosity of around 10^{21} Pa s and hence a Rayleigh number of about 10^6 (the critical Rayleigh number is about 10^3), what does this imply about the convective state of the whole mantle?

SAQ 65 (*Objective 1.58*)

For a two-scale model of shallow-mantle convection, to what do the two scales refer and what are their approximate magnitudes?

SAQ 66 (*Objective 1.58*)

What sort of geoid anomaly would you expect across Iceland in view of the pipe-like low-velocity region beneath it, and why?

ITQ ANSWERS AND COMMENTS

ITQ 1

One light year is equivalent to 9.46×10^{12} km (Section 1.2); and so 4.2 light years = $4.2 \times 9.46 \times 10^{12}$ km = 39.7×10^{12} km, or 39.7 million million kilometres.

S10

Note that in observing a star 4.2 light years away, one is seeing it as it was 4.2 years ago, because that is the time it takes light to travel from the star to the Earth. By the same token, the outermost limits of the known Universe, 10^{10} light years away, are being observed as they were before the Solar System even formed (i.e. before 4.6×10^9 years ago).

ITQ 2

The mean distance from the Sun to the Earth is about 1.5×10^8 km (Section 1.2). Light travels 3×10^5 km in one second; and so it travels 1.5×10^8 km in $\frac{1.5 \times 10^8}{3 \times 10^5}$ s = 500 s, or about 8.3 minutes.

ITQ 3

The total mass of the Sun, Moon and planets, obtained by adding all the figures in the 'mass' column of Table 1.1, is $2.002\,668\,956 \times 10^{30}$ kg. The proportion of this accounted for by the Sun is therefore: mass of Sun/mass of Sun, Moon and planets = $2 \times 10^{30}/2.002\,668\,956 \times 10^{30} = 0.998\,667\,3$, or 99.87% corrected to two decimal places.

Note that the figure of 0.998 667 3 is actually a slight overestimate, for the Sun, Moon and planets are not the only bodies in the Solar System, although the corrected figure of 99.87% is not an overestimate. In fact, this is not a very sensible calculation at all, for the masses of Mercury, Pluto, the Moon, etc. are so small compared to that of the Sun that they are not worth including in the calculation. If all planets are ignored except Jupiter, Saturn and Neptune (i.e. ignoring all planets with masses less than 1.0×10^{26} kg), the answer corrected to two decimal places turns out to be the same (99.87%).

ITQ 4

Figure 1.123 shows that for the Earth, Venus, Mars and the Moon (but not Mercury) there is a roughly linear relationship between mean density and radius.

This is really quite a remarkable result. The linearity is less important than the regularity, which could mean one of two things. The first is that the four bodies concerned are chemically identical and homogeneous, and that the only factor involved is self-compression. However, we know that the Earth is not chemically homogenous and that self-compression is not the only factor involved there; so that rules out the first option. The second is that, though not chemically homogeneous, the terrestrial planets and the Moon are indeed broadly similar in overall composition; otherwise, as the text explains, there would be greater variation in mean density, and the regularity would probably not appear. It's interesting to note that, if the points *had* been much more widely scattered, that would not necessarily have ruled out similar bulk compositions, because there could have been other disrupting factors. For example, planets with the same overall compositions could nevertheless have radically different structures; and this, too, could have introduced some variability into the mean densities. The fact that, despite all the odds, the Earth, Venus, Mars and the Moon plot as they do suggests strongly that they do indeed have similar overall compositions, although *this cannot be said to constitute conclusive proof*. Mercury is the odd body out in all this. Its mean density is clearly much higher in relation to its size than for the other bodies, implying that its chemistry differs significantly from theirs. The simplest hypothesis is that, relative to its size, it has a much bigger iron core than do its inner-Solar-System companions (iron, remember, is denser than silicates). In Section 1.3, we shall look at a possible explanation for this.

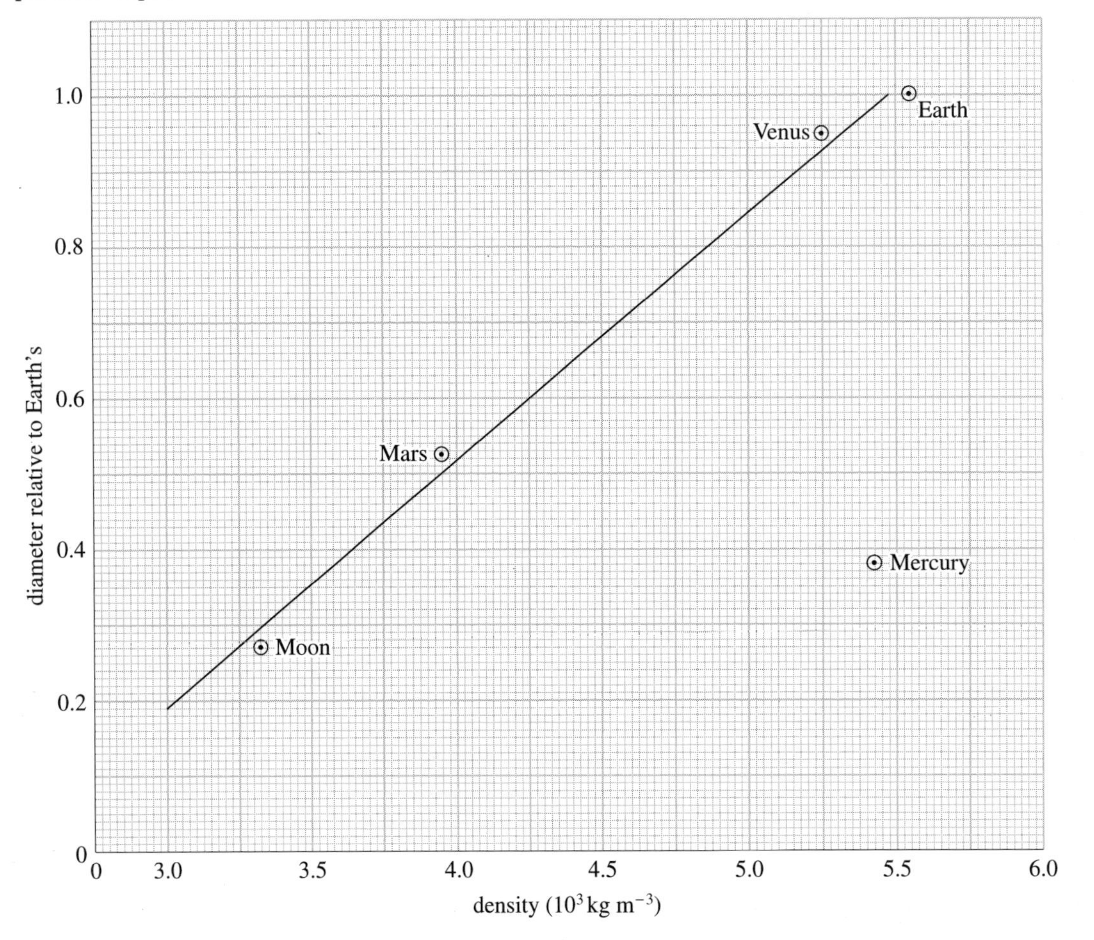

Figure 1.123 Graph relating to the answer to ITQ 4

ITQ 5

There must have been enough planetesimals to form the terrestrial planets and the Moon. The total mass of the terrestrial planets and the Moon (from Table 1.1) is 1.2×10^{25} kg. So it would have required $1.2 \times 10^{25}/10^{15} = 1.2 \times 10^{10}$ planetesimals of mass 10^{15} kg each to make up this mass. The number density is then

$$\frac{\text{no. of planetesimals}}{\text{volume (km}^3)} = \frac{1.2 \times 10^{10}}{2.8 \times 10^{18}\,\text{km}^3} = 4.3 \times 10^{-9}\,\text{km}^{-3}.$$

This figure looks a little odd, because, of course, each 10^{15} kg planetesimal has a volume much greater than 1 km^3. Perhaps it makes more sense to put it the other way round — i.e. that there is one planetesimal in every $2.8 \times 10^{18}\,\text{km}^3/1.2 \times 10^{10} = 2.3 \times 10^8\,\text{km}^3$. And as each planetesimal has a volume of about 10^4 km^3, this means that only 1/10 000 of the available space is taken up with planetesimal material. It is clear, therefore, that the planetesimals are hardly closely packed. In reality, they must have been even less closely packed, for the thickness of the disc would surely have been greater than the thickness of one planetesimal.

ITQ 6

The total mass of the four terrestrial planets is 1.2×10^{25} kg (Table 1.1). The number of 10^{23} kg planetary embryos needed to produce them is therefore $1.2 \times 10^{25}/10^{23} = 120$.

ITQ 7

You should have compared all the elements in Table 1.2 with those in Figure 1.5 to determine which are volatiles and which are refractories. You should then have spotted that all the volatiles except sulphur have Moon : Earth abundance ratios of less than 1, whereas all the refractories except iridium have ratios greater than 1. The conclusions you should have drawn are that, relative to the Earth, the Moon is *generally* depleted in volatiles and enriched in refractories, although it is neither depleted in every volatile nor enriched in every refractory. Note, however, that what are being compared here are lunar and terrestrial basalt — i.e. surface and near-surface rocks. It does not follow that these rocks are necessarily representative of the Moon and Earth as a whole (although Earth scientists often assume that they are in the case of the Moon).

ITQ 8

From Equation 1.3, Section 1.3.3,

$$\Delta T = \frac{mv^2}{2MC}$$

where $m = 10^{15}$ kg (given), $v = 10^4$ m s^{-1} (given), $C = 7.5 \times 10^2$ J kg^{-1} K^{-1} (given), and $M = 6.0 \times 10^{24}$ kg (from Table 1.1).

And so,

$$\Delta T = \frac{10^{15}\,\text{kg} \times (10^4\ \text{m s}^{-1})^2}{2 \times 6 \times 10^{24}\,\text{kg} \times 7.5 \times 10^2\ \text{J kg}^{-1}\,\text{K}^{-1}} = 1.1 \times 10^{-5}\,\text{K}.$$

This may look, and is, very small; but now do ITQ 9.

ITQ 9

The number of planetesimals of mass 10^{15} kg each required to construct the Earth is $6 \times 10^{24}/10^{15} = 6 \times 10^9$. If each of these produced the temperature rise in ITQ 8, the total temperature rise in the Earth would be $6 \times 10^9 \times 1.1 \times 10^{-5}$ K = 66 000 K.

This is a very crude, but very interesting, calculation. The temperature rise given by ITQ 8 was based on the impact of a 10^{15} kg planetesimal on an Earth of current size. In practice, most of the planetesimals going into the construction of the Earth would have impacted a *growing* Earth (i.e. one smaller than the present Earth). As Equation 1.3 shows, if M were smaller, ΔT would be bigger. In other words, the 66 000 K temperature rise obtained above can be regarded as a *minimum* temperature rise — and it is already well over the vaporization temperature of refractories! (See Figure 1.7.)

ITQ 10

Using Equation 1.2, Section 1.3.3, for a total kinetic energy of 10^{32} J/10 = 10^{31} J converted to heat,

$$\Delta T MC = 10^{31}\,\text{J}$$

and so

$$\Delta T = \frac{10^{31}}{6 \times 10^{24} \times 7.5 \times 10^2}\,\text{K} = 2\,222\,\text{K} = 2\,200\,\text{K (to two sig. figs.)}$$

In fact, this is an underestimate of the real temperature rise because it is based on an Earth mass of 6×10^{24} kg, whereas the average size of the Earth during growth would have been much less.

ITQ 11

The main features of Figure 1.12 are as follows, although you may have them in a different order:

(a) Hydrogen and helium are by far the most abundant elements in the Sun by several orders of magnitude. For every 10^6 silicon atoms, for example, there are 10^{10}–10^{11} hydrogen atoms and 10^9–10^{10} helium atoms. In fact, H and He together account for more than 98% of the Sun's mass.

(b) The abundances of the elements generally decrease with increasing atomic number, although the decrease is not smooth.

(c) Within a similar range of atomic number (e.g. around atomic number 40, etc.), elements with even atomic numbers (white circles) have greater abundances (by a factor of about 10) than those with odd atomic numbers (solid circles). Also, lithium (Li), beryllium (Be) and boron (B) are sharply depleted in comparison with other elements of similar atomic number. The reasons for these phenomena are known but are beyond the scope of this Course. We are here simply interested in the the fact that the relative abundances of the elements in the Sun are known.

ITQ 12

(a) The higher the degree of oxidation, the more the iron is combined into silicates (Fe_2SiO_4 and others) and the less it appears as either free metal (Fe) or in iron sulphide (FeS). As Figure 1.14 shows, the E-group has very little iron in silicates (but almost all as Fe or FeS), the H-group has more, the L-group has still more, the LL-group has yet more, and the C-group has most. The order of increasing oxidation is therefore E, H, L, LL, C.

(b) Along each diagonal line in Figure 1.14 the total iron content is the same, irrespective of how the iron is divided between silicates and Fe/FeS; and the higher the diagonal line the higher is the total iron content (indicated by the scale at each end of the line). This exercise is more difficult than that in (a) because the shaded boxes overlap considerably where total iron content is concerned. However, working on the basis of the average iron content in each group as represented by the mid-point of each shaded box, the order of increasing iron content is LL (about 20%), L (22%), C (24%), E (27%), H (28%).

ITQ 13

(a) Ni lies to the left of the line in Figure 1.17 and is therefore depleted in the crust compared to solar abundances, by more than 100 times. Mg is also depleted in the crust compared to solar abundances, this time by less than 10 times. You may find these figures a little difficult to determine, because the scales on Figure 1.17 are logarithmic, not linear. However, you should be able to see that for every approximately 10^2 Ni atoms in the crust, there are in excess of 10^4 in the Sun (i.e. more by a factor of at least 10^2); and that for every 10^5 Mg atoms in the crust, there are in excess of 10^5 (but fewer than 10^6) in the Sun.

(b) Figure 1.3 (caption) suggests that Mg is more than usually abundant in the mantle and the fact that Ni often alloys with iron in meteorites suggests that Ni may be abundant in the core. It is quite possible, therefore, that these elements are depleted in the crust simply because they have preferentially moved elsewhere (to the mantle in the case of Mg and to the core in the case of Ni). The crustal depletions in Ni and

Mg are thus quite consistent with an initially chondritic Earth (but do not prove the validity of the CEM).

ITQ 14

As the earthquake occurs close to the Earth's surface and within the State in which the seismometer is located, an approximate earthquake–seismometer distance will be given by Equation 1.7. The first S-waves arrive 24 s after the first P-waves, and so t = 24 s. Therefore, d = 8.65 × 24 km = 207.6 km = 208 km (to three sig. figs.)

ITQ 15

The differences in arrival times for P-waves and S-waves are:

Station A: 12.1 s
Station B: 21.9 s
Station C: 22.0 s

Using Equation 1.7 gives the following epicentral distances:

Station A: 104.7 km = 105 km (to three sig. figs.)
Station B: 189.4 km = 189 km (to three sig. figs.)
Station C: 190.3 km = 190 km (to three sig. figs.)

Arcs of circles with these radii are shown in Figure 1.124. You will notice that the arcs do not meet exactly at a point, largely as a result of experimental error in the original data. The epicentre may be presumed to lie in the small region within all three arcs.

Figure 1.124 Construction relating to the answer to ITQ 15

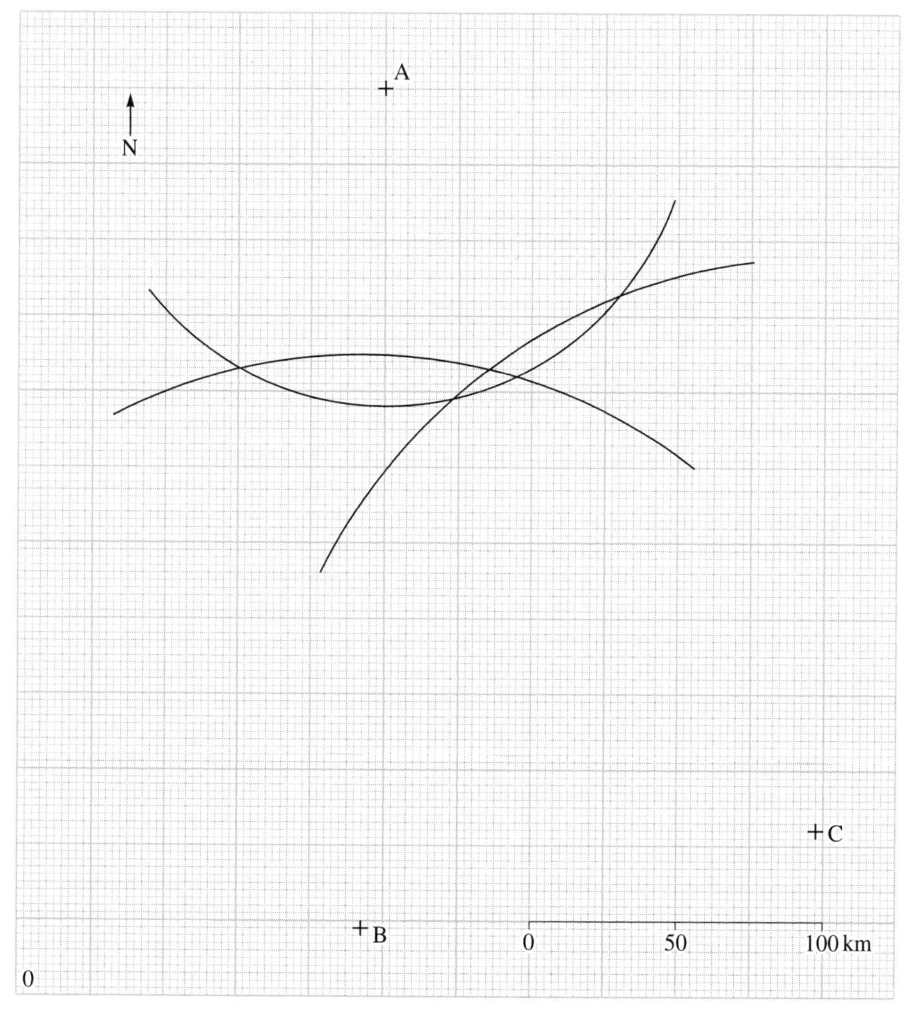

ITQ 16

The earthquake has $h < 50$ km and $\Delta > 20°$; so Equation 1.9 can be used with $A = 200\,\mu m$ (2×10^{-4} m = 200×10^{-6} m = $200\,\mu m$), $T = 20$ s, and $\Delta = 30°$. Therefore,

$$M_S = \log\left(\frac{200}{20}\right) + 1.66 \log 30 + 3.3$$

$$= 1 + (1.66 \times 1.477) + 3.3$$

$= 6.8$ (to two sig. figs., which is the maximum to which earthquake magnitudes are ever quoted).

ITQ 17

In this case, Equation 1.10 is the more appropriate, using $A = 60\,\mu m$, $T = 12$ s and $\Delta = 30°$. Therefore,

$$m_b = \log\left(\frac{60}{12}\right) + (0.01 \times 30) + 5.9$$

$$= 0.6990 + 0.3 + 5.9$$

$= 6.9$ (to two sig. figs.)

ITQ 18

Figure 1.125 shows the plot of M_S against m_b. What is immediately obvious is that no straight line could pass through all the points, for there is considerable scatter. In other words, there is no one-to-one relationship between M_S and m_b; the relationship is a statistical one. The straight line drawn through the points in Figure 1.125 is actually the best straight line that can be drawn through a worldwide set of M_S and m_b. Its equation is:

$$m_b = 2.94 + 0.55M_S.$$

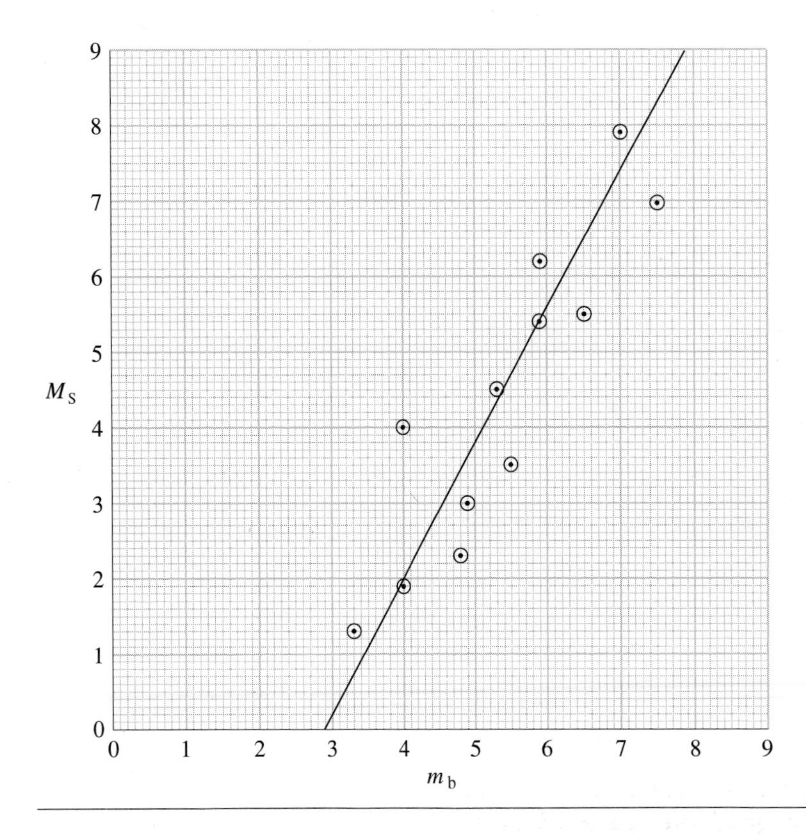

Figure 1.125 Graph relating to the answer to ITQ 18.

ITQ 19

The San Francisco earthquake had a magnitude (M_S) of 8.3. So using Equation 1.12:

$$\log E = 4.8 + 1.5(8.3) = 17.25.$$

Therefore, $E = 1.8 \times 10^{17}$ J (to two sig. figs.).

ITQ 20

Using Equation 1.12:

When M_S = (say) 5.0, $\log E = 4.8 + 1.5(5.0) = 12.3$.

Therefore $E = 2.0 \times 10^{12}$ J.

And when $M_S = 6.0$, $\log E = 4.8 + 1.5(6.0) = 13.8$.

Therefore $E = 6.3 \times 10^{13}$ J.

Thus from $M_S = 5.0$ to $M_S = 6.0$, the energy has increased by a factor of:

$$\frac{\text{energy for } M_S = 6.0}{\text{energy for } M_S = 5.0} = \frac{6.3 \times 10^{13}}{2.0 \times 10^{12}} = 31.5 = 32 \text{ (to two sig. figs.)}$$

So an earthquake of $M_S = 6.0$ releases about 32 times more energy than one of $M_S = 5.0$, $32 \times 32 = 1\,024$ times (= 1 000 times, to two sig. figs.) more energy than one of $M_S = 4.0$, and so on.

ITQ 21

Again using Equation 1.12:

For $M_S = 8.5$, $\log E = 4.8 + 1.5(8.5) = 17.55$.

Therefore $E = 3.5 \times 10^{17}$ J.

And for $M_S = 2.5$, $\log E = 4.8 + 1.5(2.5) = 8.55$.

Therefore, $E = 3.5 \times 10^{8}$ J.

And as there are 300 000 earthquakes of $M_S = 2.5$, their total energy is $3.5 \times 10^{8} \times 3 \times 10^{5}$ J $= 10.5 \times 10^{13}$ J $= 1.1 \times 10^{14}$ J (to two sig. figs.)

Thus the single $M_S = 8.5$ earthquake releases at least three orders of magnitude more energy than all 300 000 $M_S = 2.5$ earthquakes put together. In fact, if you do the calculations you will discover that the single $M_S = 8.5$ earthquake releases more energy than *all* the other earthquakes in Table 1.4 taken together.

ITQ 22

Your answer should include the following, in no particular order:

(1) Within a given layer (mantle, outer core, inner core), wave paths are generally curved rather than straight.

(2) When waves hit a layer boundary (e.g. that between mantle and outer core), they undergo a sudden change in direction (i.e. they are refracted).

(3) Waves are reflected as well as refracted at boundaries, including the Earth's surface.

(4) When P-waves are reflected or refracted, they can produce both P-waves and S-waves; likewise, when refracted and reflected, S-waves can produce both S-waves and P-waves.

ITQ 23

(a) For the 750 m source, $v_P = \dfrac{\text{distance}}{\text{time}} = \dfrac{750\,\text{m}}{0.136\,\text{s}} = 5\,515\,\text{m}\,\text{s}^{-1}$.

For the 1 000 m source, $v_P = \dfrac{1000\,\text{m}}{0.182\,\text{s}} = 5\,495\,\text{m}\,\text{s}^{-1}$.

Therefore, average $v_P = \dfrac{5\,515\,\text{m} + 5\,495\,\text{m}}{2\,\text{s}} = 5\,505\,\text{m}\,\text{s}^{-1} \sim 5.5\,\text{km}\,\text{s}^{-1}$.

For the 750 m source, $v_S = \dfrac{750\,\text{m}}{0.250\,\text{s}} = 3\,000\,\text{m}\,\text{s}^{-1}$.

For the 1 000 m source, $v_S = \dfrac{1\,000\,\text{m}}{0.330\,\text{s}} = 3\,030\,\text{m}\,\text{s}^{-1}$.

Therefore, average $v_S = \dfrac{3\,000\,\text{m} + 3\,030\,\text{m}}{2\,\text{s}} = 3\,015\,\text{m}\,\text{s}^{-1} \sim 3.0\,\text{km}\,\text{s}^{-1}$.

(b) From Equation 1.17:

$$v_S = \sqrt{\frac{\mu}{\rho}}$$

and so (by squaring each side)

$$\begin{aligned} \mu &= \rho(v_S)^2 \\ &= 2\,700\,\text{kg}\,\text{m}^{-3} \times (3\,015\,\text{m}\,\text{s}^{-1})^2 = 2.5 \times 10^{10}\,\text{N}\,\text{m}^{-2}. \end{aligned}$$

Then from Equation 1.16:

$$v_P = \sqrt{\frac{K + 4\mu/3}{\rho}}$$

and so (by squaring both sides)

$$\begin{aligned} K &= \rho v_P^2 - \frac{4\mu}{3} \\ &= 2\,700\,\text{kg}\,\text{m}^{-3} \times (5\,505\,\text{m}\,\text{s}^{-1})^2 - \frac{4 \times 2.5 \times 10^{10}\,\text{N}\,\text{m}^{-2}}{3} \\ &= 8.2 \times 10^{10} - 3.3 \times 10^{10}\,\text{N}\,\text{m}^{-2} \\ &= 4.9 \times 10^{10}\,\text{N}\,\text{m}^{-2}. \end{aligned}$$

Notice that the bulk modulus is greater than the rigidity modulus; and this is true for all Earth materials. Note also that the units of elastic moduli ($\text{N}\,\text{m}^{-2}$) are the same as those of stress (force per unit area), because strain is always a dimensionless (unitless) quantity.

ITQ 24

The critical angle is given by Equation 1.20. Thus:

$$\sin i_c = \frac{v_1}{v_2} = \frac{4}{8} = 0.5$$

and so $i_c = \sin^{-1} 0.5 = 30°$.

ITQ 25

Because the angle of incidence is less than 30°, there will be a wave refracted into the lower layer. So from Equation 1.19:

$$\sin r = \frac{v_2 \sin i}{v_1} = \frac{8 \sin 20}{4} = \frac{8 \times 0.342\,0}{4} = 0.684\,0$$

and so $r = 43°$.

ITQ 26

You first need to determine the critical distance SC. If the critical distance is $2y$ (see Figure 1.126), y is given by

$$\tan i_c = \frac{y}{h}$$

and so

$$\begin{aligned} 2y \text{ (critical distance)} &= 2h \tan i_c \\ &= 2 \times 10\,\text{km} \times \tan 30^\circ \\ &= 20 \times 0.5774\,\text{km} \\ &= 11.55\,\text{km} \\ &= 12\,\text{km (to two sig. figs.)} \end{aligned}$$

The detector is thus outside the critical distance and will thus receive both the direct wave and a refracted wave. Note that response (b) was impossible anyway; there is no point at which the detector could receive *only* a refracted wave.

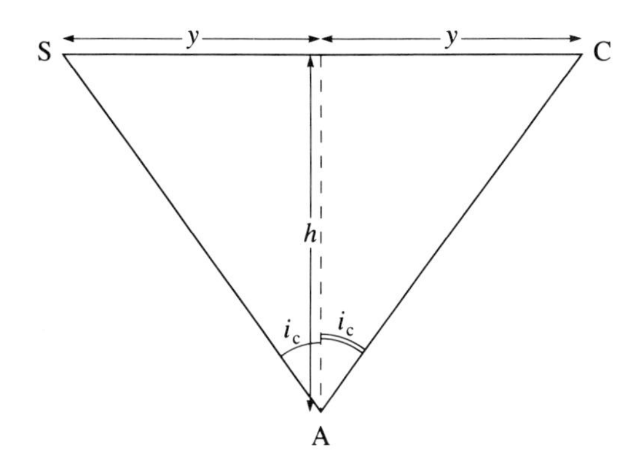

Figure 1.126 Diagram relating to the answer to ITQ 26.

ITQ 27

This is simply a matter of calculating x_d, with $v_1 = 4\,\text{km s}^{-1}$, $v_2 = 8\,\text{km s}^{-1}$ and $h = 10\,\text{km}$. Therefore, using Equation 1.24:

$$x_d = 2 \times 10 \sqrt{\frac{8+4}{8-4}}\,\text{km} = 20\sqrt{3}\,\text{km} = 35\,\text{km (to two sig. figs.)}$$

ITQ 28

The completed version of Figure 1.38 appears here as Figure 1.127.

(a) The gradient of the direct-wave line is $1/v_1$, so

$$\frac{1}{v_1} = \frac{4\,\text{s}}{12\,\text{km}};\quad \text{therefore } v_1 = 3\,\text{km}\,\text{s}^{-1}.$$

Note: The gradient is determined by counting squares at some convenient point. Here we took distance = 12 km and then counted upwards to find that time = 4 s.

Similarly, the gradient of the refracted-wave line is $1/v_2$, so

$$\frac{1}{v_2} = \frac{(8.3 - 4.9)\,\text{s}}{(40 - 16)\,\text{km}} = \frac{3.4\,\text{s}}{24\,\text{km}};\quad \text{therefore } v_2 = 7\ \text{km s}^{-1}.$$

(b) h may be obtained from Equation 1.25, which requires x_d. From the graph, $x_d = 14$ km. So:

$$h = \frac{x_d}{2}\sqrt{\frac{v_2 - v_1}{v_2 + v_1}} = \frac{14\,\text{km}}{2}\sqrt{\frac{4}{10}} = 7\,\text{km}\sqrt{0.4} = 4.4\,\text{km (to two sig. figs.)}$$

h may also be obtained from Equation 1.26, which requires t_0. From the graph, $t_0 = 2.65$ s. So:

$$h = \frac{t_0 v_1 v_2}{2\sqrt{v_2^2 - v_1^2}} = \frac{2.65 \times 3 \times 7}{2\sqrt{49 - 9}}\,\text{km} = \frac{55.65}{2\sqrt{40}}\,\text{km} = 4.4\,\text{km (to two sig. figs.)}$$

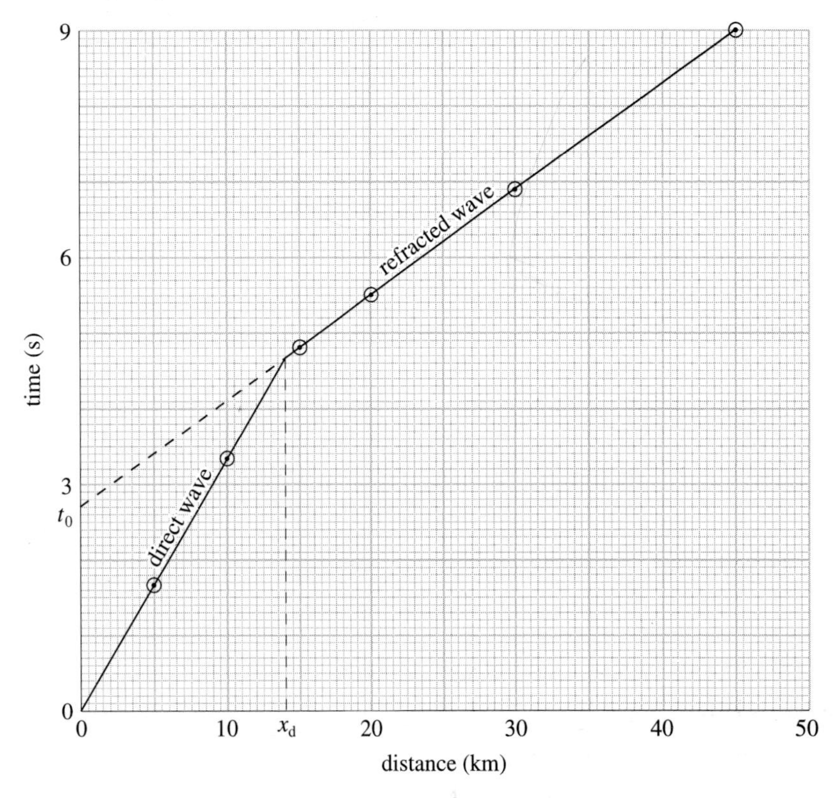

Figure 1.127 Graph relating to the answer to ITQ 28

ITQ 29

Squaring both sides of Equation 1.30 gives:

$$t^2 = \frac{x^2}{v^2} + \frac{4h^2}{v^2}$$

and so a plot of t^2 against x^2 has a gradient $1/v^2$. The plot is shown as Figure 1.128.

(a) gradient $= \frac{1}{v^2} = \frac{8.8\,\text{s}^2}{140\,\text{km}^2}$; therefore $v^2 = \frac{140\,\text{km}^2}{8.8\,\text{s}^2} = 15.9\,\text{km}^2\,\text{s}^{-2}$

(to three sig. figs.)

and so $v = 3.99\,\text{km}\,\text{s}^{-1} \sim 4\,\text{km}\,\text{s}^{-1}$, where v is the velocity above the discontinuity.

(b) Intercept on t^2 axis $= \frac{4h^2}{v^2} = 4\,\text{s}^2$; therefore $h^2 = \frac{4v^2}{4}\,\text{s}^2$

$= 16\,\text{km}^2$

and so $h = 4\,\text{km}$.

The thickness of the upper layer above the discontinuity is therefore 4 km.

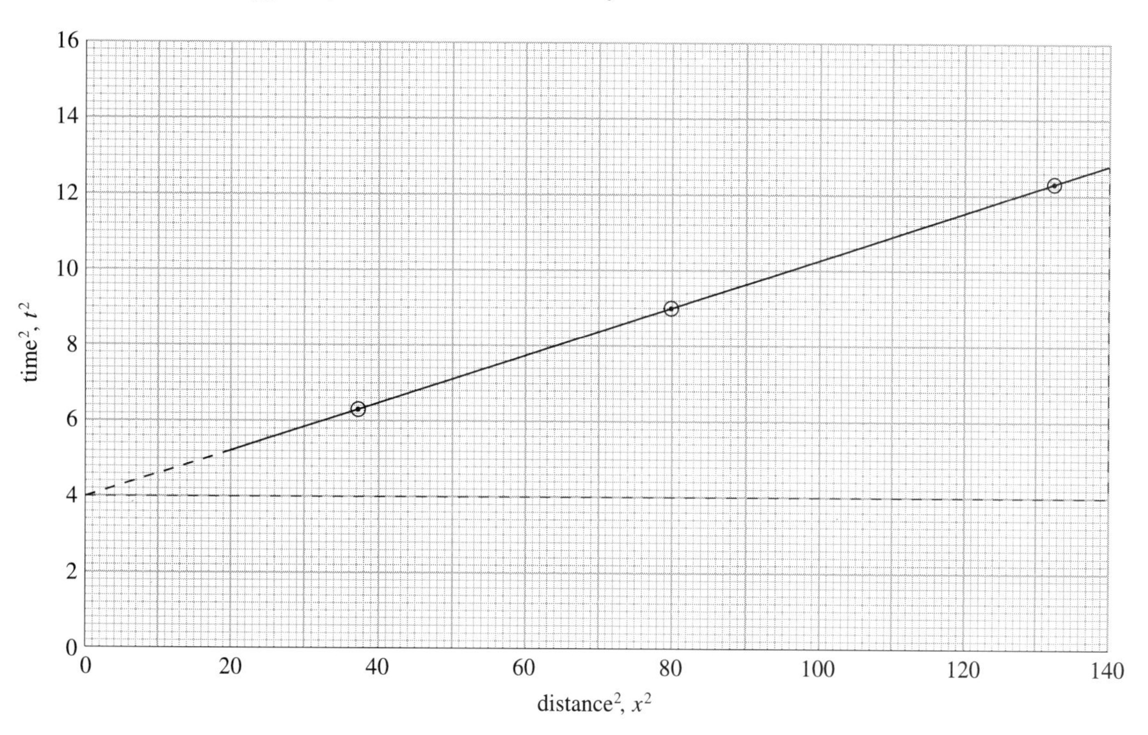

Figure 1.128 Graph relating to the answer to ITQ 29

ITQ 30

(a) The Moho reflector lies at about 10 s, which, don't forget, is the *two-way* travel time. At a velocity of $6\,\text{km}\,\text{s}^{-1}$, 10 s represents 60 km; and so the Moho lies at a depth of about 30 km.

(b) The depth covered by the Moho reflector is not very clear, but it looks to be about 0.5 s. At a velocity of $6\,\text{km}\,\text{s}^{-1}$, 0.5 s represents 3 km; so the thickness of the Moho is 1.5 km.

ITQ 31

We use exactly the same method here as we used for calculating the value of gravity at the pole, except that this time we need to use the equatorial radius instead of the polar radius:

$$g_{\text{equator}} = \frac{6.672 \times 10^{-11}\,\text{N m}^2\,\text{kg}^{-2} \times 6 \times 10^{24}\,\text{kg}}{(6\,378 \times 10^3)^2\,\text{m}^2}$$

$$g_{\text{equator}} = 9.841\,\text{N kg}^{-1}$$

$$g_{\text{equator}} = 9.841\,\text{m s}^{-2}.$$

The difference in the acceleration due to gravity between the pole and the equator is therefore:

$$9.908\,\text{m s}^{-2} - 9.841\,\text{m s}^{-2} = 6.7 \times 10^{-2}\,\text{m s}^{-2}.$$

ITQ 32

Using the IGF (Equation 1.35), the value of gravity at S is

$$\begin{aligned} g_{10.431} &= 9.780\,318\,5(1 + 5.278\,895 \times 10^{-3}\ (\sin 10.431)^2 \\ &\quad + 2.346\,2 \times 10^{-5}\ (\sin 10.431)^4)\,\text{m s}^{-2} \\ &= 9.782\,011\,1\,\text{m s}^{-2}; \end{aligned}$$

and for the value at P,

$$\begin{aligned} g_{10.590} &= 9.780\,318\,5\ (1 + 5.278\,895 \times 10^{-3}\ (\sin 10.590)^2 \\ &\quad + 2.346\,2 \times 10^{-5}\ (\sin 10.590)^4)\,\text{m s}^{-2} \\ &= 9.782\,062\,5\,\text{m s}^{-2}. \end{aligned}$$

The latitude correction is the difference between these values,

$$\begin{aligned} \Delta_1 g &= (9.782\,011\,1 - 9.782\,062\,5)\,\text{m s}^{-2} \\ &= -5.14 \times 10^{-5}\,\text{m s}^{-2}. \end{aligned}$$

These SI units are not normally used in practice since they are rather inconvenient, so we use gals. Thus

$$5.14 \times 10^{-5}\,\text{m s}^{-2} = 5.14 \times 10^{-3}\,\text{Gal}$$

and the common unit is the mGal, or thousandth of a gal. The latitude correction therefore is –5.14 mGal, and it needs to be *subtracted* from the observed value of gravity at S to make it comparable with the value at P (remembering that subtracting a negative number means that you add it).

ITQ 33

(a) We need to evaluate $\Delta_2 g$, using Equation 1.37.

If S is 10 m *above* P, then $h = 10$ m, so

$$\begin{aligned} \Delta_2 g &= -0.308\,6 \times 10\,\text{mGal} \\ &= -3.086\,\text{mGal}. \end{aligned}$$

This value is negative, but the correction is taken away from the observed gravity difference (Equation 1.38) between S and P, so in fact we add 3.086 mGal to the observed value of gravity at S to make it comparable with P.

(b) In this case, $h = -10$ m, so

$$\begin{aligned} \Delta_2 g &= -0.3086 \times (-10)\,\text{mGal} \\ &= 3.086\,\text{mGal}. \end{aligned}$$

This time the correction itself is positive, but it is again subtracted from the observed gravity difference, so that the value at S is *reduced* to make it comparable with the value at P.

(c) In this case, $h = 0$ m, so

$$\Delta_2 g = 0\,\text{mGal}.$$

There is no elevation correction if S and P are at the same height.

(d) In this case, h = 100 m, so

$$\Delta_2 g = -30.86\,\text{mGal}$$

and this value will be *added* to the value observed at S to make it comparable with the value at P.

ITQ 34

(a) You have seen already that gravity at S in Figure 1.70a will be less than at P by an amount $\Delta_2 g$, the free-air correction, because S is further from the Earth's centre than P. But the gravity at S will be *greater* than at P by an amount due to the additional downwards attraction exerted by the material (of density ρ) shaded in Figure 1.70a. Put another way, the rock column below S down at the level of P exerts more force on the point at S than would an equivalent column of air, so we must *reduce* the effect of the free-air correction to allow for this slab of rock.

(b) All other things being equal, this makes the gravity at S greater than at P, so the correction $\Delta_3 g$ must be subtracted from the measured gravity difference Δg to find the correct gravity difference $g_S - g_P$. So the correction *decreases* Δg. The Bouguer correction is *always* of the opposite sign to the free-air correction for any station.

ITQ 35

(a) Use Equation 1.35 to find the IGF value at the reference point P and the station S.

$$g_P = 9.780\,318\,5\{1 + 5.278\,895 \times 10^{-3}\,(\sin[30 - \tfrac{2}{60}])^2 + 2.346\,2 \times 10^{-5}(\sin[30 - \tfrac{2}{60}])^4\}\,\text{m s}^{-2}$$

$$= 9.780\,318\,5\{1 + 1.317\,065 \times 10^{-3} + 1.460\,5 \times 10^{-6}\}\,\text{m s}^{-2}$$
$$= 9.793\,214\,1\,\text{m s}^{-2}.$$

$$g_S = 9.780\,318\,5\{1 + 5.278\,895 \times 10^{-3}\,(\sin 30)^2 + 2.346\,2 \times 10^{-5}\,(\sin 30)^4\}\,\text{m s}^{-2}$$

$$= 9.793\,240\,2\,\text{m s}^{-2}.$$

To avoid losing precision, we keep all these significant figures for the time being. The difference between the IGF at P and S is therefore

$$\Delta_1 g = 9.793\,240\,2 - 9.793\,214\,1\,\text{m s}^{-2}$$
$$= 2.61 \times 10^{-5}\,\text{m s}^{-2}$$
$$= 2.61 \times 10^{-3}\,\text{cm s}^{-2}$$
$$= 2.61\,\text{mGal}.$$

Since S is north of P, $\Delta_1 g$ is positive.

The reference point is below the site, so the height difference is:

$$h_S - h_P = +10\,\text{m}.$$

So the free-air correction, from Equation 1.37, is:

$$\Delta_2 g = -0.308\,6 \times 10\,\text{mGal}$$
$$= -3.086\,\text{mGal}$$

and the Bouguer correction, from Equation 1.39, is:

$$\Delta_3 g = 4.192\,1 \times 10^{-5} \times 2\,670 \times 10\,\text{mGal}$$
$$= 1.119\,\text{mGal}.$$

Using Equation 1.41, and ignoring $\Delta_4 g$, we have:

$$\begin{aligned}(g_S - g_P)_{\text{corrected}} &= \Delta g - \Delta_1 g - \Delta_2 g - \Delta_3 g \\ &= 9.70 - 2.61 + 3.086 - 1.119\,\text{mGal} \\ &= 9.057\,\text{mGal}.\end{aligned}$$

This is the corrected value of the gravity difference.

(b) The most important correction is the free-air correction, followed by the latitude and Bouguer correction. Where stations are close together, the latitude correction becomes smaller and is often negligible.

(c) The corrected value of $g_S - g_P$ is 9.057 mGal, which is $9.057 \times 10^{-5}\,\text{m}\,\text{s}^{-2}$. As a percentage of the Earth's gravity, this is:

$$\frac{9.057 \times 10^{-5}}{9.81} \times 10^2 = 0.001\%.$$

The point of this exercise is to emphasize that gravity anomalies are extremely small fractions of the total field.

ITQ 36

Using the IGF once again,

$$\begin{aligned}g_P = {}& 9.780\,318\,5\{1 + 5.278\,895 \times 10^{-3}\,[\sin 30 + \tfrac{2}{60}]^2 \\ & + 2.346\,2 \times 10^{-5}\,(\sin[30 + \tfrac{2}{60}]^4\}\,\text{m}\,\text{s}^{-2}\end{aligned}$$

$$\begin{aligned}&= 9.780\,318\,5\{1 + 1.322\,384 \times 10^{-3} + 1.472\,29 \times 10^{-6}\}\,\text{m}\,\text{s}^{-2} \\ &= 9.793\,266\,2\,\text{m}\,\text{s}^{-2}.\end{aligned}$$

g_S has the same value as before, since the IGF is symmetric about the equator:

$$g_S = 9.793\,240\,2\,\text{m}\,\text{s}^{-2}.$$

So $g_P - g_S = 2.60 \times 10^{-5}\,\text{m}\,\text{s}^{-2} = 2.60\,\text{mGal}$.

This time the reference point is at the higher latitude, so $\Delta_1 g$ is negative. $\Delta_1 g = -2.60\,\text{mGal}$.

The reference point is now above the site and so $h_S - h_P = -10\,\text{m}$, and

$$\begin{aligned}\Delta_2 g &= +3.086\,\text{mGal} \\ \Delta_3 g &= -1.119\,\text{mGal}.\end{aligned}$$

$$\begin{aligned}\text{Therefore, } (g_S - g_P)_{\text{corrected}} &= 9.70 + 2.60 - 3.086 + 1.119 \\ &= 10.333\,\text{mGal}.\end{aligned}$$

You can now see how important the IGF is and how corrected gravity values depend on where in the world the survey is.

ITQ 37

Using Equation 1.43 to evaluate whether the columns are in isostatic equilibrium:

For column X: $2\,670\,\text{kg}\,\text{m}^{-3} \times 100 \times 10^3\,\text{m} = 2.67 \times 10^8\,\text{kg}\,\text{m}^{-2}$.

For column A: $2\,640\,\text{kg}\,\text{m}^{-3} \times 101 \times 10^3\,\text{m} = 2.67 \times 10^8\,\text{kg}\,\text{m}^{-2}$.

For column B: $2\,590\,\text{kg}\,\text{m}^{-3} \times 103 \times 10^3\,\text{m} = 2.67 \times 10^8\,\text{kg}\,\text{m}^{-2}$.

For column C: $2\,540\,\text{kg}\,\text{m}^{-3} \times 105 \times 10^3\,\text{m} = 2.67 \times 10^8\,\text{kg}\,\text{m}^{-2}$.

For column D: $2\,495\,\text{kg}\,\text{m}^{-3} \times 107 \times 10^3\,\text{m} = 2.67 \times 10^8\,\text{kg}\,\text{m}^{-2}$.

For column E: $2\,720\,\text{kg}\,\text{m}^{-3} \times 97 \times 10^3\,\text{m} + 1\,000\,\text{kg}\,\text{m}^{-3} \times 3 \times 10^3\,\text{m}$
$= 2.67 \times 10^8\,\text{kg}\,\text{m}^{-2}$.

For column F: $2\,735\,\text{kg}\,\text{m}^{-3} \times 96 \times 10^3\,\text{m} + 1\,000\,\text{kg}\,\text{m}^{-3} \times 4 \times 10^3\,\text{m}$
$= 2.67 \times 10^8\,\text{kg}\,\text{m}^{-2}$.

Clearly, all the columns are in isostatic equilibrium with each other.

If the density of the tallest column (column D) was increased to $2\,700\,\text{kg}\,\text{m}^{-3}$, the value from Equation 1.43 would become

For column D: $2\,700\,\text{kg}\,\text{m}^{-3} \times 107 \times 10^3\,\text{m} = 2.89 \times 10^8\,\text{kg}\,\text{m}^{-2}$.

This column would no longer be in isostatic equilibrium with the surrounding columns. Its mass would be too large, and the column would have to sink (that is become less tall) in order to compensate for this.

ITQ 38

Since the columns are in equilibrium, the masses of the columns must be equal. If the water were removed from column F, the overall mass of the column would decrease and it would no longer be in isostatic equilibrium, so it would move upwards.

ITQ 39

(a) For isostatic equilibrium, Equation 1.44 is satisfied for all columns. We only have the vertical dimensions for columns X and C, but the constant will be the same for all columns, so we really only need to do one calculation. Since you could have chosen either column X or C, we will work out both here.

For column X: $2\,670\,\text{kg}\,\text{m}^{-3} \times 30 \times 10^3\,\text{m} + 3\,270\,\text{kg}\,\text{m}^{-3} \times 27 \times 10^3\,\text{m} =$
$1.68 \times 10^8\,\text{kg}\,\text{m}^{-2}$.

For column C: $2\,670\,\text{kg}\,\text{m}^{-3} \times (30 + 27 + 6) \times 10^3\,\text{m} = 1.68 \times 10^8\,\text{kg}\,\text{m}^{-2}$.

The value of the constant for this model is therefore $1.68 \times 10^8\,\text{kg}\,\text{m}^{-2}$.

(b) The effect of doubling the width of column C is the same as inserting a new column between columns C and D. The thickness and density of the new column will be the same as those for column C, and since C is in isostatic equilibrium with its surroundings, the new column will also be in isostatic equilibrium.

ITQ 40

A super-adiabatic temperature gradient means that temperature increases more rapidly with depth than it would under self-compression alone. At any particular depth within a super-adiabatic layer, material will be more expanded (because of higher temperatures) than it would be under an adiabatic gradient. Therefore, its density will be lower than under an adiabatic gradient.

ITQ 41

The atomic masses of Fe and Si are, respectively, 55.85 and 28.09. An atom of Fe therefore has a mass 55.85/28.09 = 1.988 times greater than that of an atom of Si. The relative abundance of Fe compared to Si in terms of mass is therefore $1.2 \times 1.988 = 2.3856 \sim 2.4$.

ITQ 42

The density can be read directly from the Nafe–Drake curve, from which it turns out to be about $2\,700\,\text{kg}\,\text{m}^{-3}$. Note that this is the value read from the curve. Because of the scatter of points on the graph, the value is probably uncertain to within at least $\pm 100\,\text{kg}\,\text{m}^{-3}$.

ITQ 43

You should have realized by now what the problem is and why it was that in the paragraph immediately above this pair of ITQs we used such phrases as 'on the face of it' and 'in principle'. The fact is that the density ranges of rocks overlap so much that density alone can seldom be used to determine the nature of the rock at any point in the Earth. Thus, for example, the density of 2 700 kg m^{-3} could indicate sandstone, shale, limestone, rhyolite, andesite, granite, basalt, gabbro, quartzite, schist, granulite, marble, slate or gneiss! Similar problems would also have arisen if we had chosen to compare not densities but seismic velocities. In practice, some of the rocks above could probably be excluded by other (e.g. geological) criteria, depending on the context in which the particular study is being carried out; but generally, neither the seismic-velocity method nor the density method will give a unique solution.

ITQ 44

(a) The completed table should read (figures in mass per cent):

SiO_2	29.8
TiO_2	0.13
Al_2O_3	2.18
Cr_2O_3	0.26
MgO	25.1
FeO	5.3
MnO	0.09
CaO	2.05
Na_2O	0.26
K_2O	0.02

(b) See Figure 1.129.

(c) The oxides of Si, Mg, Al and Ca fall on, or almost on, the line, thus indicating that they are present in the bulk Earth in chondritic, or near-chondritic, proportions. In short, the results provide strong support for the chondritic Earth model.

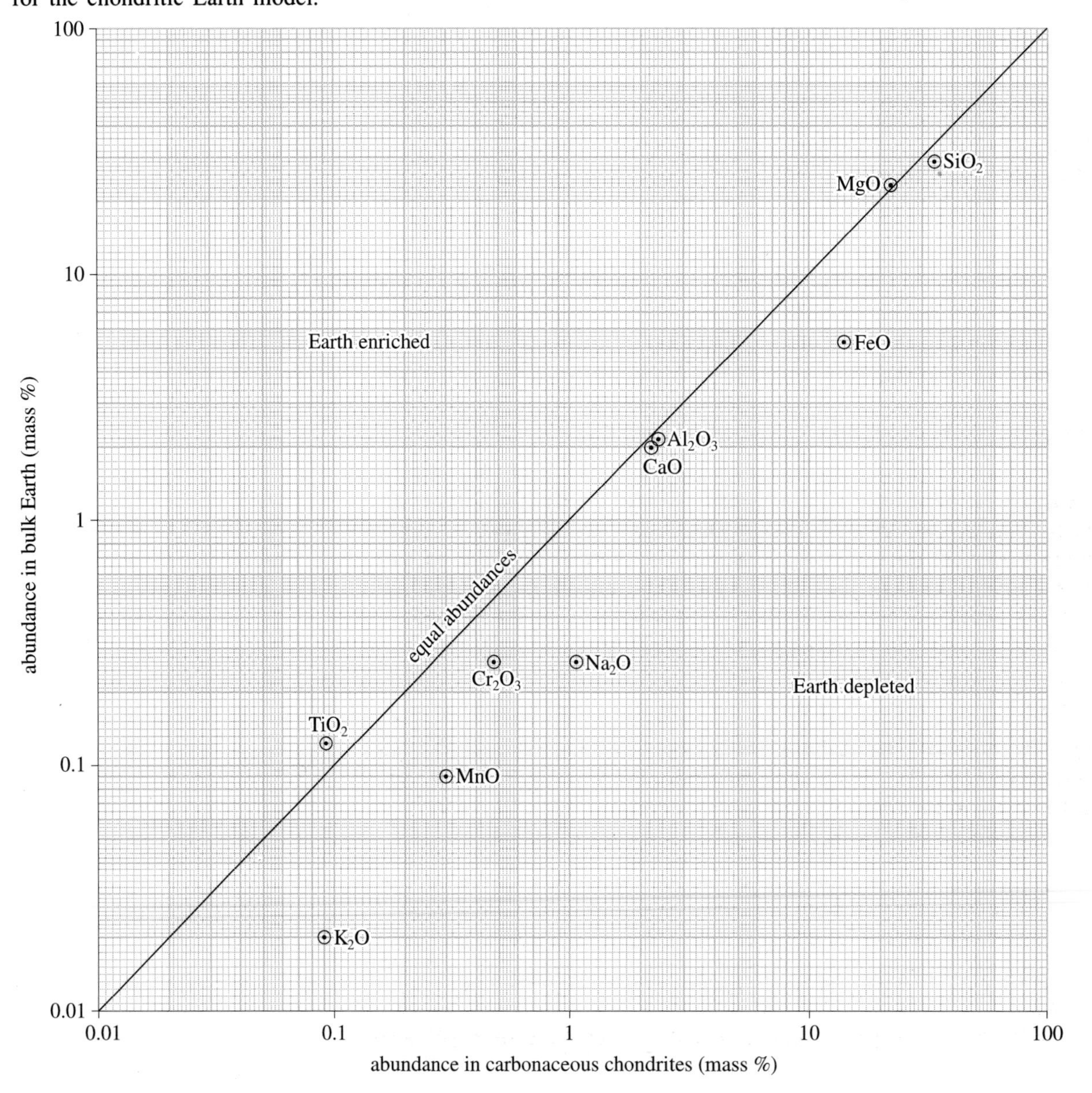

Figure 1.129 Graph relating to the answer to ITQ 44.

ITQ 45

Ocean tides (a) result principally from a combination of the gravitational influence of the Moon and Sun (A ii) and the Earth's rotation and gravity field (B ii). Atmospheric circulation (b), ocean currents (d) and weathering, erosion and sedimentation (e) result mainly from the combined effects of solar energy (A i) and the Earth's rotation and gravity field (B ii). Biological activity (c) owes its origin principally to solar energy (A i), but it could also be argued that life would be less successful under zero gravity conditions on a non-rotating Earth; so the Earth's rotation and gravity field (B ii) might also be relevant. Volcanism (f), metamorphism (g) and earthquakes (h) are all manifestations of the Earth's internal heat (B i), all being related to the plate-tectonic cycle.

ITQ 46

The football field has an area of $100\,\text{m} \times 70\,\text{m} = 7\,000\,\text{m}^2$. For a heat flow of $80\,\text{mW m}^{-2}$, or $0.08\,\text{W m}^{-2}$, this would yield a total heat energy of $0.08 \times 7\,000 = 560\,\text{W}$, sufficient to illuminate just 5.6 100 W electric light bulbs.

ITQ 47

Taking the granodiorite, we see from Table 1.12 that 1 kilogram will contain $3.97 \times 10^{-6}\,\text{kg}\ ^{238}\text{U}$, $0.03 \times 10^{-6}\,\text{kg}\ ^{235}\text{U}$, $18 \times 10^{-6}\,\text{kg}\ ^{232}\text{Th}$, and $3.5 \times 10^{-6}\,\text{kg}\ ^{40}\text{K}$.

By multiplying each of these masses by the appropriate heat generation value (in mW kg^{-1}) from Table 1.11, you should have calculated the heat generation values listed below (all to two sig. figs.):

^{238}U: $3.97 \times 10^{-6} \times 9.6 \times 10^{-2}\,\text{mW kg}^{-1} = 38 \times 10^{-8}\,\text{mW kg}^{-1}$;

^{235}U: $0.03 \times 10^{-6} \times 56 \times 10^{-2}\,\text{mW kg}^{-1} = 1.7 \times 10^{-8}\,\text{mW kg}^{-1}$;

^{232}Th: $18 \times 10^{-6} \times 2.6 \times 10^{-2}\,\text{mW kg}^{-1} = 47 \times 10^{-8}\,\text{mW kg}^{-1}$;

^{40}K: $3.5 \times 10^{-6} \times 2.8 \times 10^{-2}\,\text{mW kg}^{-1} = 9.8 \times 10^{-8}\,\text{mW kg}^{-1}$.

Summing the heat generations in the last column gives a value of $96 \times 10^{-8}\,\text{mW kg}^{-1}$ for the heat generation of the granodiorite.

The results in your Table 1.12 indicate very clearly where the Earth's heat generation is concentrated. Heat generation per kilogram of upper continental crust (granodiorite) is about five times that of the oceanic crust (gabbro), and the heat generation of the mantle (peridotite) is at least two orders of magnitude lower. (But remember that the mantle is many times more massive that the crust — see SAQ 55.)

ITQ 48

(a) The question should be answered by using the same method as applied in ITQ 47. Taking the chondritic meteorite, 1 kilogram will contain $0.01 \times 10^{-6}\,\text{kg}\ ^{238}\text{U}$, $7 \times 10^{-5} \times 10^{-6}\,\text{kg}\ ^{235}\text{U}$, $0.04 \times 10^{-6}\,\text{kg}\ ^{232}\text{Th}$, and $0.10 \times 10^{-6}\,\text{kg}\ ^{40}\text{K}$.

By multiplying each of these masses by the appropriate heat generation value (in mW kg^{-1}) from Table 1.11, you should have calculated the heat generations listed below (to one sig. fig.):

^{238}U: $0.01 \times 10^{-6} \times 9.6 \times 10^{-2}\,\text{mW kg}^{-1} = 0.1 \times 10^{-8}\,\text{mW kg}^{-1}$;

^{235}U: $7 \times 10^{-5} \times 10^{-6} \times 56 \times 10^{-2}\,\text{mW kg}^{-1} = 0.004 \times 10^{-8}\,\text{mW kg}^{-1}$;

^{232}Th: $0.04 \times 10^{-6} \times 2.6 \times 10^{-2}\,\text{mW}\,\text{kg}^{-1} = 0.1 \times 10^{-8}\,\text{mW}\,\text{kg}^{-1}$;

^{40}K: $0.10 \times 10^{-6} \times 2.8 \times 10^{-2}\,\text{mW}\,\text{kg}^{-1} = 0.3 \times 10^{-8}\,\text{mW}\,\text{kg}^{-1}$.

Summing the above heat generation values gives a value of $\sim 0.5 \times 10^{-8}\,\text{mW}\,\text{kg}^{-1}$.

(b) The total heat generation in each of the various parts of the Earth is as follows (to one sig. fig.):

upper continental crust: $96 \times 10^{-8} \times 8 \times 10^{21}\,\text{mW} = 8 \times 10^{15}\,\text{mW}$
lower continental crust: $40 \times 10^{-8} \times 8 \times 10^{21}\,\text{mW} = 3 \times 10^{15}\,\text{mW}$;

oceanic crust: $19 \times 10^{-8} \times 7 \times 10^{21}\,\text{mW} = 1 \times 10^{15}\,\text{mW}$;

mantle: $0.26 \times 10^{-8} \times 4080 \times 10^{21}\,\text{mW} = 1 \times 10^{16}\,\text{mW}$;

core: $0 \times 10^{-8} \times 1880 \times 10^{21}\,\text{mW} = 0\,\text{mW}$.

The total heat generation in the Earth is therefore $22 \times 10^{15}\,\text{mW}$, which (because the mass of the Earth is $6 \times 10^{24}\,\text{kg}$) is equivalent to

$$\frac{22 \times 10^{15}}{6 \times 10^{24}} = 0.4 \times 10^{-8}\,\text{mW}\,\text{kg}^{-1}.$$

(c) The heat generation estimated for the bulk Earth is a very high proportion (about 80%) of that based on the chondritic Earth model. In fact, by juggling estimates for heat-producing elements in different parts of the Earth only very slightly, it would be possible to make the agreement exact!

ITQ 49

(a) The temperature gradients may be determined from the graph, with the following results:

$$\text{top layer: gradient} = \frac{175\,\text{K}}{10\,\text{km}} = 17.5\,\text{K}\,\text{km}^{-1} = 0.0175\,\text{K}\,\text{m}^{-1};$$

$$\text{middle layer: gradient} = \frac{700\,\text{K} - 175\,\text{K}}{20\,\text{km}} = 26.25\,\text{K}\,\text{km}^{-1} = 0.02625\,\text{K}\,\text{m}^{-1};$$

$$\text{bottom layer: gradient} = \frac{865\,\text{K} - 700\,\text{K}}{15\,\text{km}} = 11.0\,\text{K}\,\text{km}^{-1} = 0.011\,\text{K}\,\text{m}^{-1}.$$

The middle layer therefore has the steepest temperature gradient.

(b) Heat flow can then be calculated using Equation 1.51 with the above values of temperature gradient and the values of thermal conductivity shown on the graph.

For the top layer:

$$q = 2.05 \times 0.0175\,\text{W}\,\text{m}^{-2} = 0.036\,\text{W}\,\text{m}^{-2} = 36\,\text{mW}\,\text{m}^{-2}.$$

For the bottom layer:

$$q = 3.30 \times 0.011\,\text{W}\,\text{m}^{-2} = 0.36\,\text{W}\,\text{m}^{-2} = 36\,\text{mW}\,\text{m}^{-2}.$$

The heat flow is therefore the same through the top layer as through the bottom layer. In fact, if you had calculated the heat flow through the middle layer, you would have found it also to be $36\,\text{mW}\,\text{m}^{-2}$. In other words, the system is in equilibrium, with no heat building up within it. Note, however, that this is a somewhat artificial situation because, in practice, the Earth's crust contains heat sources.

ITQ 50

(a) In both the North Atlantic and the North Pacific Oceans, the depth evidently varies with the square root of the age of the lithosphere — i.e. there is a positive correlation between depth and age.

(b) The departure from linearity occurs at a little over 8 on the horizontal scale, which is equivalent to an age of about 70 Ma.

(c) For crust older than about 70 Ma, the depth increases less rapidly than predicted by the simple linear relationship. This suggests that the older lithosphere has contracted less than predicted, which implies that it is warmer than predicted.

ITQ 51

The heat flow, q, for oceanic lithosphere 36 Ma old is given by (Equation 1.55)

$$q = \frac{473}{(36)^{0.5}} \text{ mW m}^{-2} = \frac{473}{6} \text{ mW m}^{-2} = 79 \text{ mW m}^{-2}.$$

The mean heat flow for oceanic lithosphere 35 Ma in age on the Reykjanes Ridge (south of Iceland) has been determined as $83 \pm 10 \text{ mW m}^{-2}$. The geophysicists who made the measurements comment that 'The mean value ... is in excellent agreement with the relationship between heat flow and the reciprocal of the square root of age predicted for oceanic crust less than 100 Ma old'.

ITQ 52

(a) The obvious similarity is that there is a decrease in heat flow with increasing age in both oceanic and continental crust.

(b) The rate of that decrease is very different in the two kinds of crust: in oceanic areas, heat flow decreases from about 150 mW m^{-2} to a few tens of mW m^{-2} in something like 100 Ma; in continental areas, it decreases from less than 100 mW m^{-2} to less than 50 mW m^{-2} in something like 2 000 Ma.

ITQ 53

(a) Equation 1.56 is of the form $y = mx + c$, where m is the gradient and c is the intercept on the y axis. The gradient of the line in Figure 1.104 is therefore b.

(b) By the same token, the intercept on the heat-flow axis is q^*.

(c) From Figure 1.104, the intercept on the q axis (q^*) is about 33 mW m^{-2}. The gradient (b) is

$$\frac{108 - 33}{10} \text{ km} = 7.5 \text{ km}.$$

(*Note*: You should pay particular attention to the units in Equation 1.56. In any equation, the units of each term — not each quantity! — must be the same. Thus in Equation 1.56 the terms q, bA and q^* must all have the same units. Thus if q and q^* were in mW m^{-2}, bA would be in mW m^{-2}, which means that if A were in mW m^{-3}, b would be in m. However, whereas the units of q and q^* in Equation 1.56 as stated above are actually in mW m^{-2}, the units of A in Figure 1.104 are in $\mu\text{W m}^{-3}$. This automatically gives b in $10^3 \text{ m} = \text{km}$.)

ITQ 54

The simplest way to do this is to look at the point where the mean heat flow, $\bar{q}$, is $100\,\mathrm{mW\,m^{-2}}$, and find the corresponding value of q^*, which is $60\,\mathrm{mW\,m^{-2}}$. The slope of the graph is therefore 0.6, and we conclude that this is also approximately the proportion of the total heat flow provided by the reduced heat flow q^*. (Because the graph of q^* against $\bar{q}$ is a straight line passing through the origin, the line has the equation $q^* = c\bar{q}$, where c is a constant equal to the gradient of the line.)

ITQ 55

The range of heat flows, estimated from Plate 11 and characteristic of the different geological provinces, is shown below.

Geological province	**Heat flow range ($\mathrm{mW\,m^{-2}}$)**
(a) ocean ridges	generally 120–320, but parts of some ocean ridges (for example, the Mid-Atlantic Ridge) have heat flow of 80–120
(b) old ocean floor (for example, in the western Pacific Ocean, southeast of Japan)	<60
(c) ancient continental crust (for example, the Canadian Shield in central Canada	<60
(d) island arcs and continental destructive margins	generally 60–80 but some destructive continental margins (for example, the central Andes) and island arcs (for example, Sumatra) have heat flow <60

ITQ 56

(1) In Section 1.7.4, we saw that seismic-velocity models (i.e. graphs or tables that show in detail how seismic velocity varies with depth) place the top of the asthenosphere (as defined by the LVZ) at a depth of 40–60 km. As, by definition, the lithosphere is what overlies the asthenosphere, on this account the lithosphere must be 40–60 km thick.

(2) The seismic-wave dispersion data uncorrected for anisotropy, as shown in Figure 1.66, indicate the thickness of the lithosphere to be about 125 km.

(3) The seismic-wave dispersion data corrected for anisotropy, as also shown in Figure 1.66, suggest that the base of the lithosphere lies at a depth of about 60 km.

(4) The elastic-loading data illustrated in Figure 1.66 appear to place the base of the lithosphere at more like 30–40 km.

(5) Irrespective of whether the oceanic lithosphere is modelled as a layer of uniform thickness (the plate model) or as a layer of gradually increasing thickness (the boundary-layer model), the thermal models in Figure 1.96 conclude that the lithosphere has a thickness of about 125 km. However, the thermal data are less consistent than this would suggest. We deliberately omitted the dimensions from Figure 1.98; in fact, whereas the boundary-layer model therein indicates a lithosphere thickness of about 125 km, the version of the plate model illustrated (it was devised by researchers different from those responsible for the models in Figure 1.96) proposes a lithospheric thickness of only about 75 km.

ITQ 57

The Rayleigh number (Ra) is given by

$$\text{Ra} = \frac{\alpha \Delta T g d^3 \rho}{\kappa \eta}$$

and the units are

$$\text{Ra} = \frac{\alpha\,(\text{K}^{-1})\,\Delta T\,(\text{K})\,g\,(\text{m s}^{-2})\,d^3\,(\text{m}^3)\,\rho\,(\text{kg m}^{-3})}{\kappa\,(\text{m}^2\,\text{s}^{-1})\,\eta\,(\text{Pa s})}.$$

Remembering that $1\,\text{Pa s} = 1\,\text{kg m}^{-1}\,\text{s}^{-1}$, the units cancel out. So the Ra has no units and is therefore referred to as a dimensionless number.

ITQ 58

(a) At the time when the ice sheet began to melt, a peripheral bulge of either about 10 m (deep-flow model) or about 30 m (channel-flow model) surrounded the ice-loaded area. The bulge sloped steeply from its crest towards the edge of the ice sheet, and more gradually from the crest away from the ice.

(b) After deglaciation, the peripheral bulge collapsed quickly in the case of the channel-flow model and more slowly in the case of the deep-flow model. In the deep-flow model, uplift preceded the sinking, whereas in the channel-flow model, deglaciation is associated only with peripheral *sinking*.

(c) As a consequence of (b), in the channel-flow model the contours that define zero uplift migrate *outwards* during deglaciation, whereas in the deep-flow model the zero uplift contours migrate *inwards*.

ITQ 59

(a) Figure 1.113 clearly shows an *uplift* of about 85 m from 12 000 to 4 000 years ago, followed by more recent *subsidence* of 3 m.

(b) The pattern described in (a) is similar to that predicted by the deep-flow model for mantle flow below Fennoscandia (Figure 1.111b) rather than to that predicted by the channel-flow model (Figure 1.112b) for that region. The deep-flow model is therefore more appropriate for explaining the mantle deformation below the east coast of the USA than the channel-flow model.

ITQ 60

The Rayleigh number is given by:

$$\text{Ra} = \frac{\alpha \Delta T g d^3 \rho}{\kappa \eta}.$$

Substituting the values given:

$$\text{Ra} = \frac{2 \times 10^{-5}\,\text{K}^{-1} \times 1\,\text{K} \times 10\,\text{m s}^{-2} \times (3 \times 10^6)^3\,\text{m}^3 \times 3\,300\,\text{kg m}^{-3}}{10^{-6}\,\text{m}^2\,\text{s}^{-1} \times 10^{21}\,\text{kg m}^{-1}\,\text{s}^{-1}}$$

$$= 1.8 \times 10^4\,\frac{\text{K K}^{-1}\,\text{m s}^{-2}\,\text{m}^3\,\text{kg m}^{-3}}{\text{m}^2\,\text{s}^{-1}\,\text{kg m}^{-1}\,\text{s}^{-1}}.$$

The units cancel out, proving that Ra is a dimensionless number. (It is always a good idea to check equations by making sure that the units on each side of the equals sign balance. This is called dimensional analysis.)

So, for $\Delta T = 1$ K and remembering that for liquid in a spherical shell, $\text{Ra}_{cr} \sim 10^3$, $\text{Ra} > \text{Ra}_{cr}$, and convection will occur.

For $\Delta T = 10^2$ K, we multiply our answer for $\Delta T = 1$ K by 10^2:

$$\text{Ra} = 1.8 \times 10^4 \times 10^2 = 1.8 \times 10^6.$$

Again Ra > Ra_{cr} which shows that convection will certainly occur within the mantle. Since our estimated value of Ra is orders of magnitude greater than Ra_{cr}, not only with the mantle convect, but it will do so turbulently.

ITQ 61

This question will be rather easier to answer when you have studied the Smithsonian Map in detail in Block 2. Nevertheless, a quick glance shows the belts of earthquakes and volcanoes which mark the edges of tectonic plates. There is a crude correlation between slow Rayleigh wave velocity and young volcanic and earthquake zones. There is also a correlation between faster Rayleigh wave velocities and older lithosphere. Fast regions represent cold material and slow regions are hotter with a degree of partial melting. If you don't remember much about plate tectonics, don't worry because this subject will be revised and developed in Block 2.

ITQ 62

Iceland is characterized by a broad geoid high. The anomaly is much wider than the island. A geoid high means that there is an excess mass beneath the Earth's surface. To compensate for this extra mass, the equipotential surface is displaced upwards above the spheroid surface, so that the value of gravity at the equipotential (geoid) surface is equal to the value on the spheroid at that point in the absence of the anomalous mass.

ITQ 63

A region of partial melt would be buoyant and therefore less dense than its surroundings. This would imply that in a given volume, there would be less mass in the partially molten region and this would cause a geoid low. Since the region of partial melt is vertically extensive but laterally confined, the geoid anomaly would also be restricted laterally and would not be broad. The deeper parts would give rise to a laterally extensive anomaly, albeit quite small at the margins of the main anomaly. So in summary, a narrow geoid low would be expected above a region of laterally confined partial melt

---

SAQ ANSWERS AND COMMENTS

SAQ 1

(a) True. This ordering should be clear from the text (Sections 1.2 and 1.2.1).

(b) False. *All* the planets orbit the Sun in the same direction (Section 1.2).

(c) True. The most inclined orbit is that of Pluto at 17.15° (Table 1.1).

(d) False. There are indeed three planets with retrograde rotation, but they are Venus, Uranus and Pluto (Section 1.2.1).

(e) False. The equator of Uranus lies at 97.86° to the planet's orbital plane, which means that the angle between the axis and the orbital plane is only 7.86° (Table 1.1).

(f) True. This should be clear from the text (Section 1.2.1).

(g) False. Pluto is an outer planet, but is unique among the outer planets in not being a gaseous giant (Section 1.2.1).

(h) True. It probably has a much higher proportion of iron than does the Earth (Section 1.2.1 and ITQ 4).

(i) False. Neptune is a gaseous planet, and so its surface couldn't possibly record meteorite impacts (Section 1.2.1).

(j) False. Although the mass and diameter of Venus are close to those of the Earth, there are major differences in, for example, the atmospheres and in that there is no significant plate tectonics on Venus (Section 1.2.1).

(k) True. The Moon is the largest satellite in the relation to its primary (Section 1.2.1).

(l) True. This point is made in the text (Section 1.2.1).

(m) True. Self-compression is a general principle to which there can be no exceptions, although clearly the smaller the planet the less will be the effect of self-compression (Section 1.2.2).

SAQ 2

Pluto's mean distance from the Sun is $5\,913.5 \times 10^6$ km (Table 1.1), and the speed of light is $3 \times 10^5\,\text{km s}^{-1}$ (Section 1.2). As speed = distance/time, we have: time taken for light to travel from Pluto to the Earth = $5\,913.5 \times 10^6\,\text{km} / 3 \times 10^5\,\text{km s}^{-1} = 19\,712\,\text{s} \sim 5.5$ hours.

SAQ 3

The approximate distance to the edge of the observable Universe is 10^{10} light years (Figure 1.1). As one light year is 9.46×10^{12} km (Section 1.2), the distance is $9.46 \times 10^{12} \times 10^{10}\,\text{km} = 9.46 \times 10^{22}$ km, or 94.6 thousand million million million kilometres.

SAQ 4

The total mass of the planets (Table 1.1) is 2.7×10^{27} kg and the mass of Jupiter is 1.9×10^{27} kg. The proportion accounted for by Jupiter is therefore $1.9/2.7 = 0.70 = 70\%$.

SAQ 5

If the planet lies to the left of the line, its density must be lower in relation to its size than for the Earth, Venus, Mars and the Moon. This means that the ratio of lower-density silicates to higher-density iron is probably higher, which suggests that, relatively, the planet has a smaller iron core than have the Earth, Venus, Mars and the Moon (cf. Mercury, which lies to the right of the line and probably has a relatively large iron core).

SAQ 6

You should have at least five items on your list, of which the most important are probably the following: (1) the disc-like shape of the Solar System; (2) the fact that all the planets orbit the Sun in the same direction; (3) the near-circularity of most planetary orbits; (4) the great diversity of the Solar System, both in general (planets, planetary satellites, asteroids, comets, meteoroids, interplanetary dust and gas) and in detail (rocky bodies, liquid/gaseous bodies); (5) the rotation of the planets (three retrograde), with axes not perpendicular to the ecliptic.

SAQ 7

(a) True. The solar nebula only contained a very small proportion of the total mass going into the Solar System (Section 1.3), which is why the Sun now accounts for most of the Solar System's mass (see Section 1.2 and ITQ 3).

(b) False. The order of increasing size is planetesimals $\rightarrow$ planetary embryos $\rightarrow$ planets (Section 1.3).

(c) False. Planetary embryos would have formed in 10^5 years, but it would have taken much longer than 10^6 years — actually 10^7–10^8 years — for the planets to form fully (Section 1.3).

(d) True. But note that the solar-nebula gas might not have been present during the later stages of the Earth's formation (cf. the relatively rapid dissipation of gas clouds around new stars), in which case there would have been no primordial atmosphere (Section 1.3).

(e) True. This is so by definition of the two terms (Section 1.3).

(f) False. Inspection of Figure 1.5, Section 1.3, shows that sulphur is a volatile with a vaporization temperature in the range 1 300–600 K, whereas iodine is a volatile with a vaporization temperature lower than 600 K.

(g) True. Inspection of Figure 1.5, Section 1.3, shows that titanium is a refractory, whereas sulphur is as in (f) above.

(h) True. This is one of the main distinguishing tenets of the current standard model (Section 1.3).

(i) True. However, note that, while not necessarily false, this hypothesis is currently a minority view.

(j) False. On the contrary, the giant-impact hypothesis is the only one of many hypotheses for the origin of the Moon that does explain the Earth–Moon system's dynamic characteristics, although there could be geochemical problems with it (Section 1.3.1).

(k) False. Inspection of Figure 1.7, Section 1.3.2, shows that Zr is one of the early-condensing refractories, whereas F condenses out below 1 100 K.

(l) True. Inspection of Figure 1.7, Section 1.3.2, shows that $CaTiO_3$ is an early-condensing refractory, whereas $MgSiO_3$ is a major silicate condensing out later.

(m) True. In fact, all material would be in vapour form at an even lower temperature (Section 1.3.2 and Figure 1.7).

(n) False. This is unlikely to have been true because conduction is a very slow way of transferring heat in the Earth (because silicates are very poor conductors). The mass movement of rock in the Earth, by contrast, can transfer heat (by convection) very much more quickly (Section 1.3.3).

(o) False. By the definition of original heat given in the text (Section 1.3.3), planetesimals would have none — i.e. they would be cold. If the planetesimals had no original heat, nor could an Earth constructed from them.

(p) True. This is because the Earth would have become heated during accretion (see text, Section 1.3.3, for definition of initial heat).

(q) True. The heat from all three sources arises from gravitational effects (Section 1.3.3).

(r) False. In fact, the proportion of incoming kinetic energy likely to have been converted into heat is only 5–15%, the rest of the energy going, for example, to form a crater (Section 1.3.3).

(s) False. Some have such long half-lives and are in such low abundance that their heat production can be ignored (Section 1.3.3).

SAQ 8

The mass of the Earth is 6×10^{24} kg (Table 1.1) and so, at 15× this, the mass of Jupiter's core would be $15 \times 6 \times 10^{24}\,\text{kg} = 9 \times 10^{25}\,\text{kg}$. The number of planetary embryos is therefore simply $9 \times 10^{25}\,\text{kg}/10^{23}\,\text{kg} = 900$.

SAQ 9

(1) The anomalously small silicate mantle of Mercury, (2) the origin of the Moon, (3) the tilts of planetary equators away from their respective orbital planes, and (4) the removal of the Earth's primordial atmosphere (Section 1.3.1).

SAQ 10

The homogeneous accretion model assumes that the Earth accreted from cold, rocky fragments and thus formed as a homogeneous mass. The iron core would therefore not have been a primary feature, although it would have begun to form by the fall of liquid iron towards the Earth's centre even as the Earth was heating up during accretion. The heterogeneous accretion model, by contrast, assumes that the Earth formed in embryo before the solar nebula cooled. The core would therefore have been a primary feature, forming by condensation from the solar nebula before the silicate mantle (Section 1.3.2).

SAQ 11

The blob model involves (1) iron blobs up to several kilometres in size (2) falling through a solid silicate matrix. The rainout model, by contrast, involves (i) small droplets of iron (ii) falling through an iron/silicate emulsion (Section 1.3.2).

SAQ 12

(1) Accretion, (2) self-compression, (3) core formation, (4) decay of short-lived radioactive isotopes, (5) decay of long-lived radioactive isotopes, and (6) tidal dissipation.

SAQ 13

Using Equation 1.3, Section 1.3.3, we have

$$\Delta T = \frac{mv^2}{2MC}$$

where $m = 10^{17}\,\text{kg}$, $v = 1\,\text{km s}^{-1} = 10^3\,\text{m s}^{-1}$, $M = 6 \times 10^{23}\,\text{kg}/6 = 10^{23}\,\text{kg}$, and $C = 7.5 \times 10^2\,\text{J kg}^{-1}\,\text{K}^{-1}$.

And so,

$$\Delta T = \frac{10^{17} \times (10^3)^2}{2 \times 10^{23} \times 7.5 \times 10^2}\,\text{K} = 6.7 \times 10^{-4}\,\text{K}.$$

SAQ 14

Kinetic energy (Section 1.3.3) is given by $E = \frac{1}{2}mv^2$.

So for episode 1, $E_1 = 2 \times \frac{1}{2}mv^2 = mv^2$,

and for episode 2, $E_2 = \frac{1}{2} \times 2m \times \left(\frac{v}{2}\right)^2 = \frac{mv^2}{4}$.

Episode 1 therefore imparts the greater energy by a factor of 4.

SAQ 15

In four ways: (1) by introducing a large amount of kinetic energy, (2) by burying the resulting heat deeply, (3) by generating penetrating seismic waves — and hence spreading heat, and (4) by introducing 'original heat' from a new source (Section 1.3.3).

SAQ 16

(a) True. In its early stages, the Universe cooled and expanded so rapidly that no fusion reactions were able to proceed beyond the production of helium (Section 1.4).

(b) False. The 'burning' of light elements to form heavier ones always generates heat and is therefore exothermic (Section 1.4).

(c) False. It is true that the interiors of stars are important production factories for elements (by fusion reactions and slow neutron capture), but elements are also generated by rapid neutron capture during supernova explosions (Section 1.4).

(d) True. Note, however, that the constituents of white dwarves differ, depending on just how far fusion reactions had proceeded in the stars that formed the original red giants (Section 1.4).

(e) False. Fusion reactions can produce elements up to only iron (atomic number = 26) and nickel (atomic number = 28). Elements up to bismuth are generated by slow neutron capture — the s-process (Section 1.4).

(f) True. The Sun is too small to allow helium burning to take place, producing elements heavier than helium (Section 1.4)

(g) True. Indeed, heavier-than-helium elements would not otherwise have been incorporated into the Solar System (Section 1.4).

(h) False. Quite the contrary. Heavy elements are never destroyed; so as more and more supernova explosions occur, the proportion of such elements in the Universe increases (Section 1.4).

(i) True. See the beginning of Section 1.4.1.

(j) True. This was one of the features of Figure 1.12 identified in ITQ 11 (Section 1.4.1).

(k) False. On the contrary, they are the most common (Section 1.4.1).

(l) False. The order is: carbonaceous, enstatite, ordinary (Section 1.4.1).

(m) True. This is quite clearly shown by Figure 1.13.

(n) True. This is quite clearly shown by Figure 1.14.

(o) False. These are volatile elements. They plot to the left of the line in Figure 1.16 and are thus clearly relatively the more abundant in the Sun than in carbonaceous chondrites.

(p) True. That this is so is immediately apparent from Figure 1.15. Most L-chondrites are of petrological type 6 (most highly altered), whereas most C-chondrites are of petrological types 1–3 (least altered).

(q) True. You can read this directly from Table 1.3. However, this is by no means the element with the widest range; Ru has a ratio of 500!

(r) False. Ge is depleted in the crust relative to the Sun and, accordingly, lies to the left of the line in Figure 1.17. However, Zr lies to the right of the line; it is enriched in the crust relative to the Sun.

(s) False. Siderophile elements prefer to exist as metals and thus preferentially migrate to the metallic core.

SAQ 17

The Fraunhofer lines are shown in Figure 1.11. First, you need to convert the frequencies to wavelengths (as shown on the Figure 1.11 scale) using the equation in the caption to Figure 1.11. Thus, for example,

$$\lambda = \frac{3 \times 10^8\,\mathrm{m\,s^{-1}}}{4.580 \times 10^{14}\,\mathrm{Hz}} = 655 \times 10^{-9}\,\mathrm{m} = 655\,\mathrm{nm}$$

and, similarly, the other two frequencies give wavelengths of 485 nm and 430 nm. Comparison with Figure 1.11 shows that these three wavelengths correspond to the prominent lines C, F and G. The first two indicate *hydrogen* and the last *iron*.

SAQ 18

Examination of Figure 1.12 shows that C (about 10^7 atoms) is enriched in the Sun compared to Si (10^6 atoms, by definition) and that Pb (1–10 atoms) and Nd (about 1 atom) are depleted. There is thus about 10 times more C than Si in the Sun; there is less Pb by a factor of 10^5–10^6; and there is less Nd by a factor of about 10^6.

SAQ 19

(a) B, D, G; (b) B, E; (c) C; (d) A, F. These answers follow from the descriptions of meteorites in Section 1.4.1

SAQ 20

This may be answered by reference to Figure 1.13. Reading from the vertical scale, for every atom of Cd (fourth element from right) in the carbonaceous chondrite there is 0.1 atom in the enstatite chondrite. Thus for every $100x$ atoms of Cd in the carbonaceous chondrite there are $10x$ atoms in the enstatite chondrite.

SAQ 21

$$\text{mass of gold in core} = \frac{1.5 \times 2 \times 10^{24}\,\text{kg}}{10^6} = 3 \times 10^{18}\,\text{kg}$$

$$\text{mass of gold in mantle} = \frac{0.005 \times 4 \times 10^{24}\,\text{kg}}{10^6} = 0.02 \times 10^{18}\,\text{kg}$$

therefore, total mass of gold in the core + mantle = $3.02 \times 10^{18}\,\text{kg}$

$$\text{therefore, proportion of gold in core} = \frac{3 \times 10^{18}}{3.02 \times 10^{18}}$$
$$= 0.993 \text{ or } 99.3\%.$$

SAQ 22

(a) A; (b) B; (c) C; (d) A; (e) C; (f) B. These follow from the definitions of lithophile, chalcophile and siderophile (Section 1.4.2).

SAQ 23

(a) False. An earthquake is a sudden release of *strain* in a local region of the Earth's crust or upper mantle (Section 1.5.1). The strain is the distortion produced by an applied stress; and the stress can continue acting even as the earthquake takes place to reduce or eliminate the strain that built up prior to the earthquake.

(b) True. To measure the complete motion it is necessary to measure the three components of motion at right angles (Section 1.5.1).

(c) True. This is really the definition of a P-wave (Section 1.5.1).

(d) False. In Rayleigh waves the particles travel in *vertical* ellipses (Section 1.5.1).

(e) True. S-waves travel through the interior, or body, of the Earth (Section 1.5.1).

(f) False. Love waves are *surface* waves, travelling along or close to the Earth's surface (Section 1.5.1).

(g) True. This is a matter of observation (Section 1.5.1).

(h) True. All seismic waves are elastic waves (Section 1.5.1).

(i) True. This is simply the definition of earthquake focus (Section 1.5.1).

(j) True. An epicentre is, by definition, the point on the Earth's surface directly above the focus (Section 1.5.1).

(k) False. P-waves travel faster than S-waves. The order of decreasing velocity is therefore: P-waves, S-waves, Love waves, Rayleigh waves (Section 1.5.1).

(l) True. This is a matter of observation (Section 1.5.2).

(m) True. Both Wadati–Benioff zones and oceanic trenches are the consequence of subduction of oceanic lithosphere (Section 1.5.2).

(n) False. During an earthquake the rocks on either side of a fault slip to new relative positions (Section 1.5.4).

(o) False. Non-earthquake sources are poor generators of S-waves (Section 1.5.5).

SAQ 24

According to the equation given in the caption to Figure 1.20, the velocity (v) of a wave is related to the wave's period (T) and wavelength (λ) by $v = \lambda/T$. So for $v = 6\,\text{km}\,\text{s}^{-1} = 6\,000\,\text{m}\,\text{s}^{-1}$ and $T = 20\,\text{s}$,

$$\lambda = 6\,000\,\text{m}\,\text{s}^{-1} \times 20\,\text{s} = 120\,000\,\text{m} = 120\,\text{km}.$$

Similarly, frequency (f) = v/λ; so

$$f = 6\,000\,\text{m}\,\text{s}^{-1}/120\,000\,\text{m} = 0.05\,\text{Hz}.$$

Incidentally, as $6\,\text{km}\,\text{s}^{-1}$ is a typical P-wave velocity in the upper continental crust, this calculation gives you an idea of just how long near-surface seismic waves are.

SAQ 25

According to the equation given in the caption to Figure 1.20, $f = v/\lambda$, where f is frequency. As the unit of v is $\text{m}\,\text{s}^{-1}$ and the unit of λ is m,

$$\text{unit of } f = \frac{\text{unit of } v}{\text{unit of } \lambda} = \frac{\text{m}\,\text{s}^{-1}}{\text{m}} = \text{s}^{-1}.$$

In other words, frequency is the reciprocal of time (i.e. 1/time).

SAQ 26

As the earthquake occurred within a few hundred kilometres of the seismometer, the approximate distance between the two will be given by Equation 1.7, with $t = 43.9\,\text{s}$. Thus

$$d = 8.65t = 8.65\,\text{km}\,\text{s}^{-1} \times 43.9\,\text{s} = 380\,\text{km (to three sig. figs.)}$$

SAQ 27

(a) Using the data, the time intervals between P-wave and S-wave arrivals are:

Station A: 3.5 s

Station B: 2.9 s

Station C: 4.1 s

Then using Equation 1.7, the distances from the stations to the epicentre are:

Station A: 30.3 km = 30 km (to two sig. figs.)

Station B: 25.1 km = 25 km (to two sig. figs.)

Station C: 35.5 km = 35 km (to two sig. figs.)

You could, and might, havc plotted these data on a graph to determine the position of the epicentre relative to B (as in ITQ 15). There's nothing wrong with that. However, you may have spotted that as the distance from B to the epicentre is about 25 km, the distance from C to the epicentre is about 35 km and the distance CB is 10.5 km, C, B and the epicentre must be roughly in a straight line with the epicentre on the opposite side of B to C (see Figure 1.130). In other words, the epicentre is 25 km from B and approximately due east of B. You should have obtained the same answer if you used the graphical method.

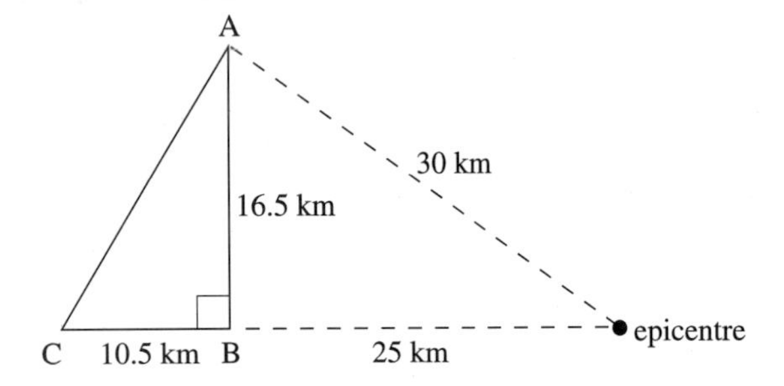

Figure 1.130 Diagram relating to the answer to SAQ 27

(b) You should recall that Equation 1.7 is based on the seismic velocities $v_P = 5.60\,\text{km}\,\text{s}^{-1}$ and $v_S = 3.40\,\text{km}\,\text{s}^{-1}$. Using the distance data in (a) and bearing in mind that velocity = distance/time, it is possible to calculate six seismic-wave travel times as follows:

	for P-waves	for S-waves
epicentre to station A	5.4 s	8.8 s
epicentre to station B	4.5 s	7.4 s
epicentre to station C	6.3 s	10.3 s

Subtracting these travel times from the six times given in the question makes it quite clear that the earthquake occurred at 12 hours, 30 minutes, 50 seconds (GMT).

SAQ 28

Because the depth of the earthquake is less than 50 km and the epicentral angle is in excess of 20°, Equation 1.9 can be used, with $A = 2 \times 10^{-4}\,\text{m} = 200\,\mu\text{m}$, $T = 20\,\text{s}$ and $\Delta = 41.7°$. So

$$\begin{aligned} M_S &= \log\left(\frac{200}{20}\right) + 1.66\log 41.7° + 3.3 \\ &= 1 + 2.7 + 3.3 \\ &= 7.0. \end{aligned}$$

SAQ 29

Using Equation 1.10, with $A = 6.0 \times 10^{-5}\,\text{m} = 60\,\mu\text{m}$, and $T = 12\,\text{s}$ and $\Delta = 41.7°$,

$$\begin{aligned} m_b &= \log\left(\frac{60}{12}\right) + (0.01 \times 41.7) + 5.9 \\ &= 0.7 + 0.4 + 5.9 \\ &= 7.0. \end{aligned}$$

For this particular earthquake, therefore, M_S and m_b turn out to be the same.

SAQ 30

The simplest way of determining the energy is to use Equation 1.12 with $M_S = 7.0$. Thus

$$\log E = 4.8 + 1.5(7.0) = 4.8 + 10.5 = 15.3$$
$$\text{and } E = 2.0 \times 10^{15}\,\text{J}.$$

A less direct way would be to use m_b to determine M_S from Equation 1.11. Thus

$$M_S = \frac{7.0 - 2.94}{0.55} = 7.4.$$

And then again from Equation 1.12

$\log E = 4.8 + 1.5(7.4) = 4.8 + 11.1 = 15.9$
and $E = 7.9 \times 10^{15}$ J.

The first of these results is the more valid, because it was determined by the more direct method. You should recall that the relationship between m_b and M_S (Equation 1.11) is only a statistical relationship. The value of M_S calculated using it will thus generally be less accurate than the value of M_S determined more directly.

SAQ 31

Using Equation 1.12 for $M_S = 8$,

$\log E = 4.8 + 1.5(8) = 4.8 + 12.0 = 16.8$
and $E = 6.3 \times 10^{16}$ J.

Using Equation 1.12 for $M_S = 4$,

$\log E = 4.8 + 1.5(4) = 4.8 + 6.0 = 10.8$
and $E = 6.3 \times 10^{10}$ J.

And so for 10^6 earthquakes of $M_S = 4$, $E = 6.3 \times 10^{10}\,\text{J} \times 10^6 = 6.3 \times 10^{16}$ J. A single earthquake of $M_S = 8$ thus releases the *same* amount of energy as a million earthquakes of $M_S = 4$.

SAQ 32

The answers to this series of questions are all to be found in Section 1.5, as follows: (1) (c); (2) (d); (3) (h); (4) (a); (5) (g); (6) (j); (7) (e); (8) (i); (9) (f); (10) (b).

SAQ 33

(a) True. As elastic modulus = stress/strain, for a given stress the smaller the elastic modulus the greater will be the strain, i.e. deformation (Section 1.6.1).

(b) False. As wave velocity $\propto \sqrt{\text{elastic modulus}}$, the smaller the elastic modulus the lower will be the wave velocity (Section 1.6.1).

(c) True. This is the definition of angle of incidence (Section 1.6.2).

(d) True. This is the definition of critical distance (Section 1.6.2).

(e) False. There is *no* critical distance for the direct wave (Section 1.6.2).

(f) True. The critical distance relates only to refracted waves (Section 1.6.2).

SAQ 34

This involves a straightforward application of Equation 1.17. Squaring both sides of the equation gives

$$\rho = \frac{\mu}{v_S^2}.$$

So for $\mu = 3.5 \times 10^{10}\,\text{N m}^{-2}$ and $v_S = 3.0\,\text{km s}^{-1} = 3\,000\,\text{m s}^{-1}$,

$$\rho = \frac{3.5 \times 10^{10}\,\text{N m}^{-2}}{(3\,000\,\text{m s}^{-1})^2} = \frac{3.5 \times 10^{10}\,\text{N m}^{-2}}{9 \times 10^6\,\text{m}^2\,\text{s}^{-2}} = 3\,889\,\text{kg m}^{-3}$$
$$= 3\,900\,\text{kg m}^{-3} \text{ (to two sig. figs.)}$$

SAQ 35

Squaring both sides of Equation 1.16 gives

$$v_P^2 = \frac{K + 4\mu/3}{\rho} \quad \text{and so} \quad K = \rho v_P^2 - 4\mu/3.$$

So for $\rho = 3\,900\,\text{kg m}^{-3}$, $\mu = 3.5 \times 10^{10}\,\text{N m}^{-2}$ and $v_P = 6.0\,\text{km s}^{-1} = 6\,000\,\text{m s}^{-1}$,

$$K = 3\,900\,\text{kg m}^{-3} \times (6\,000\,\text{m s}^{-1})^2 - \frac{4 \times 3.5 \times 10^{10}\,\text{N m}^{-2}}{3}$$

$$= 14 \times 10^{10} - 4.7 \times 10^{10}\,\text{N m}^{-2}$$

$$= 9.3 \times 10^{10}\,\text{N m}^{-2}.$$

SAQ 36

The axial modulus (Equation 1.18) is given by $K + 4\mu/3$

$$= 9.3 \times 10^{10} + \frac{4 \times 3.5 \times 10^{10}}{3}\,\text{N m}^{-2}$$

$$= 9.3 \times 10^{10} + 4.7 \times 10^{10}\,\text{N m}^{-2}$$

$$= 14 \times 10^{10}\,\text{N m}^{-2}.$$

SAQ 37

This is a straightforward application of Snell's law (Equation 1.19), with $v_1 = 4.0\,\text{km s}^{-1}$, $v_2 = 6.0\,\text{km s}^{-1}$ and $i = 30°$. So

$$\sin r = \frac{v_2 \sin i}{v_1} = \frac{6 \sin 30}{4} = 0.75$$

whence $r = 48.5° \sim 49°$.

SAQ 38

The critical angle occurs when $r = 90°$ (and hence $\sin r = 1$). Then

$$\sin i = \frac{v_1}{v_2} = \frac{4}{6} = 0.67$$

whence $i = 42.1° \sim 42°$.

The angle of incidence is thus greater than the critical angle, and so the wave will be reflected back into the first layer.

SAQ 39

(a) It should be clear from Figure 1.36c that

$$\tan i_c = \frac{\text{half the critical distance}}{\text{thickness of upper layer}}$$

and so

$$\tan i_c = \frac{100}{100} = 1, \text{ and } i_c = 45°.$$

Then from Equation 1.20

$$v_2 = \frac{v_1}{\sin i_c} = \frac{4.24\,\text{km s}^{-1}}{\sin 45°} = \frac{4.24\,\text{km s}^{-1}}{0.707} = 6.00\,\text{km s}^{-1}.$$

(b) From Equation 1.16 (squaring both sides)

$$\rho = \frac{K + 4\mu/3}{v_P^2}$$

$$= \frac{(6 \times 10^{10}\,\mathrm{N\,m^{-2}}) + (4 \times 3 \times 10^{10}\,\mathrm{N\,m^{-2}})/3}{(6\,000\,\mathrm{m\,s^{-1}})^2}$$

$$= \frac{10^{11}\,\mathrm{N\,m^{-2}}}{36 \times 10^6\,\mathrm{m^2\,s^{-2}}}$$

$$= 2\,778\,\mathrm{kg\,m^{-3}}$$

$$= 2\,780\,\mathrm{kg\,m^{-3}}\ \text{(to three sig. figs.)}$$

SAQ 40

The thickness of the upper layer (h) and the distance (x_d) at which the direct and refracted waves are received at the same time are related in Equation 1.25. Using that equation,

$$h = \frac{x_d}{2}\sqrt{\frac{v_2 - v_1}{v_2 + v_1}}$$

$$= \frac{150}{2}\sqrt{\frac{6-4}{6+4}}\,\mathrm{km}$$

$$= 75\sqrt{0.2}\,\mathrm{km}$$

$$= 33.5\,\mathrm{km} \sim 34\,\mathrm{km}.$$

SAQ 41

(a) The time–distance graph is shown in Figure 1.131.

(b) The graph has three straight-line segments, indicating three distinct layers.

(c) The P-wave velocity (v_1) in the uppermost layer is given by the gradient of the direct-wave curve, the left-hand segment of the graph.

$$\text{gradient} = \frac{1}{v_1} = \frac{20\,\mathrm{s}}{40\,\mathrm{km}} = 0.5\,\mathrm{s\,km^{-1}}, \text{ and so } v_1 = 2.0\,\mathrm{km\,s^{-1}}.$$

The P-wave velocity (v_3) in the lowermost (third) layer is given by the gradient of the second refracted-wave curve, the right-hand segment of the graph.

$$\text{gradient} = \frac{1}{v_3} = \frac{(61.4 - 52)\,\mathrm{s}}{(240 - 160)\,\mathrm{km}} = \frac{9.4\,\mathrm{s}}{80\,\mathrm{km}}, \text{ and so } v_3 = 8.5\,\mathrm{km\,s^{-1}}.$$

Note that the figures for determining the gradients were read at convenient points on the graph. You may have used different points, but the results should be same.

(d) The thickness of the upper layer is given by Equation 1.25 in terms of v_1 (= $2.0\,\mathrm{km\,s^{-1}}$), v_2 (the P-wave velocity in the second layer) and x_d (the distance at which the direct-wave and first-refracted-wave curves intersect). Reading from the graph, $x_d = 60\,\mathrm{km}$, but v_2 is not yet known.

Alternatively, the thickness of the upper layer is given by Equation 1.26 in terms of v_1, v_2 and t_0 (the intercept of the first-refracted-wave curve on the time axis). Reading from the graph, $t_0 = 15\,\mathrm{s}$, but again v_2 is not yet known.

v_2 must be obtained from the gradient of the second-refracted-wave curve, the middle segment of graph.

$$\text{gradient} = \frac{1}{v_2} = \frac{(45 - 35)\,\mathrm{s}}{(120 - 80)\,\mathrm{km}} = \frac{10\,\mathrm{s}}{40\,\mathrm{km}}, \text{ and so } v_2 = 4.0\,\mathrm{km\,s^{-1}}.$$

Then,

$$h = \frac{60}{2}\sqrt{\frac{4-2}{4+2}}\,\text{km} = \frac{30}{\sqrt{3}}\,\text{km} = 17.3\,\text{km}$$

or

$$h = \frac{15 \times 2 \times 4}{2\sqrt{16-4}}\,\text{km} = \frac{60}{\sqrt{12}}\,\text{km} = 17.3\,\text{km}.$$

(e) Reading from the graph, (i) the second arrival at 48 km arrives at 26.8 s, and (ii) the second arrival at 88 km arrives at 44.0 s.

(f) The critical angle is given simply by Equation 1.20

$$\sin i_c = \frac{v_1}{v_2} = \frac{2}{4} = 0.5,$$

and so $i_c = 30°$.

Figure 1.131 Graph relating to the answer to SAQ 41.

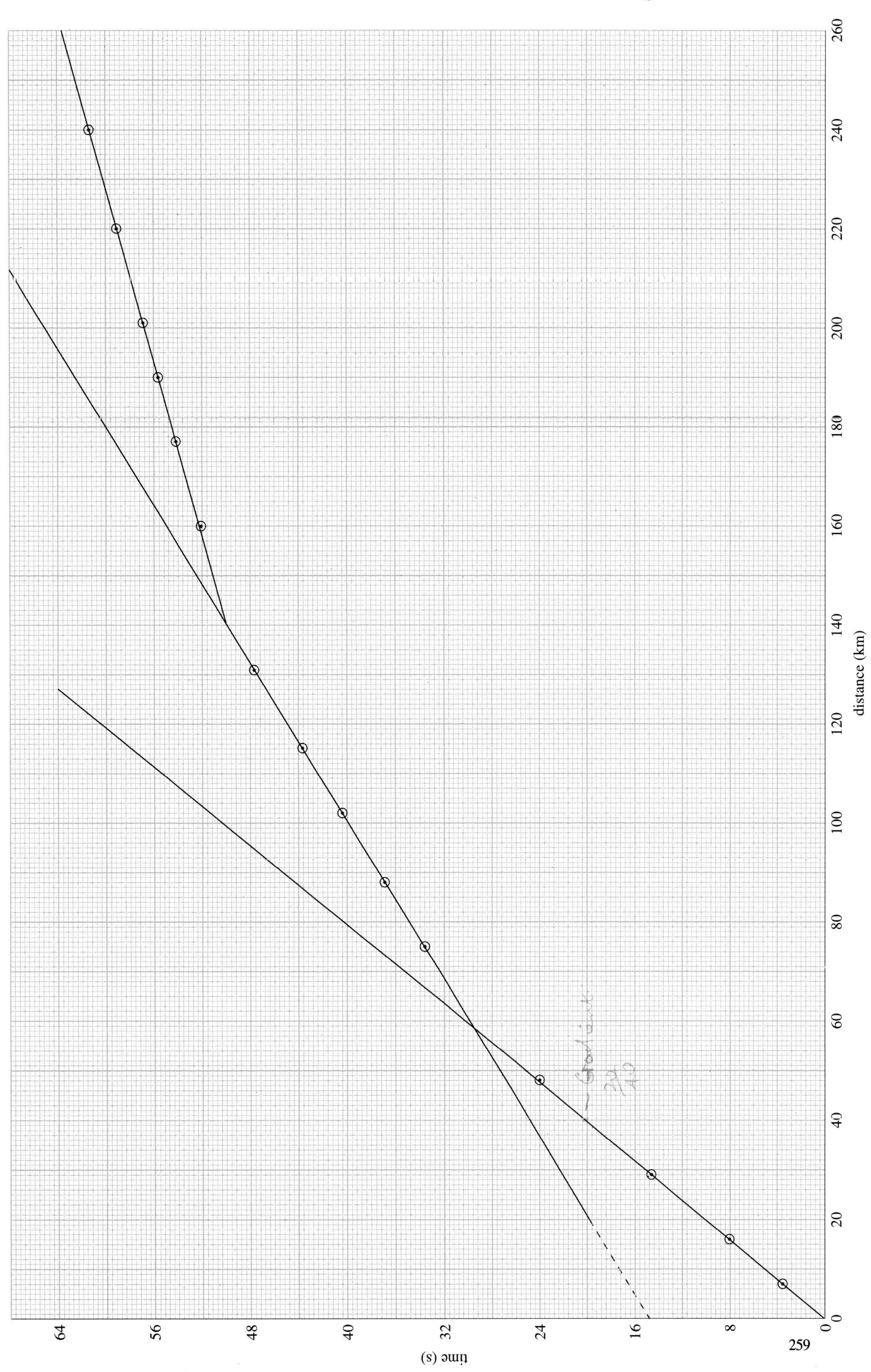
time (s)
0
8
16
24
32
40
48
56
64
distance (km)
0
20
40
60
80
100
120
140
160
180
200
220
240
260

SAQ 42

(a) True. Both the phenomena and the reason for them are true (Section 1.7.1).

(b) True. S-waves do not pass through fluids, and so if the Earth has a fluid outer core there is bound to be a zone in which refracted S-waves cannot emerge. The fact that this zone begins at a particular epicentral angle of 103° is not really relevant to the issue of the outer core's state, but it does enable the radius of the core to be determined (Section 1.7.1).

(c) False. The shadow zone refers to *refracted* waves. Reflected S-waves can be received within it (Section 1.7.1).

(d) False. The P-wave shadow zone lies between 103° and 142° (Section 1.7.1).

(e) False. Figure 1.52 (Section 1.7.1) clearly shows that the P-wave velocity is lower.

(f) True. As Figure 1.52 shows, there is some variation with depth, but not much.

(g) True. However you manipulate the figures of 35–40 km for continental crust and 5–7 km for oceanic crust, the ratio of the two is always 5 or greater (Section 1.7.2).

(h) False. Figure 1.54 (Section 1.7.2) clearly shows that in parts of southern England the crustal thickness exceeds 35 km.

(i) False. There is certainly a general increase with depth; but as Figure 1.57 (Section 1.7.3) shows, the increase could not be considered smooth.

(j) False. Layer 1 is the sedimentary surface layer of the oceanic crust (Section 1.7.4).

(k) True. This is evident from a reading of Figure 1.62 (Section 1.7.5).

(l) True. The LVZ is clearly shown in Figure 1.63 (Section 1.7.5).

(m) True. This is really the definition of lithosphere (Section 1.7.5).

(n) True. For 220 km is the depth of the base of the LVZ as indicated by seismic models (Section 1.7.5).

(o) True. This is clearly so if the base of the asthenosphere is placed at 670 km, because this would mean that the asthenosphere includes the partially molten LVZ and the region between 220 km and 670 km in which solid-state creep is thought to occur. However, even if the asthenosphere includes only the partially molten zone, some of the flow there could be due to solid-state creep. There is apparently nothing to prevent creep in the LVZ (Section 1.7.5).

(p) True. This is clearly evident from Figure 1.66 (Section 1.7.5) as long as we use the 'corrected' seismic dispersion data. Note, however, that we have not yet considered the thermal data.

(q) False. For reasons explained in Section 1.7.6, higher temperature means lower density and lower seismic velocity.

(r) True. The jump is from ~ 10.4 km s^{-1} to ~ 11.0 km s^{-1} (about 5%).

SAQ 43

For a wave travelling towards the Earth's centre (i.e. vertically downwards) the angle of incidence at any boundary (e.g. at the core–mantle boundary when entering or leaving the core) will be zero. So from Snell's law (Equation 1.19),

$$\frac{\sin i}{\sin r} = \frac{v_1}{v_2}$$

and for any values of v_1 and v_2,

$$\sin r = \frac{v_2 \sin i}{v_1} = \frac{v_2 \sin 0°}{v_1} = 0 \text{ (as } \sin 0° = 0).$$

In other words, any wave travelling vertically downwards will never deviate from its original direction and will therefore emerge at an epicentral angle of 180°.

SAQ 44

(a) True. (Section 1.8.3 and Figure 1.68.)

(b) False. The free-air anomaly is calculated after only the height and latitude difference between stations (S) and the reference (P) have been accounted for. Bouguer anomalies are calculated after corrections for latitude, elevation, density and terrain have been made (Section 1.8.4). However, it is true that free-air anomalies are usually computed for marine gravity surveys and Bouguer anomalies for land surveys.

(c) False. Isostatic rebound is the process by which the lithosphere rises or falls in response to loading. The models of Pratt and Airy describe the lithosphere in terms of columns of equal cross-sectional area and variable density (Pratt) or depth to level of compensation (Airy). For isostatic equilibrium, the weight of each column (for either Pratt or Airy models) must be the same. If the columns are not of equal weight, then the region is not in isostatic equilibrium.

SAQ 45

(a) The reference point and station are the same latitude, so $\Delta_1 g = 0$. From Equation 1.37, since $h_S - h_P = +20$, the free-air correction is:

$$\begin{aligned} \Delta_2 g &= -0.308\,6 \times 20\,\text{mGal} \\ &= -6.172\,\text{mGal} \end{aligned}$$

and it is negative because the site is *above* the reference point, where gravity is less:

$$(g_S - g_P)_{\text{corrected}} = \Delta g - \Delta_1 g - \Delta_2 g.$$

The expression $(-\Delta_1 g - \Delta_2 g)$ is what the question asks for,

This is $-(0) - (-6.172) = +6.172\,\text{mGal}$.

Therefore we must add 6.172 mGal to the observed gravity anomaly.

(b) From Equation 1.39, the Bouguer correction is:

$$\begin{aligned} \Delta_3 g &= 4.192\,1 \times 10^{-5} \rho h\,\text{mGal} \\ &= 4.192\,1 \times 10^{-5} \times 2\,670 \times 20\,\text{mGal} \\ &= 2.238\,\text{mGal} \end{aligned}$$

and, as Equation 1.40 shows:

$$(g_S - g_P)_{\text{corrected}} = \Delta g - \Delta_1 g - \Delta_2 g - \Delta_3 g.$$

The Bouguer correction must be subtracted from the observed gravity anomaly because the site lies *above* the reference point.

The total correction to Δg, therefore, becomes:

$$-(0) - (-6.172) - (+2.238)\,\text{mGal} = +3.934\,\text{mGal}.$$

SAQ 46

The matching pairs of text are: (i) A (Section 1.8.4); (ii) C (Section 1.8.2); (iii) D (Section 1.8.4); (iv) B (Section 1.8.2).

SAQ 47

(a) From Equation 1.44, if the columns are in isostatic equilibrium according to Airy's hypothesis:

$$\rho_c h + \rho_s l + \rho_w d = \text{constant}.$$

Using densities in $\text{kg}\,\text{m}^{-3}$ and heights in m:

For column A: $(2\,670 \times 38.6 \times 10^3)\,\text{kg}\,\text{m}^{-2} + (3\,270 \times 20 \times 10^3)\,\text{kg}\,\text{m}^{-2}$
$= 1.685 \times 10^8\,\text{kg}\,\text{m}^{-2}$.

For column B: $(2\,670 \times 30.0 \times 10^3)\,\text{kg}\,\text{m}^{-2} + (3\,270 \times 27.0 \times 10^3)\,\text{kg}\,\text{m}^{-2}$
$= 1.684 \times 10^8\,\text{kg}\,\text{m}^{-2}$.

Therefore the two columns have roughly equal masses per unit cross-section down to the level of compensation and are in isostatic equilibrium.

(b) The height of column A above sea-level must be the difference in the heights of the two columns, since B reaches sea-level. This is:

$$58.6\,\text{km} - 57.0\,\text{km} = 1.6\,\text{km}.$$

SAQ 48

These words should be inserted in the following order:

velocity, low, Bouguer, isostatic, isostatic, partially molten, trench, island arc, negative, Bouguer, deficiency, excess.

SAQ 49

(a) True. See Equation 1.50 and the discussion in Section 1.10.

(b) False. The Monte Carlo inversion method described in Section 1.10 uses a random number generation method for guessing the Earth's *density* at depth; the observed mass and moment of inertia are used as checks on plausible density distributions.

(c) True. See Figure 1.80.

SAQ 50

(a) True. Working down from the top right-hand corner of Figure 1.16, it becomes clear that, after the group of elements considered right at the beginning of Section 1.11, the next four most abundant elements are Al, Ca, Na and Ni. However, as Figure 1.5 shows, Na is a volatile. The fifth, sixth and seventh most refractory elements in the Earth (the precise order is unimportant here), assuming the Earth to be chondritic, are therefore Al, Ca and Ni.

(b) False. As we determined in Section 1.11, the four most abundant refractory elements are O, Fe, Si and Mg.

(c) True. See Section 1.11.

(d) False. The Nafe–Drake curve does indeed enable density to be obtained from seismic velocity, but the rock cannot then be specified uniquely because rocks have overlapping density ranges (Section 1.11.1).

(e) True. Reading from the Nafe–Drake curve (Figure 1.82), a P-wave velocity of $6\,km\,s^{-1}$ is equivalent to a density of about $2\,700\,kg\,m^{-3}$. This is outside the usual density range of clay (Table 1.5). The rock is therefore unlikely to be clay (although given the uncertainty in the Nafe–Drake curve, clay might just be possible).

(f) True. All the estimates of SiO_2 in Table 1.6 are in excess of 55%.

(g) False. The seismic data from the lower crust are incompatible with the lower crust's being of the same composition as the upper crust (Section 1.11.1).

(h) True. See Section 1.11.2.

(i) True. This becomes clear by comparing the relevant figures in Table 1.7 with those in Table 1.6.

(j) True. This can be checked by adding the figures in each column of Table 1.7.

(k) True. As explained in Section 1.11.2, the iron in the mantle (on the assumption that the proportion in the lower mantle is the same as in the upper mantle), in the crust and in the core is apparently insufficient to add up to an Earth fully chondritic in iron.

(l) False. The phase changes take place in peridotite (Section 1.11.2).

(m) False. The huge amount of iron in the core is also required (Section 1.11.3).

(n) True. See Section 1.11.3.

(o) True. See Section 1.11.3.

(p) False. It is clear from Figure 1.83 that the opposite is the case (i.e. carbon is the more depleted).

(q) True. See Section 1.11.3.

(r) True. This becomes clear from an examination of Figure 1.129 (see answer to ITQ 44).

SAQ 51

According to Figure 1.16, in a chondritic Earth there would be about 10^4 atoms of Cr for every 10^6 atoms of Si. As the atomic mass of Si is 28.09, an atom of Cr has a mass of 52.00/28.09 = 1.851 times that of Si. The relative abundance of Cr compared to Si in terms of mass is therefore $(1.851 \times 10^4)/10^6 = 0.018\,5$.

SAQ 52

Reading from the Nafe–Drake curve (Figure 1.82), a P-wave velocity of about $8.5\,km\,s^{-1}$ translates into a density of about $3\,500\,kg\,m^{-3}$. The only rock listed in Table 1.5 with such a high density is gabbro, which therefore becomes the most likely choice. (Perhaps you should regard this question as an exercise rather than as representing a realistic situation. A P-wave velocity as high as $8.5\,km\,s^{-1}$ is a pretty unlikely phenomenon to observe in the uppermost crust, although it is presumably theoretically possible for a form of gabbro at the higher end of its density range.)

SAQ 53

(a) True. This is a matter of observation (Section 1.12).

(b) True. The unit of heat flow (q) is $\mathrm{W\,m^{-2}}$ (Table 1.10) and the unit of heat generation (A) is $\mathrm{W\,m^{-3}}$. So the unit of q/A is

$$\frac{\mathrm{W\,m^{-2}}}{\mathrm{W\,m^{-3}}} = \mathrm{m}.$$

(c) False. This should be clear from Table 1.12, by comparing the figure you have filled in for granodiorite with the corresponding figure for gabbro (extreme right-hand column).

(d) True. See Section 1.12.2.

(e) False. Using Equation 1.51,

$$q = k\frac{\Delta T}{\Delta z} \text{ where } k = 1.50\,\mathrm{W\,m^{-1}\,K^{-1}}$$

$$\text{and } \frac{\Delta T}{\Delta z} = 27.3\,\mathrm{K\,km^{-1}} = 0.027\,3\,\mathrm{K\,m^{-1}}.$$

So $q = 1.50 \times 0.027\,3 = 0.041\,\mathrm{W\,m^{-2}} = 41\,\mathrm{mW\,m^{-2}}$.

(f) True. Using Equation 1.52,

$$d = 2\,500 + 350\sqrt{t}\,\mathrm{m} = 2\,500 + 350\sqrt{49}\,\mathrm{m} = 4\,950\,\mathrm{m}.$$

(g) False. Equation 1.52 will give a depth of 6700 m. Remember, however, that this equation does not apply to lithosphere older than about 70 Ma (Section 1.12.4).

(h) False. It is the plate model that assumes constant lithosphere thickness (Section 1.12.4).

(i) False. Precisely the opposite, in fact; the predicted heat flow agrees least well in the 0–50 Ma range (Section 1.12.4).

(j) True. Using Equation 1.55,

$$q = \frac{473}{t^{0.5}} = \frac{473}{(100)^{0.5}}\,\mathrm{mW\,m^{-2}} = 47.3\,\mathrm{mW\,m^{-2}}.$$

(k) True. See Section 1.12.5.

(l) False. Using Equation 1.56,

for $A = 8\,\mu\mathrm{W\,m^{-3}}$ and $q = 30\,\mathrm{mW\,m^{-2}}$, $30 = 8b + q^*$; (Equation A)

for $A = 4\,\mu\mathrm{W\,m^{-3}}$ and $q = 20\,\mathrm{mW\,m^{-2}}$, $20 = 4b + q^*$. (Equation B)

If Equation B is multiplied throughout by 2, it becomes $40 = 8b + 2q^*$, and from Equation A, $8b = 30 - q^*$.

So $40 = 30 - q^* + 2q^*$, and so $q^* = 10\,\mathrm{mW\,m^{-2}}$.

Feeding this value back into Equation B gives $20 = 4b + 10$, and so $b = 2.5\,\mathrm{km}$ (i.e. not 5 km).

(m) True. This is a matter of observation — or, rather, calculations based on observation (Section 1.12.6).

SAQ 54

Gabbro contains 0.96 ppm of ^{40}K (Table 1.12), and so 1 kg contains 0.96×10^{-6} kg of ^{40}K. 1 kg of ^{40}K generates heat at the rate of 2.8×10^{-2} mW (Table 1.11), and so 0.96×10^{-6} kg will generate heat at the rate of

$$0.96 \times 10^{-6} \times 2.8 \times 10^{-2}\,\text{mW kg}^{-1} = 2.7 \times 10^{-8}\,\text{mW kg}^{-1}$$

(to two sig. figs.)

SAQ 55

1 kg of granodiorite generates heat at the rate of 96×10^{-8} mW kg^{-1} (ITQ 47), and the mass of the upper continental crust is 8×10^{21} kg (ITQ 48). The upper continental crust therefore generates heat at the rate of

$$96 \times 10^{-8} \times 8 \times 10^{21}\,\text{mW} = 8 \times 10^{15}\,\text{mW (to one sig. fig.)}$$

1 kg of peridotite generates heat at the rate of 0.26×10^{-8} mW kg^{-1} (Table 1.12), and the mass of the mantle is 4×10^{24} kg (ITQ 48). The mantle therefore generates heat at the rate of

$$0.26 \times 10^{-8} \times 4 \times 10^{24}\,\text{mW} = 1 \times 10^{16}\,\text{mW (to one sig. fig.)}$$

The mantle therefore generates heat at the greater rate. Note, however, that, despite its much greater mass (500 times that of the upper continental crust), the mantle generates heat at a rate only 1.25 times greater than does the upper continental crust.

SAQ 56

The situation is as shown in Figure 1.132. The temperature drop along the cooler brass rod is 8 °C − 7 °C = 1 °C; and assuming that no heat is lost from the sides of the apparatus, this must also be the temperature drop along the hotter brass rod, which is identical. The temperature at the inner end of the hotter brass rod is therefore 27 °C − 1 °C = 26 °C, and hence the temperature drop across the rock sample is 26 °C − 8 °C = 18 °C. Then using the appropriate formula in the AV 04 notes (which is simply derived using Equation 1.51)

$$k_r = k_b \left(\frac{\Delta T}{\Delta z}\right)_b \left(\frac{\Delta z}{\Delta T}\right)_r$$

$$= 113 \times \frac{1}{0.05} \times \frac{0.005}{18}\,\text{W m}^{-1}\,\text{K}^{-1}$$

$$= 0.63\,\text{W m}^{-1}\,\text{K}^{-1}.$$

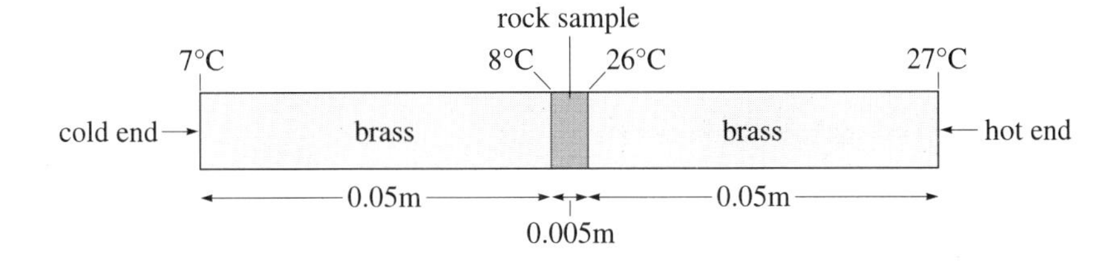

Figure 1.132 Diagram relating to the answer to SAQ 56.

SAQ 57

Using the appropriate formula in the AV 04 notes, the temperature (T) rises with time (t) according to

$$T = \frac{Q}{4\pi k} \ln t + C$$

where Q is the heat applied to the needle per unit length (m) per second, k is the thermal conductivity, and C is a constant.

Q = 36 J per minute per 60 mm. Therefore

$$Q = \frac{36 \times 1\,000}{60 \times 60} = 10\,\text{J per second per metre (J s}^{-1}\,\text{m}^{-1}\text{)}.$$

When t = 10 s, T = 30.3 °C and so $30.3 = \frac{10}{4\pi k} \ln 10 + C$.

When t = 1 000 s, T = 34.9 °C and so $34.9 = \frac{10}{4\pi k} \ln 1\,000 + C$.

Subtracting the two,

$$34.9 - 30.3 = 4.6 = \frac{10}{4\pi k} \ln 1\,000 + C - \frac{10}{4\pi k} \ln 10 - C$$
$$= \frac{10}{4\pi k}(\ln 1\,000 - \ln 10) = \frac{10}{4\pi k}(6.9 - 2.3) = \frac{10}{4\pi k}(4.6)$$

and so

$$k = \frac{10}{4\pi} = 0.8\,\text{W m}^{-1}\,\text{K}^{-1}.$$

SAQ 58

Using Equation 1.56 and substituting the values of q and A given in the question:

For province X: $96 = 5b + q^*$ and $36 = 1b + q^*$, so by subtraction, $96 - 36 = 4b$ and so $b = \frac{60}{4} = 15$ km.

For province Y: $80 = 7.7b + q^*$ and $50 = 1.7b + q^*$, so by subtraction, $80 - 50 = 6b$ and so $b = \frac{30}{6} = 5$ km.

SAQ 59

Substituting the values of b back into one of the corresponding equations:

For province X: $36 = 15 + q^*$ and so $q^* = 21\,\text{mW m}^{-2}$.

For province Y: $50 = (5 \times 1.7) + q^*$ and so $q^* = 41.5\,\text{mW m}^{-2}$.

SAQ 60

As q^* represents the heat-flow contribution from the lower crust and mantle (i.e. that not generated in the upper crust), the proportion required will be

$$\frac{q^*}{q} = \frac{41.5}{59.3} = 0.70 \text{ or } 70\%.$$

SAQ 61

(a) cannot be true because, if it were, q^* would be zero, which is clearly not the case. (b) cannot be true because, if it were, all the bA values would be zero, which is equally clearly not the case. From SAQ 60, q^*/q = 70%, whereas for the Earth's continents as a whole q^*/q = 60% (Section 1.12.5). Thus for province Y, q^*/q is greater than that for the Earth's continents as a whole by 70/60 = 1.17 or 117%. (c) is therefore a valid conclusion, which means that (d) cannot be.

That's the easy part. What follows is very much trickier, and you should not be discouraged if you find you have failed to work it out. Let's suppose that, sometime during the previous few hundred million years, province Y had experienced a magmatic/metamorphic episode in the upper crust, thereby introducing an extra component of heat into the province.

The average heat flow in the region would therefore be higher than it would be on the basis of upper crustal radioactivity and lower crustal/mantle background heat alone. Individual heat flow values (q) would also be higher, making it appear that the ($bA + q^*$) values in Equation 1.56 were higher. However, the bA values represent radioactive heat generation in the upper crust, which would presumably remain unaltered by a magmatic/metamorphic episode. The effect of the episode, therefore, would be to make q^* appear bigger than it really is. Moreover, q^* would rise by a greater proportion than q. Suppose, for example, that the effect of the episode was to raise q by $x\%$. Then ($bA + q^*$) would appear to increase by $x\%$; but because bA remains the same, q^* would appear to rise by more than $x\%$, and the ratio q^*/q would therefore appear to rise above the normal value of 0.6. Accordingly, a q^*/q value greater than 0.6, as in province Y, would strongly imply that a magmatic/metamorphic episode had indeed taken place in the province's upper crust. (e) is therefore a reasonable conclusion, which means that (f) cannot be. (Of course, all this rather assumes that the episode introduced extra heat uniformly throughout the province. If it did not do so, individual q values would have risen by different amounts and a plot of q against A would no longer have been a straight line.)

Finally, as Figure 1.107 suggests, the heat from a magmatic/metamorphic episode largely dissipates within a few hundred million years. In 500 million years' time, therefore, the heat flow should have reduced to the observed value of 40–50 mW m^{-2} for continental regions greater than about 1 500 Ma old (Figure 1.107) In short, (g) is a valid conclusion, in which case (h) cannot be. In summary, the correct answer to SAQ 61 is (c), (e) and (g).

SAQ 62

For the variables involved in the Rayleigh number, the completed table below indicates whether each promotes or inhibits convection.

Variable	Promotes convection	Inhibits convection
(a) volume coefficient of thermal expansion (α)	✓	
(b) temperature difference (ΔT)	✓	
(c) gravitational acceleration (g)	✓	
(d) depth of fluid layer (d)	✓	
(e) thermal diffusivity (κ)		✓
(f) dynamic viscosity (η)		✓

SAQ 63

(a) According to the channel-flow model, mantle deformation associated with the growth of large ice sheets is restricted to a channel at relatively shallow depth below the ice sheet. By contrast, the deep-flow model predicts that such mantle deformation occurs over a wide depth range, up to a depth equivalent to about half the horizontal extent of the ice sheet.

(b) The major predicted distinction for the deformation of deglaciated areas is that at the start of deglaciation the channel-flow model predicts the existence of a significant peripheral bulge around the ice-sheet, as a result of mantle material being squeezed outwards and upwards by the pressure of the ice above and rigid mantle below. Following deglaciation, this bulge would collapse quickly so that areas around the ice sheet would show only *sinking*, on the channel-flow model. In contrast, the deep-flow model predicts a small *uplift* followed by sinking.

SAQ 64

The high estimated Rayleigh number for the mantle (about 10^6) in comparison with the critical Rayleigh number (about 10^3) implies that the mantle as a whole is in convective motion, and that the whole-mantle convection model may be valid.

SAQ 65

In the two-scale model of shallow-mantle convection, the large-scale convection represents the horizontal motions of the lithospheric plates themselves. This scale has a range of around 2 000–10 000 km. The small-scale convection extends from the lithosphere to a depth of 700 km, and the convection cells have similar widths and heights. Both large-scale and small-scale models are thought to stop below about 700 km.

SAQ 66

A low-velocity region such as revealed in Figure 1.121 represents relatively hot buoyant material. This material is less dense than its surroundings, so we would expect there to be a negative Bouguer anomaly and a negative geoid anomaly (the surface is depressed below the spheroid in order for the IGF value to be attained).

SUGGESTIONS FOR FURTHER READING

Much of Block 1 is concerned with basic geophysics. Unfortunately, although a number of books advertise themselves as 'introductions' to geophysics, few, if any, can be regarded as truly basic. In particular, they usually assume a competence in mathematics way beyond the level of that in S267. Nevertheless, you should be able to understand parts of them. One of the best and most recent of such texts is

Fowler, C. M. R. (1990) *The Solid Earth: An Introduction to Global Geophysics* Cambridge University Press.

Two older, but rather less mathematical, texts, which also give some attention to the geochemistry of the Earth, are

Bott, M. H. P. (1982) *The Interior of the Earth: Its Structure, Constitution and Evolution*, 2nd edn, Edward Arnold

and

Brown, G. C. and Mussett, A. E. (1981) *The Inaccessible Earth* George Allen and Unwin. [A revised version of this is due from Chapman and Hall in mid-1993.]

Curiously, basic geophysical topics are often better covered in books on applied geophysics (i.e. geophysics applied to resource exploration), one of the best and most recent of which is

Kearey, P. and Brooks, M. (1991) *An Introduction to Geophysical Exploration*, 2nd edn, Blackwell Scientific Publications.

Less wide-ranging books that we would recommend, but again with the warning that you may find parts of them difficult, are

Bolt, B. A. (1988) *Earthquakes*, 2nd edn, W. H. Freeman,

Bolt, B. A. (1982) *Inside the Earth: Evidence from Earthquakes* W. H. Freeman

and

Tsuboi, C. (1983) *Gravity* George Allen & Unwin.

The 'standard' model of the origin of the Solar System is so new that, as far as we know, it has not yet (in late 1992) emerged from original scientific papers, reviews and research monographs. We are therefore unable to recommend a book on that topic. However, a recent text that covers the origin of the elements in the Solar System and the terrestrial distribution of elements is

Cox, P. A. (1989) *The Elements: Their Origin, Abundance and Distribution* Oxford University Press.

Finally, an excellent general introduction to a wide variety of topics, some of which relate to Block 1 (and, indeed, to other parts of this Course), is

Brown, G. C., Hawkesworth, C. J. & Wilson, R. C. L. (eds) (1992) *Understanding the Earth: A New Synthesis* Cambridge University Press.

ACKNOWLEDGEMENTS

Thanks are due to the Course Team for improving early drafts of this Block; thanks are also due to Philip Kearey, David Darbishire, Colin Hayes and Ray Macdonald.

Grateful acknowledgement is made to the following sources for permission to reproduce material in this Block:

Figure 1.1: Moore, Hunt, Nicolson and Cattermole *The Atlas of the Solar System*, 1983, published by Mitchell Beazley International Ltd; *Figure 1.3*: J. Kelly Beatty and Andrew Chaikin *The New Solar System*, 3rd edition 1990, Cambridge University Press; *Figures 1.4 and 1.74*. D. J. Levin 'The Upper Mantle', in M. H. P. Bott (cd.) *The Interior of the Earth*, © 1972, 1982 Elsevier Science Publishers BV; *Figures 1.5 and 1.7*: J. W. Morgan and E. Anders *Proc. Nat. Acad. Sci.*, 77, 6973, © 1980 National Academy of Science; *Figure 1.6*: Boss *Science*, 231, pp. 341–345, 1986, © American Association for the Advancement of Science; *Figure 1.8*: Stevenson 'Fluid dynamics of core formation', in Newson and Jones (eds) *Origin of the World*, Oxford University Press, © 1990 Lunar and Planetary Institute, Huston; *Figure 1.9*: Don L. Anderson *Theory of the Earth*, © 1989 Blackwell Scientific Publications Limited; *Figure 1.10*: *New Scientist*, 3 February 1990, p. 4, © IPC Magazines 1990; *Figure 1.12*: R. C. O. Gill, *Chemical Fundamentals of Geology*, 1982, Chapman & Hall; *Figures 1.13, 1.15 and 1.16*: H. Y. McSween. *Meteorites and their Parent Planets*, 1987, p. 44, Cambridge University Press; *Figure 1.14*: adapted from K. Keil 'Classification of Chondrites', in K. H. Wedepohl (ed.) *Handbook of Geochemistry*, 4th edition 1974, Springer-Verlag; *Figure 1.19*: From *Nuclear Expolosions and Earthquakes*, by Bruce A. Bolt, copyright © 1976 by W. H. Freeman and Company. Reprinted by permission; *Figure 1.21*: D. Davies 'A comprehensive test ban', in *Science Journal*, vol. 4, no. 11, November 1968, © 1968 ILIFFE Industrial Publications Ltd; *Figure 1.24*: From *Earthquakes*, by Bruce A. Bolt, copyright © 1978, 1988 by W. H. Freeman and Company. Reprinted by permission; *Figure 1.25*: © National Geophysical Data Center, Boulder, Colorado, USA; *Figure 1.27*: From B. Isacks, J. Oliver and L. R. Sykes 'Seismology and new global tectonics', in *Journal of Geophysical Research*, vol. 73, © 1968 American Geophysical Union; *Figure 1.28*: A. F. Espinosa et al. 'The San Fernando, California, earthquake of February 9, 1971', *USGS*, © 1971, US Geological Survey; *Figure 1.34*: J. A. Jacobs *Deep Interior of the Earth*, 1992, Chapman & Hall, © 1992 J. A. Jacobs; *Figure 1.41*: J. P. Eaton 'Crustal structure from San Francisco, California, to Eureka, Nevada, from seismic refraction measurements', in *Journal of Geophysical Research*, vol. 68, © American Geophysical Union; *Figure 1.42*: G. Muller and R. Kind 'Observed and computed seismogram sections for the whole Earth', in *Geophys. J. R. astr. Soc.* vol. 44, 1976, pp. 699–716, reproduced by permission of the Royal Astronomical Society and the authors; *Figure 1.45*: A. D. Gibbs, 'Structural evolution of extensional basin margins', in *Journal of the Geological Society*, vol. 141, no. 4, 1984, © Geological Society 1984; *Figures 1.46 and 1.47*: Brewer and Oliver, 'Seismic reflection studies of deep crustal structure', reproduced, with permission, from the *Annual Review of Earth and Planetary Sciences*' vol. 8, © 1980 by Annual Reviews Inc.; *Figure 1.49*: K. E. Bullen and B. A. Bolt *An Introduction to the Theory of Seismology*, © Cambridge University Press 1963, 1985; *Figure 1.52*: A. M. Dziewonski and D. L. Anderson 'Preliminary reference Earth model', in *Physics of the Earth and Planetary Interiors*, vol. 25, no. 4, June 1981, © 1981 Elsevier Science Publishers BV; *Figures 1.53 and 1.56*: From *Inside the Earth* by Bruce A. Bolt, copyright © 1982 by W. H. Freeman and Company. Reprinted by permission; *Figure 1.54*: R. J. Allenby and

C. C. Schnetzler, 'United States crustal thickness', in *Tectonophysics*, vol. 93, © 1983 Elsevier Science Publishers BV; *Figure 1.55*: L. C. Pakiser 'Structure of the crust and upper mantle in the Western United States', in *Journal of Geophysical Research*, vol. 68, © 1963 American Geophysical Union; *Figure 1.57*: J. A. Brewer et al. 'The Laramide orogeny: evidence from COCORPS deep crustal seismic profile in the Wind River Mountains', in *Tectonophysics*, vol. 62, © 1979 Elsevier Scientific Publishers BV; *Figures 1.58 and 1.80*: G. C. Brown and A. E. Mussett *The Inaccessible Earth*, 1981, Chapman & Hall; *Figure 1.60*: M. Ewing 'The sediments of the Argentine Basin', in *The Quarterly Journal of the Royal Astronomical Society*, vol. 6, March 1965, © The Royal Astronomical Society; *Figure 1.62*: E. Herrin, 'Regional variations of P-wave velocity in the upper mantle beneath North America', in P. J. Hart (ed) *The Earth's Crust and Upper Mantle*, © 1969 American Geophysical Union; *Figure 1.64*: Reprinted with permission from K. C. Condie *Plate Tectonics and Crustal Evolution*, © 1982 Pergamon Press Ltd; *Figure 1.65*: W. A. Nierenberg, *Encyclopedia of Earth System Science*, vol. 3, © 1992 Academic Press Inc.; *Figures 1.68 and 1.69*: D. G. King-Hele 'The shape of the Earth', in *Royal Aircraft Establishment Technical Memo Space 130*, September 1969, © D. G. King-Hele; *Figure 1.71*: H. Rymer and G. C. Brown 'Gravity fields and the interpretation of volcanic structures: Geological discrimination and temporal evolution', in the *Journal of Volcanology and Geothermal Research*, vol. 27, © 1986 Elsevier Science Publishers BV; *Figure 1.72:* C. M. Fowler, *The Solid Earth*, 1990, Cambridge University Press; *Figure 1.77*: F. A. Vening Meinesz *Developments in Solid Earth Geophysics*, © 1964 Elsevier Science Publishers BV; *Figure 1.78*: M. Talwani, G. H. Sutton and J. L. Worzel 'A crustal section across the Puerto Rican Trench', in *Journal of Geophysical Research*, vol. 64, © 1959 American Geophysical Union; *Figure 1.82*: J. E. Nafe and C. L. Drake 'Physical properties of marine sediments', in M. N. Hill (ed) *The Sea*, vol. 3, © 1963 Interscience Publishers (Division of John Wiley & Sons Inc.); *Figure 1.83*: V. Rama Murthy 'Composition of the core and the early chemical history of the Earth', in B. F. Windley (ed.) *The Early History of the Earth*, © 1976 John Wiley & Sons Ltd. Reprinted by permission of John Wiley & Sons Ltd.; *Figures 1.91 and 1.97*: J. G. Sclater et al. 'The heat flow through the oceanic and continental crust and the heat loss of the Earth', in *Reviews of Geophysics and Space Physics*, vol. 18, no. 1, American Geophysical Union, © authors and Dept. of Earth & Planetary Science, MIT; *Figures 1.94 and 1.99*: B. Parsons and J. G. Sclater 'An analysis of the variation of ocean floor bathymetry and heat flow with age', in the *Journal of Geophysical Research*, vol. 82, no. 5, February 1977, © 1977 American Geophysical Union; *Figure 1.95*: P. Kearey and F. Vine *Global Tectonics*, © 1990 Blackwell Scientific Publications Ltd; *Figure 1.100*: M. G. Langseth Jr and R. P. Von Herzen 'Heat flow through the floor of the world oceans', in A. E. Maxwell (ed) *The Sea*, vol. 4, part 1, © 1971 John Wiley and Sons Inc.; *Figure 1.104*: R. F. Roy et al. 'Heat generation of plutonic rocks and continental heat flow sources', in *Earth and Planetary Science Letters*, vol. 5, © 1968 Elsevier Science Publishers BV; *Figures 1.106 and 1.107*: I. Vitorello and H. N. Pollack 'On the variation of continental heat flow with age and the thermal evolution of continents', in *Journal of Geophysical Research*, vol. 85 B2, © 1980 American Geophysical Union; *Figures 1.111 and 1.112*: L. M. Cathles *The Viscosity of the Earth's Mantle*, © 1975 Princeton University Press; *Figure 1.113*: Adaption of C. A. Kaye and E. S. Barghoorn, *Bull. Geological Society of America*, vol. 75, p. 76, © 1964 Geological Society of America; *Figures 1.114 and 1.115*: J. M. Hewitt, D. P. McKenzie and N. O. Weiss 'Large aspect ratio cells in two dimensional thermal convection', in *Earth and Planetary Science Letters*, vol. 51, © 1980 Elsevier Science Publishers BV; *Figures 1.117 and*

1.120: D. L. Anderson *Theory of the Earth*, © 1989 Blackwell Scientific Publications Ltd.; *Figures 1.118 and 1.119*: T. Tanimoto and D. L. Anderson 'Lateral heterogeneity and azimuthal anisotropy of the upper mantle: Love and Rayleigh waves 100–250 s', in *Journal of Geophysical Research*, © 1985 American Geophysical Union; *Figures 1.121 and 1.222*: Michael P. Ryan *Magna Transport and Storage*, © 1990 John Wiley and Sons Ltd. Reproduced by permission of John Wiley and Sons Ltd.

INDEX FOR BLOCK 1

(**bold** entries are to key terms; *italic* entries are to tables and figures)

BLOCK 1 COLOUR PLATE SECTION

Plate 1.1 A false-colour optical photograph of the spiral Andromeda Galaxy (M31), about 2.2 million light years away. One of the largest members of the Local Group of galaxies, Andromeda is about 170 000 light years across and is visible to the naked eye as a faint patch. Also shown are Andromeda's two dwarf elliptical satellite galaxies, NGC 205 (top right) and NGC 221 (the bright spot on the lower edge of Andromeda itself).

Plate 1.2 An image of the asteroid Gaspra, taken by the passing Galileo spacecraft on 20 October 1991. Gaspra has a maximum dimension of about 15 km.

Plate 1.3 Comet West as photographed on 9 March 1976, showing its dust tail (white) and gas tail (blue).

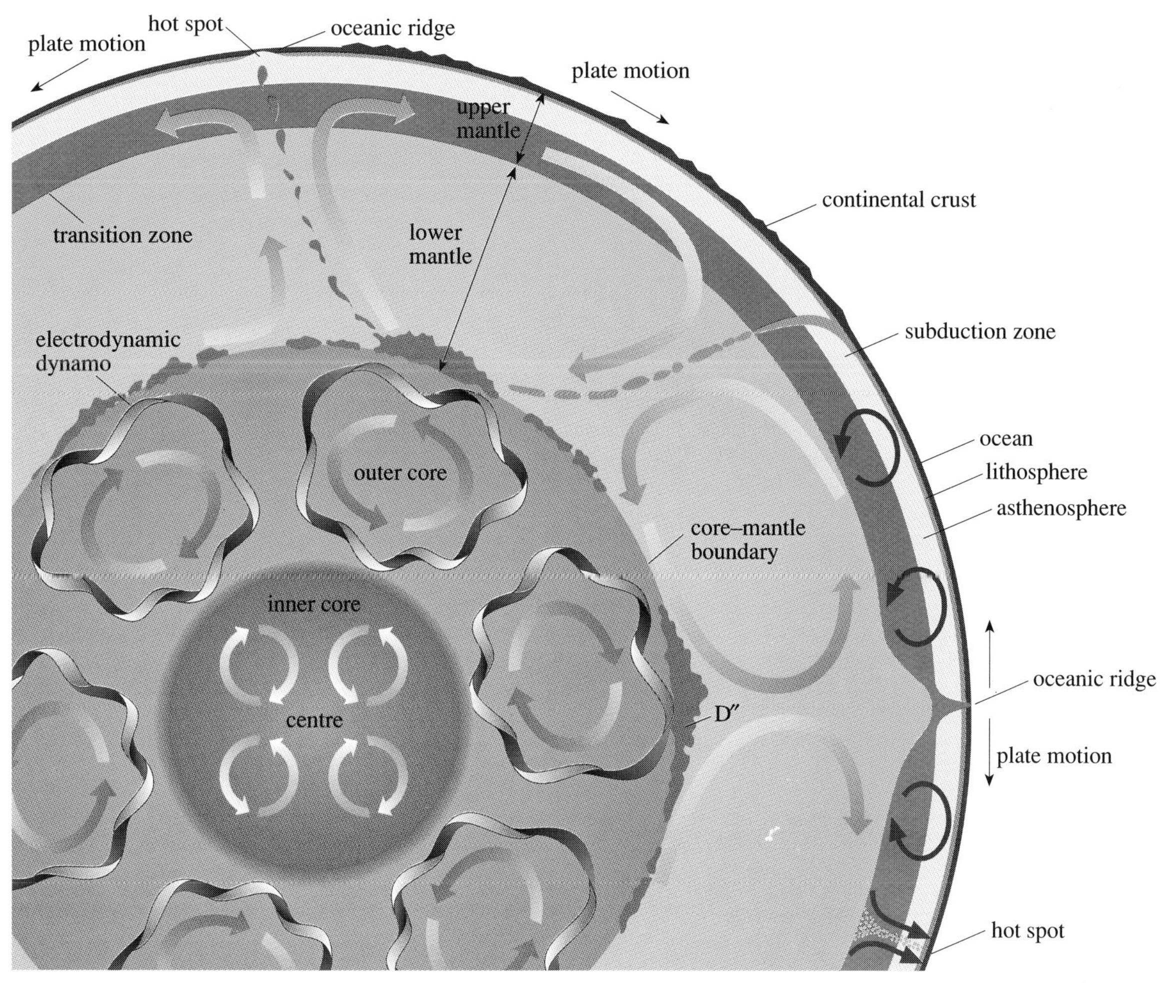

Plate 1.4 An artist's impression of how the escape of heat from the Earth's interior produces ceaseless churning of core and mantle.

Plate 1.5 Supernova SN 1987a (lower right), the brightest supernova observed since 1607, resulted from the explosion of a star in the Large Magellanic Cloud, a satellite galaxy of the Milky Way. It was first seen on 24 February 1987 (this photograph was taken on 7 March 1987). The larger image on the upper left is of the Tarantula Nebula, a giant nebula in the Large Magellanic Cloud.

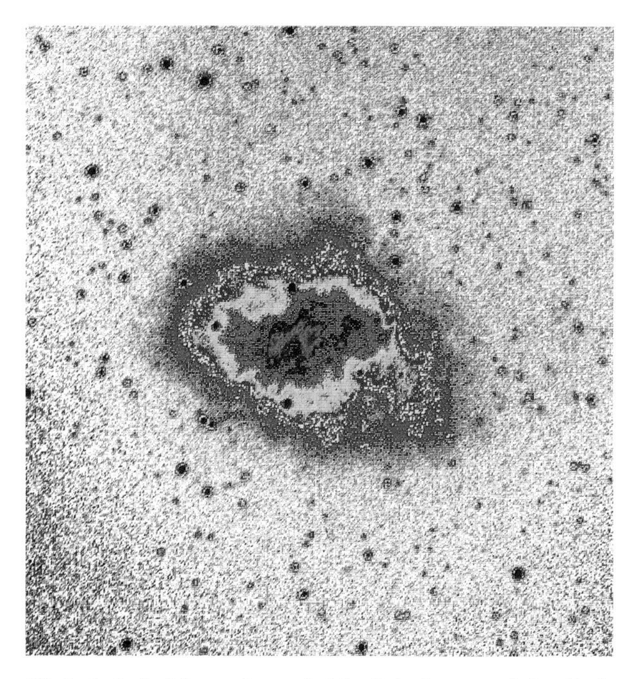

Plate 1.6 A false-colour visible-light image of the Crab Nebula (M1). The colours have been added by computer to represent the intensity of the image, from green (lowest intensity) to dark blue (highest). The Crab Nebula is the remnant of a supernova explosion observed by Chinese astronomers in 1054, and is about 6 500 light years away in the constellation Tauras.

Plate 1.7 Part of the solar spectrum, showing dark absorption lines known as Fraunhofer lines. (Note that here the wavelength decreases from left to right, whereas in Figure 1.11 of Block 1 it increases from left to right.)

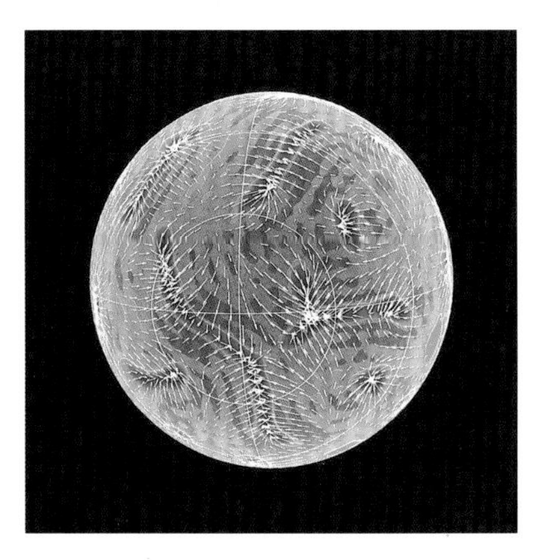

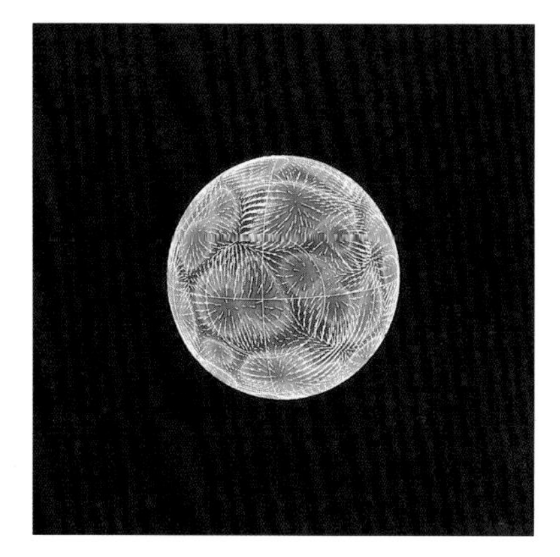

Plate 1.8 Seismic tomographic images of the Earth's mantle at three different depths. These are 'snapshots' of a three-dimensional model at depths of (a) 430 km, (b) 2 020 km, and (c) 2 600 km. The blues represent material in which the seismic velocity is higher than average (higher density, lower temperature) and the orange and yellow zones have lower-than-average seismic velocity (lower density, higher temperature). These images, which demonstrate marked lateral inhomogeneities in the mantle, can be related to convection in the mantle. The 'cold' (blue) regions are where denser material is sinking, and the 'hot' (orange/yellow) regions are where less dense material is rising. Tomographic images of the uppermost mantle can be closely related to plate-tectonic processes (to be covered in Block 2); for example, oceanic ridges are associated with hotter rising material, whereas subduction zones are associated with colder sinking material. As the depth increases, however, the correlation with surface tectonic features decreases. The white arrows indicate the directions of flow at the respective surfaces.

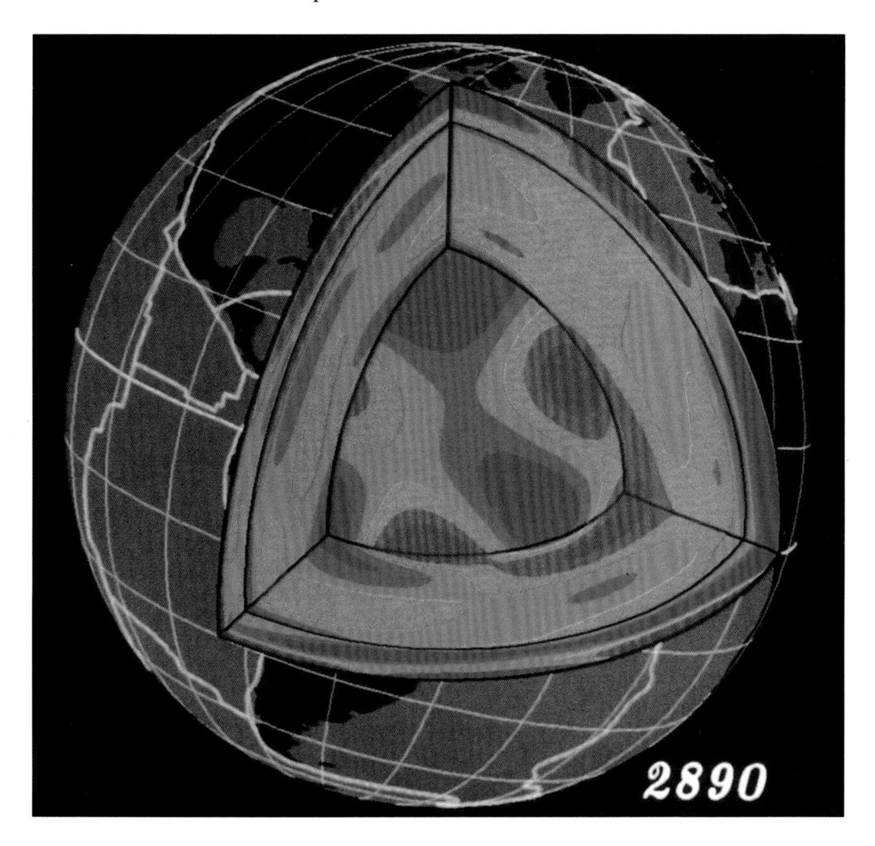

Plate 1.9 Tomographic data of the same type as in Plate 1.8 but here displayed as a 'three-dimensional' cut-away of the Earth down to the core–mantle boundary. The colour convention is the same as for Plate 1.8.

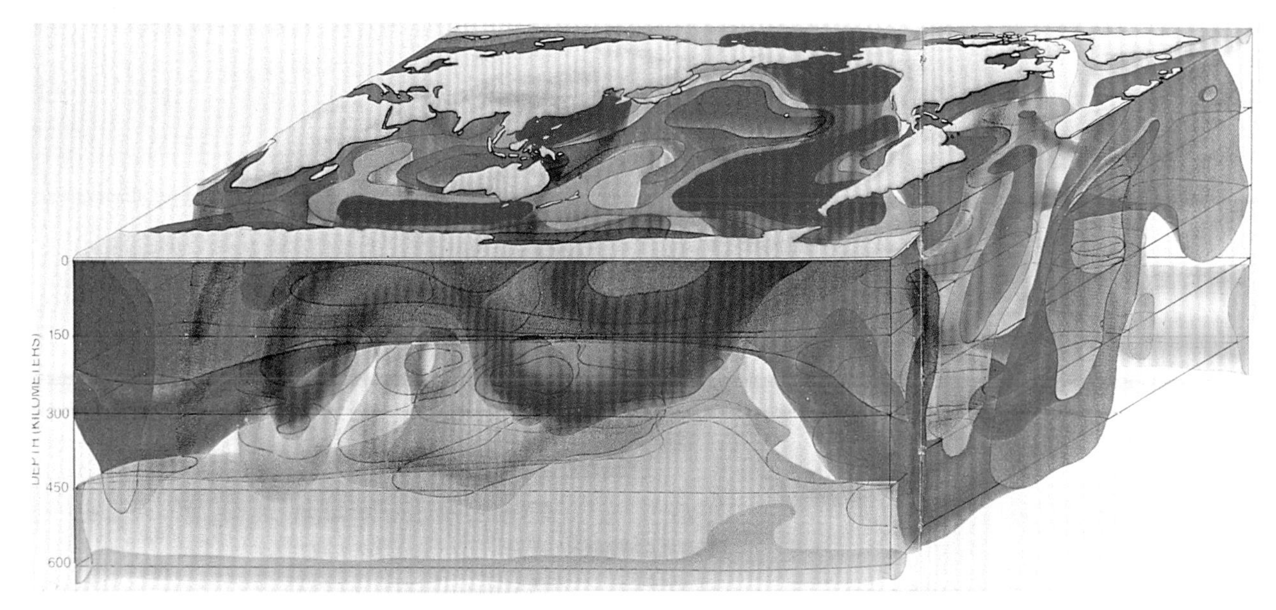

Plate 1.10 A tomographic image of the Earth's upper mantle. The colour convention is broadly the same as that in Plates 1.8 and 1.9 but with the addition of green, which marks regions where a cold anomaly (blue) lies in front of or behind a hot one (red). Note that the hot regions (red and yellow) associated with oceanic ridges (Block 2) persist to considerable depth, but that as the depth increases they become more and more offset from the surface features. This suggests that the lateral movement of both hot and cold material in the upper mantle is extensive.

GLOBAL HEAT FLOW

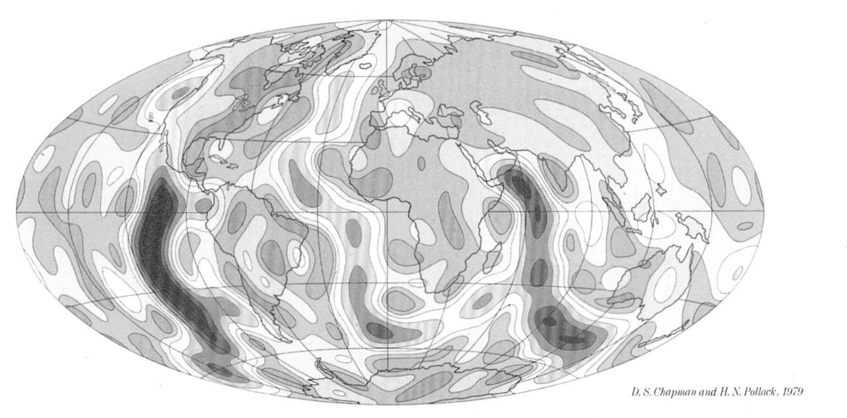

45 60 80 120 160 200 240 280

mW m^{-2}

Plate 1.11 The pattern of global heat flow, based on measurements and predictions in 5° × 5° squares. For those squares in which no measurements are available, values were predicted from the relationship between heat flow and age in oceanic and continental provinces. The variations in heat flow are represented by contour lines determined by the techniques of spherical harmonic analysis.

Acknowledgements

We are grateful to the following institutions and people outside the Course Team who provided the photographic material from which these plates were made:

Plate 1.1 Tony Ward, Tetbury, Science Photo Library, London. *Plate 1.2* NASA. *Plate 1.3* Rev. Ronald Royer, Science Photo Library, London. *Plate 1.4* Artwork from *Scientific American* June 1991 © Scientific American Inc., George V. Kelvin. *Plate 1.5* National Optical Astronomy Observatories, Science Photo Library, London. *Plate 1.6* George Fowler, Science Photo Library, London. *Plate 1.7* National Solar Observatory, Sunspot, New Mexico 88348. *Plate 1.8* *Scientific American* June 1991. © Scientific American Inc., Gary A. Glatzmaier, Los Alamos. *Plate 1.9* Don L. Anderson and Adam M. Dziewonski *Seismic tomography* Copyright © October 1984 by Scientific American Inc. All rights reserved. *Plate 1.10* Prof. J. H. Woodhouse, Oxford University. *Plate 1.11* H. N. Pollack in *Proceedings of the NATO Advanced Study Institute* on 'The mechanics of continental drift and plate tectonics'.